# Design of
# URBAN RUNOFF
# QUALITY CONTROLS

Proceedings of an
Engineering Foundation Conference on
Current Practice and Design Criteria
for Urban Quality Control

Trout Lodge
Potosi, Missouri
July 10-15, 1988

Organized by the
Urban Water Resources Research Council of the
Technical Council on Research of the
American Society of Civil Engineers

co-sponsored and funded by
the U.S. Environmental Protection Agency

Conference Committee
Larry A. Roesner, Chairman
Ben Urbonas, co-Chairman
Thomas O. Barnwell, Jr.
Jonathan E. Jones
Michael L. Testriep

Edited by
Larry A. Roesner, Ben Urbonas, and Michael B. Sonnen

Published by the
American Society of Civil Engineers
345 East 47th Street
New York, New York 10017-2398

ENVIRONMENTAL S/E
933 W. LIBERTY DRIVE
WHEATON, ILLINOIS 60187
708/690-4090
FAX 708/690-4167

## ABSTRACT

This book describes current practice in the design of polution controls for urban runoff. The contents comprise the proceedings of an Engineering Foundation Conference held in July, 1988, titled "Current Practice and Design Criteria for Urban Runoff Water Quality Control." The papers are concerned with the pragmatic, functional design and maintenance of devices that have been demonstrated to work in the field. These include: wet and dry detentrion ponds, infiltration devices, sedimentation tanks, swirl concentrators, wetlands, and source controls. Three related sessions address receiving water responses to urban runoff, the Water Quality Act of 1987, and institutional issues related to implementation of urban runoff quality control. These proceedings bring together much of the collective knowledge of American and European technology in the subject area. Many of the authors are internationally recognized in their field.

### Library of Congress Cataloging-in-Publication Data

Engineering Foundation Conference on Current Practice and Design Criteria for Urban Quality Control (1988: Potosi, Mo.)
    Design of urban runoff quality controls; proceedings of an Engineering Foundation Conference on Current Practice and Design Criteria for Urban Quality Control; Trout Lodge, Potosi, Missouri, July 10-15, 1988/organized by the Urban Water Resources Research Council of the Technical Council on Research of the American Society of Civil Engineers; co-sponsored and funded by the U.S. Environmental Protection Agency; edited by Larry A. Roesner, Ben Urbonas, and Michael B. Sonnen.
    p. cm.
    Includes indexes.
    ISBN 0-87262-695-4
    1. Urban runoff-Congresses. 2. Water quality management–Congresses. I. Roesner, Larry A. II. Urbonas, Ben. III. Sonner, Michael B. IV. American Society of Civil Engineers. Urban Water Resources Research Council, V. United States. Environmental Protection Agency. VI. Title.
TD657.E64 1988                                        89-6498
628'.212—dc19                                                CIP

The Society is not responsible for any statements made or opinions expressed in its publications.

Authorization to photocopy material for internal or personal use under circumstances not falling within the fair use provisions of the Copyright Act is granted by ASCE to libraries and other users registered with the Copyright Clearance Center (CCC) Transactional Reporting Service, provided that the base fee of $1.00 per article plus $.15 per page is paid directly to CCC, 27 Congress Street, Salem, MA 01970. The identification for ASCE Books is 0-87262/88. $1 + .15. Requests for special permission or bulk copying should be addressed to Reprints/Permissions Department.

Copyright © 1989 by The American Society of Civil Engineers
All Rights Reserved
Library of Congress Catalog Card No: 89-6498
ISBN 0-87262-695-4
Manufactured in the United States of America

# FOREWORD

For more than 20 years, the Urban Water Resources Research Council of the American Society of Civil Engineers has strongly promoted the transfer of urban runoff technology among researchers, practitioners, and administrators. The primary vehicles used by the Council for the transfer are Engineering Foundation Conferences, International Symposia, and technical sessions at professional meetings. In each case, a proceedings or series of papers has been published.

This Engineering Foundation Conference on Urban Runoff Quality is one of a series of more than ten organized by the Council on the subject of urban stormwater management. This conference builds in particular on the Engineering Foundation Conference on Urban Runoff Quality held in Henniker, New Hampshire, June 23-27, 1986.

All papers presented at this conference were by invitation. Papers were reviewed prior to acceptance for publication by the appropriate session chairman, and by the editors of this proceedings. Each paper was presented by the author and subjected to open forum discussion and review by conference participants. These proceedings contain papers presented at the conference. The papers are grouped by session. Supplemental papers are also included at the end of some sessions. All papers are eligible for discussion in the appropriate ASCE Journal, and all papers are eligible for ASCE awards.

The conference-organizing committee expresses its thanks to the Engineering Foundation and to its Director, Mr. Harold A. Commerer, for providing administrative support in organizing this conference. Thanks also go to the support staff at Trout Lodge, whose day-to-day management of the conference facilities provided a comfortable working conference atmosphere. The Council is especially grateful to Mr. Dennis W. Athayde and to Mr. James D. Gallup for arranging funding support by the U.S. Environmental Protection Agency for this conference.

# CONTENTS

Foreword

**CONFERENCE OVERVIEW AND SUMMARY**
    by Larry A. Roesner, Ben Urbonas, and Michael B. Sonnen ............... 1

**SESSION 1: RECEIVING WATER RESPONSES TO URBAN RUNOFF**
    Chairman: Thomas O. Barnwell

The Development of Environmental Criteria for Urban Detention Pond
Design in the UK
    by J. Bryan Ellis ................................................. 14

Habitat and Water Quality Considerations in Receiving Waters
    by Lewis L. Osborne and Edwin E. Herricks ......................... 29

State Perspectives on Water Quality Criteria
    by Eric H. Livingston ............................................ 49

**SESSION 2: POLLUTION CONTROL**
    Chairman: Dr. Wayne C. Huber

Technological, Hydrological and BMP Basis for Water Quality Criteria
and Goals
    by Wayne C. Huber ............................................... 68

Cost-Effectiveness and Urban Storm-Water Quality Criteria
    by James P. Heaney .............................................. 84

**SESSION 3: WATER QUALITY ACT OF 1987 —
PROBLEMS AND POSSIBILITIES**
    Chairman: Dr. Stuart G. Walesh

Federal Requirements for Storm Water Management Programs
    by James D. Gallup and Kevin Weiss .............................. 100

A View From The Middle — Water Quality Act of 1987: Problems and
Possibilities for the Design Professional
    by Michael T. Llewelyn .......................................... 109

A View from the "Bottom": Challenges and Prospects
    by L. Scott Tucker .............................................. 111

## SESSION 4A: DETENTION AND RETENTION
### Chairman: Ben Urbonas

Basis for Design of Wet Detention Basis BMP's
  by *John P. Hartigan* ................................................. 122

Long Term Performance of Water Quality Ponds
  by *Eugene D. Driscoll* ............................................... 145

Mixing and Residence Times of Stormwater Runoff in a
Detention System
  by *Edward H. Martin* ................................................ 164

Design of Extended Detention Wet Pond Systems
  by *Thomas R. Schueler and Mike Helfrich* .......................... 180

Supplemental Paper: Water Quality Ponds — Are They the Answer?
  by *Patrick F. Mulhern and Timothy D. Steele* ...................... 203

## SESSION 4B: DETENTION AND RETENTION
### Chairman: Marshall E. Jennings

Rainfall Analysis for Efficient Detention Ponds
  by *T. Hvitved-Jacobsen, Y.A. Yousef, and M.P. Wanielista* .............. 214

Monitoring and Design of Stormwater Control Basins
  by *J.E. Veenhuis, J.H. Parrish, and M.E. Jennings* .................... 224

Multiple Treatment System for Phosphorus Removal
  by *James T. Wulliman, Mark Maxwell, William E. Wenk, and
  Ben Urbonas* ...................................................... 239

Supplemental Papers: Load-Detention Efficiencies in a Dry-Pond Basin
  by *Larry M. Pope and Larry G. Hess* ............................... 258

Simulated Water-Quality Changes in Detention Basins
  by *Phillip J. Zariello* ............................................ 268

Water Quality Study on Urban Wet Detention Ponds
  by *Jy S. Wu, Bob Holman, and John Dorney* ......................... 280

## SESSION 5: INFILTRATION DEVICES
### Chairman: Thomas R. Schueler

Design and Construction of Infiltration Trenches
  by *Bruce W. Harrington* ............................................ 290

Swedish Approach to Infiltration and Percolation Design

    by *Peter Stahre and Ben Urbonas* ................................. 307

### SESSION 6: INSTITUTIONAL ISSUES
Chairman: Jonathan E. Jones

Institutional Aspects of Stormwater Quality Planning
    by *Nancy U. Schultz and Ronald L. Wycoff* ......................... 324

Institutional Stormwater Management Issues
    by *H. Earl Shaver* ............................................... 340

Urban Runoff Management — An Industry Perspective: An Opinion
    by *Neil G. Jaquet* ............................................... 349

Summary of Institutional Issues
    by *D. Earl Jones* ................................................ 356

### SESSION 7: RETROFITTING STORM SEWERS
Chairman: Dennis W. Athayde

Combined-Sewer-Overflow Control in West Germany — History, Practice and Experience
    by *Hansjoerg Brombach* ........................................... 359

Nonpoint Pollution First Step in Control
    by *James Murray* ................................................ 378

Swirl Concentrators Revisited: The American Experience and New German Technology
    by *William C. Pisano* ........................................... 390

### SESSION 8: SOURCE CONTROLS
Chairman: Michael L. Terstriep

Source Control of Oil and Grease in an Urban Area
    by *Gary S. Silverman and Michael K. Stenstrom* ..................... 403

Program to Reduce Deicing Chemical Use
    by *Byron N. Lord* ............................................... 421

Source Tracing of Toxicants in Storm Drains
    by *Thomas P. Hubbard and Timothy E. Sample* ....................... 436

### SESSION 9: WETLANDS AS TREATMENT SYSTEMS
Chairman: Dr. Kenneth D. Jenkins

Long-Term Effects of Urban Stormwater on Wetlands
    by *Richard R. Horner* .......................................... 451

The Use of Wetlands for Urban Stormwater Management
 by *Eric H. Livingston* ............................................. 467

Subject Index ....................................................... 491

Author Index ....................................................... 493

# CONFERENCE OVERVIEW AND SUMMARY

Larry A. Roesner, Ph.D., P.E. [1]
Ben Urbonas, P.E. [2]
Michael B. Sonnen, Ph.D., P.E. [3]

## INTRODUCTION

In June, 1986, an Engineering Foundation Research Conference was held in Henniker, New Hampshire, to review current knowledge and experience concerning the quality aspects of urban stormwater runoff. One of the major findings of that conference was that, while much work had been done across the United States in measuring the efficiency of treatment methods for urban runoff, the literature contained very little in the way of design parameters for these devices. Moreover, in the face of anticipated promulgation by the U.S. Environmental Protection Agency of rules and regulations for control of urban runoff pollution, it was obvious that better design guidance was necessary if the engineering profession was to achieve success in designing, building, and operating urban runoff water quality control devices. So, it was recommended by the participants of that conference that another Engineering Foundation Conference be convened in 1988 whose purpose would be to present the state-of-the-art design criteria for urban runoff water quality control devices.

The follow-up Engineering Foundation Research Conference, on "Current Practices and Design Criteria for Urban Runoff Water Quality Control," was indeed convened on July 10-15, 1988 at Trout Lodge, Potosi, Missouri. The conference was sponsored and organized by the Urban Water Resources Research Council of the American Society of Civil Engineers with funding provided by the Enforcement Division and the Nonpoint Source Division of the Office of Water, U.S. EPA. These proceedings of that conference bring together much of the collective knowledge of American and European technology in this subject area. Many of the authors are internationally recognized experts in their fields. Their papers are concerned with pragmatic, functional designs of devices that have been demonstrated in the field. Most of the designs presented consider the maintenance of the devices, an aspect which is often ignored. It is the hope of all the

---

[1] Regional Director of Water Resources, Camp Dresser & McKee, Inc., 555 Winderley Place, Suite 200, Maitland, FL 32751, and Conference Chairman.

[2] Chief, Master Planning Program. Urban Drainage and Flood Control District, 2780 West 26th Avenue, Suite 156 B, Denver, CO 80211, and Conference Co-Chairman.

[3] Director, Technical Services, Disposal Control Service, Inc., 1369 West 9th Street, Upland, CA 91786, and Conference Reporter.

attendees at that conference that these proceedings will provide some guidance to those who will be responsible for implementing urban runoff controls under the Water Quality Act of 1987.

## WHAT WAS LEARNED?

From the responses received at the conferences and from the review of conference critique forms filled out by the majority of those attending, the conference was an unqualified success. Not only was the conference format received by the conferees with enthusiasm, its content and the quality of the presentations and papers provided a cornucopia of technical, institutional, and practical state-of-the-art information on the topic of enhancing the quality urban stormwater facilities design. The overall lessons learned were many and included:

o    Urban runoff quality problems that are identified as threatening receiving waters in various communities in the United States are, for the most part, being addressed by those communities.

o    A variety of engineering techniques and management practices are becoming understood, and meaningful design guidance for them is emerging.

o    Unlike many of the past conferences on this topic, this one included reports of many design and implementation successes.

o    A nation-wide, federally mandated program regulating urban stormwater discharge permits is now the law of the land.

Urban Runoff Quality Control Devices.

We saw many different techniques for water quality treatment being described and critiqued. Examples of devices for which design guidance was presented by the speakers included:

o    Ponds
o    Detention basins
o    Extended detention basins
o    Sand filter beds
o    Infiltration basins
o    Percolation trenches
o    Swales
o    Oil and grease traps
o    Wetlands
o    Mechanical/structural devices

All of these treatment systems remove from stormwater a portion of the suspended solids and the pollutants that are attached to them. Some of these devices also remove a portion of the dissolved constituents found in stormwater. Many of these devices can be used individually or in combination with each other.

Among all of these devices the most promising and best understood are detention and extended detention basins and ponds. Less reliable in terms of predicting performance, but showing promise, are sand filter beds, wetlands, infiltration basins, and percolation trenches. All of the latter appear to be in their infancy and lack the necessary long-term field testing that would provide data for the development of sound design practices.

Of particular concern to the authors are infiltration and percolation devices which can be installed at the source of runoff. These so-called "local disposal" devices rely on dispersion of stormwater into the ground, and the levels of pollutant removal they can achieve appear to be good. However, if a site of this kind is not regularly maintained, the soil pores can become clogged or permanently saturated, so that it is no longer capable of performing its functions. Are there remedies for this situation, or has the device unfortunately become no more than a permanent nuisance? Such questions and others concerning long-term reliability and potential groundwater contamination persist.

For all of these devices, long-term performance, which always depends on regular maintenance, should be the goal of all regulators and stormwater management professionals. Advances in design continue to be made, and confidence in the application of some of these "innovative" methodologies is growing. However, it is clear from these proceedings that caution must still be observed, and that these devices should not be placed in wide-spread use until certain minimum requirements are met. First, many of these devices need to be extensively field-tested under the direct control of field engineers. Such testing will allow present design parameters to be either verified or adjusted. And second, before any such device is installed, a thorough site investigation must be taken, conservative safety factors need to be built into the application, and careful field monitoring of construction should be arranged.

One last message offered to the conferees by those involved in or consulting to local governments is that there needs to be centralized regulation of privately owned quality control devices. Without such oversight, institutional and environmental problems inevitably arise relating to inspection and maintenance; in the long run, the devices eventually lose their design performance and even their reputations as effective control methodologies.

<u>Runoff-Independent Controls.</u>

In addition to devices and best management practices used to directly control urban runoff, other activities were reported which reduce or eliminate the pollutants entering receiving waters through stormwater outfalls. These include:

- o  Devices for combined sewer overflow control;
- o  Detention and removal of illicit connections ;
- o  Detection of illegal dumpings; and
- o  Changes in road de-icing practices.

Combined sewer overflow controls presented included the swirl concentrator being used in the United States and on-line and off-line storage and flow control devices being used in West Germany. The degree to which these devices can be applied directly to separate stormwater remains unclear. The potential is there, but questions regarding how they should be modified for such applications and how cost-effective they may be have yet to be answered.

The descriptions of the attack on illicit connections in Wayne County, Michigan, and on illegal dumping in Seattle, Washington, were impressive. The amounts of pollutants reaching the receiving waters in these two areas were reduced significantly through local efforts in pursuit of observed problems. It may be that many other stormwater quality-related problems in the United States are directly related to illicit connections and illicit discharges of pollutants into receiving waters. This insight could help prioritize resources when it comes to attacking observed problems in our receiving waters.

Problem Definition and Resolution.

Much formal and informal discussion was directed at the identification and quantification of the impacts on receiving waters and on the benefits of stormwater quality control. It is clear that urban runoff quality is and will continue to be controlled by local agencies. This has been done mostly in response to the views and complaints of local constituents. It is the value local citizens place on the receiving waters and whether or not they perceive urban runoff as causing the problems that drive local agency efforts and budgets in that regard. Locally perceived values are likely to be protected or enhanced in direct proportion to the pressures applied at the local level and in much less direct proportion to the pressures applied at more distant state and federal levels.

Problem identification and quantification needs much more emphasis before urban runoff quality solutions are mandated. Such data-based information can be used to educate local constituents as to the effects of urban runoff on the receiving waters. If the evidence of existing or projected water-quality degradation is clear, and so far in many instances it may not be, the observed trends are that local action will result, as has happened in many communities throughout the United States, where various technologies have been applied and are working well. Thus, when the need arises, engineers and other scientists can look at existing installations in the field and determine for themselves what is most likely to work best for their particular environs.

## SYNOPSIS OF THE SESSIONS

The Conference consisted of ten sessions. The first three sessions (I-III) of the conference addressed receiving water responses to urban runoff, pollution control approaches, and a panel discussion of the Water Quality Act of 1987 as it relates to the regulation of pollution from urban runoff. These sessions set the stage for three more sessions (IV A, IV B, and V), addressing design criteria for retention, detention, and infiltration methods. In Session VI, the institutional issues associated with implementing urban runoff controls and with permitting storm drainage systems were discussed. The last three sessions (VII, VIII, and IX), dealt with retrofitting existing systems, source controls and methods, or treatment systems.

The range of subjects covered in many of the papers is broad. Thus, to enable the reader to locate pertinent material easily in these proceedings, we have identified for each paper the significant subject areas discussed by the authors. That information is presented below by session.

### Session I--Receiving Water Responses to Urban Runoff

**Ellis: The Development of Environmental Criteria for Urban Detention and Design in the United Kingdom.**

- o Institutional view and organizational framework for urban runoff management in the UK was presented.

- o Detention pond coliform concentration resulting from stormwater inflow was discussed.

- o Guidelines were given for design of multiple-use facilities for stormwater detention and recreation.

**Osborne and Herricks: Habitat and Water Quality Considerations in Receiving Waters.**

- o Stream habitat and water quality control measures must take into account both temporal and spatial dynamic aspects of the watershed.

- o Since stream habitat and ecosystems vary with location in a watershed, the effects of urban runoff on stream ecosystems are also dependent on the order of the stream into which the runoff discharges.

- o An integration of physical and chemical habitat analysis is proposed to develop management programs for instream habitat.

**Livingston: State Perspective on Water Quality Criteria.**

- o The scientific and regulatory framework for Florida's Urban Runoff Quality Control Program is presented.

- o The State's urban runoff quality control regulations are described and explained.

- o Design guidelines for swales, retention, detention with filtration, and distribution and wetland treatment are presented.

- o The challenge of retrofitting existing systems is discussed.

Session II--Pollution Control.

**Huber: Technological, Hydrological, and BMP Bases for Water Quality Criteria and Goals.**

o   The linkages among various approaches to control of urban runoff quality are discussed.

o   Technological controls are shown to be not currently applicable to treatment of urban runoff.

o   Hydrologically based water quality standards are perceived as most scientifically attractive, but setting of wet-weather water quality criteria and determination of benefits remains difficult.

**Heaney: Cost-Effectiveness and Urban Stormwater Quality Criteria.**

o   Alternative methods of evaluating the effectiveness of investments in urban stormwater quality improvements are examined.

o   Methods of risk assessment and their potential applicability to evaluating urban runoff control programs are discussed.

o   Benefit-cost analysis methods applicable to urban stormwater problems are summarized.

Session III--Water Quality Act of 1987.

**Gallup and Weiss: Federal Requirements for Stormwater Management Programs.**

o   The history of EPA regulation of stormwater quality since the passage of the Clean Water Act by Congress in 1972 is presented.

o   Requirements of the Water Quality Act passed by Congress in 1987, with respect to stormwater discharges, are identified and discussed.

o   The current status of rulemaking by EPA to establish NPDES permit application requirements for stormwater discharges is presented.

**Llewelyn: View from the Middle: Problems and Possibilities.**

o   State's views of the 1987 Water Quality Act regarding the stormwater discharge NPDES permit program are discussed.

o   Wisconsin's program, in existence since 1978, is discussed.

o   Problems and opportunities facing the design engineers are identified.

# OVERVIEW AND SUMMARY

**Tucker: A Local Agency's View.**

- o   Local government's view of the 1987 Water Qualty Act is presented regarding stormwater discharge NPDES permit program.

- o   Paper stresses that cost of program should be done totally by local governments.

- o   The author is apprehensive about "problems" being defined in Washington with each local government having to respond, instead of allowing each metropolitan area or state to identify problems and respond accordingly.

<u>Session IV A--Detention and Retention.</u>

**Hartigan: Design of Wet Detention BMPs.**

- o   Two approaches for the design of wet detention ponds are presented and compared: one is based on solids settling theory; the second on eutrophicalian principles.

- o   Recommended design criteria for wet detention are presented and discussed.

- o   Cost associated with alternative design criteria are provided for a typical regional BMP Master Plan.

**Driscoll: Long-Term Performance of Water Quality Ponds.**

- o   Design guidelines based on sedimentation are developed from review of data collected in the National Urban Runoff Program for infiltration devices and wet detention ponds.

- o   Effects of the coefficient of variation for runoff on the removal of suspended solids is illustrated.

**Martin: Mixing and Residence Times of Stormwater Runoff in a Detention System.**

- o   Results of dye tracer studies in a wet detention pond are used to examine the applicability of plug flow and completely mixed flow theory for pollutant removal.

- o   The phenomenon of short-circuiting observed in a particular pond is discussed.

**Schueler and Helfrich: Design of Extended Detention Wet Pond Systems.**

- o   Extended detention pond design parameters are presented for several variables.

- o   Extended detention with a permanent wet pool is recommended and design criteria are given.

o   Maintenance requirements, safety and aesthetics, and aspects of designs are discussed.

**Mulhern and Steele:  Water Quality Ponds--Are They the Answer?**

o   Results of a study of phosphorus removal by an aesthetic golf course pond are presented.

o   Data indicate a net outflow of phosphorus during low flow and small storm periods, but this is fairly minor compared to load removals during storms.

o   The net removal of annual phosphorus load was only 20%, which was lower than expected.

<u>Session IV B --Detention and Retention.</u>

**Hvitved-Jacobsen, et al.:   Rainfall Analysis for Efficient Detention Ponds.**

o   Storage volume provided in ponds should be based, at least in part, on the inter-event dry period from the rainfall record, to minimize cumulative effects of pollutants discharged to receiving streams.

o   Inter-event times of 72 to 96 hours may be necessary for detention ponds to be effective treatment devices.

**Veenhuis, et al.: Monitoring and Design of Stormwater Control Basins.**

o   Numerous quality constituents were monitored for Austin, Texas, ponds during 22 storms in 1982-84.

o   Main outlet for pond storage volume is through a sand and gravel filter in the pond bottom.

o   Monitoring suggested that basins should be placed off-line to provide sufficient time and low enough turbulence to realize high removals.

**Wulliman, et al.: Multiple Treatment System for Phosphorous Removal.**

o   Phosphorus loads in urban areas were shown to be related directly to the amount of disturbed land in the upstream watershed.

o   A non-mechanical, low-maintenance combination of wet detention and downstream wetland/infiltration areas was designed to optimize post-development phosphorus control.

o   Although the devices are designed, they are not yet built; so removal efficiencies are not yet available.  However, a number of design suggestions are given for ponds and their appurtenances as well as wet lands and infiltration areas.

**Pope and Hess: Load-Detention Efficiencies in a Dry-Pond Basin.**

- o  A dry pond in Topeka, Illinois, was monitored by USGS for 19 storms over 14 months. Eleven constituents were sampled (automatically).

- o  Approximately 65% of the metals were removed; dissolved salts, including organic nitrogen and carbon, actually increased; suspended solids removal, though positive, was very low.

**Zarriello: Simulated Water-Quality Changes in Detention Basins.**

- o  Four potential basin sites near Rochester, New York, were simulated. Suspended sediment, phosphorus, lead, and zinc removals were predicted.

- o  Temporary (small) and maximum-storage (larger) basins were tried at each site.

- o  Sediment retention averaged about 50% more in the larger basins, although their annual solids' retention efficiency only ranged from 33 to 60%.

- o  Dissolved constituents had predicted removals from 16 to 59%; larger basins were 80% more efficient than smaller basins, however.

**Wu, et al.: Performance of Urban Wet Detention Ponds.**

- o  Three wet detention ponds built for storm peak-flow reduction were studied for their pollutant removal efficiencies.

- o  Larger ponds proved more efficient than the smaller ponds for suspended solids removal, although results were not as good and were mixed for dissolved constituents (phosphorus and total Kjeldahl nitrogen).

Session V--Infiltration Devices.

**Harrington: Design and Construction of Infiltration Trenches.**

- o  Two hundred infiltration "practices" have been surveyed; most were infiltration trenches, and 80% of these were working as designed.

- o  Drainage area for a single trench typically is less than 10 acres.

- o  Grass buffer strips ahead of the trenches and very careful (non-clogging, non-smearing) construction techniques markedly improve trench performance.

- o  Numerous design criteria are presented and described.

**Stahre and Urbonas: Swedish Approach to Infiltration and Percolation Design.**

o   A point system for evaluating infiltration sites is presented (given six soil, cover, and traffic criteria, a score of 20 or less is poor, 30 or more is excellent).

o   Criteria for percolation basins are presented.

o   Both volume and rate-of-outflow (infiltration) design criteria are described and related to theoretical considerations.

## Session VI--Institutional Issues.

**Shultz and Wycoff: Institutional Aspects of Stormwater Quality Planning.**

o   Six institutional frameworks for stormwater runoff quality control programs are presented (Chicago, Seattle, Milwaukee, St. Louis, Santa Clara Valley, CA, and Boston).

o   Local perceptions of threats to human health or water-use impairments are the most successful motivational forces for getting runoff quality problems solved.

o   Combined sewer overflow problems receive far earlier attention and more intensive control than do separate stormwater or storm sewer problems.

**Shaver: Institutional Stormwater Management Issues.**

o   Maryland administrator for stormwater management outlines programmatic and "nitty-gritty" implementation issues that make construction and maintenance of stormwater controls effective.

o   Issues include problem-by-problem legislation (flood control versus water pollution control requirements (responsibility, regulation guidance, etc.); regulation content; manpower (for design reviewers, inspectors, etc.); education and training; maintenance; and utilities as a management concept.

**Jacquet: Urban Runoff Management--An Industry Perspective.**

o   Paper describes a particular private-landowner's problems with increased runoff and constituent loads discharged onto his property by continuing development(s) upstream.

o   The landowner reported becoming a quasi-regulator through a design-review process in which his company participates, which concerns its neighbors' plans for more development upstream.

o   This private, individual, industrial, downstream, beneficial water user came to the independent judgment that upstream retardation and detention of runoff flow and pollutants make the most economical and logical sense among all the runoff control alternatives.

**Jones: Summary of Institutional Issues.**

Important issues, stressed by earlier speakers were:

1. Local motivation and local wherewithal to implement runoff controls are crucial; without them, implementation will not occur.

2. Regulations, ordinances, design criteria, construction specifications, and inspection and maintenance procedures must be positive, explicit, single-minded, and definitive. Otherwise, implementation and continued performance of design controls are bound to break down.

Session VII--Retrofitting Storm Sewers.

**Brombach: Combined Sewer Overflow Control in West Germany--History, Practice, and Experience.**

o  West Germany now has more than 8,000 storm overflow storage tanks on combined systems; total volume is almost 5 million cubic meters.

o  Design criteria are given for smaller "first-flush" tanks and for larger "storm overflow" tanks; the former are for temporary storage; the latter provide both storage and sedimentation treatment.

o  Outflow throttling, tank cleaning, and performance monitoring are also discussed.

**Murray: Nonpoint Pollution: First Step in Control.**

o  A survey of illicit connections to separate storm sewers in Ann Arbor, Michigan, revealed that many automobile-related industrial and commercial establishments had illegal connections; very few, if any, fecal coliform sources were identified.

o  The mean concentration of the outfall sample for a variety of chemicals exceeded the upstream control station mean concentration (on the same sewer) 87% of the time--before improper or illegal connections were found.

o  Automobile-related facilities were improperly connected at 60% of such facilities; other chemical users had a 25% frequency of improper connections.

o  Concentrations of various chemicals (of 38 constituents measured) were 2 to 10 times higher at the outfall than in the control section of the sampled storm sewer.

**Pisano: Swirl Concentrators Revisited.**

o  Case studies of previously installed swirl concentrators based on a U.S. design are presented to show that solids removal is effective only at flow ranges substantially below the original "design" flow level.

o  Intended installations based on a recent West German design innovation are described, although no performance data are yet available.

Session VIII--Source Controls.

**Silverman and Stenstrom: Source Control of Oil and Grease in an Urban Area.**

o  Conventional separators and other waste treatment devices are ineffective and inordinately expensive for stormwater oil-and-grease treatment.

o  Commercial (rather than residential or other) land uses appear to provide the greatest potential for pollution reduction via source control.

o  Porous pavements, greenbelts, absorbents in storm drain inlets (catch basins), and pavement surface cleaning with wet scrubbing were evaluated. No field performance data are presented.

**Lord: Program to Reduce De-icing Chemical Usage.**

o  While environmental and other problems from use of salt as road de-ices have been known for decades, almost 10 million tons are applied to U.S. roads each year.

o  Alternatives are being sought, but the more promising ones are still 10 to 20 times as expensive as salt.

o  An $8 million research effort is continuing to find new de-icing methods or improvements.

**Hubbard and Sample: Source Tracing of Toxicants in Storm Drains.**

o  Seattle-area research demonstrated that sampling of receiving water sediments for toxic chemicals was more effective at identifying specific commercial and industrial sources than were measurements made of (high volume, dilute concentration) storm flows.

o  Several Seattle-area industrial sources of lead and other pollutants were found for which prodigious quantities of pollutants were removed at the source (e.g. 20 cu. yds of sediment high enough in lead (25-36%) that it could be recovered by smelting).

Session IX--Wetlands as Treatment Systems

**Horner: Long-Term Effects of Urban Stormwater on Wetlands.**

o  A four-part research program is in progress in Puget Sound, King

County, Washington, to discover long-term effects of urban stormwater discharges on wetlands.

o The first part, a synoptic survey, found mainly aesthetic problems in wetlands receiving urban runoff and suggested pre-settling before urban stormwater discharge to a natural wetland.

o Bacterial counts in urban wetlands, though higher than in non-urban counterparts, did not exceed existing or proposed standards.

o Study is continuing of remaining three components: a long-term study of urban runoff impacts; an investigation of water quality benefits provided by wetland treatment; and a laboratory study.

**Livingston: The Use of Wetlands for Urban Stormwater Management**

o In using wetlands for pollutant removal in stormwater management systems, care must be taken to avoid ecological damage.

o Design variables which will avoid these problems are discussed, as well as the many unknown factors which remain.

o A Florida natural wetland treatment system and a Maryland constructed wetland system are discussed.

The Development of Environmental Criteria for Urban
Detention Pond Design in the UK

J Bryan Ellis*

## Abstract

Working guidelines for water quality, amenity and ecological design
of urban detention ponds in the UK are identified based on best
engineering and management practise. The organisational framework for
and attitudes towards urban runoff control and surface stormwater
storage in the UK are briefly reviewed together with those operational
and performance features which have influenced the development of
working design criteria and methods. The emergence of ecological
pressures and parameters for urban wetland habitat management to be
incorporated into the design procedures are also described.

## THE ORGANISATIONAL FRAMEWORK FOR URBAN RUNOFF CONTROL

The duty of a UK Water Authority to drain an urban catchment area
by providing public sewers and dealing with the contents of those
sewers derives from Section 14 of the 1973 Water Act. Under other
sections of the Act however, as well as under specific provisions
contained in the 1983 Water Act, there is a clear intention that there
should be a division of duties in respect of sewerage and drainage
between the Water Authorities and Local (District) Authorities. Local
and County Authorities under the Land Drainage Act 1976 also retain
permissive statutory powers and responsibilities in respect of
'non-main' watercourses which receive surface water discharges and such
non-main rivers are also subject to riparian and prescriptive rights.
It is therefore not open to Water Authorities to exercise their powers
relative to public sewer discharges so as to undertake improvements to
existing urban watercourses unless agreements and contributions are
reached with the District Council or unless the watercourse can be
redesignated as a 'main' river. Hence there are many areas of doubt
regarding the status of and responsibilities for urban drainage systems
where local watercourses have been pregressively built-over and piped.

An additional complication arises from the conveyance of surface
water from highways and roads as reflected in the 1936 Public Health
Act and the 1980 Highways Act which vest dual responsibilities shared
between Highway and Water Authorities. Either type of authority may
provide and operate drains to carry, in part, flows for which the other
type of authority has responsibility although highway authorities are
not required to contribute to renewals or replacements arising from
improvements, diversions etc, and which result in increased intake of
highways water. The drainage of trunk roads and motorways are the

* Head of the Urban Pollution Research Centre, Middlesex Polytechnic,
Queensway, Enfield, EN3 4SF, UK.

# DETENTION POND DESIGN

separate responsibility of the Department of Transport with whom the Water Authorities must negotiate directly when seeking to undertake surface drainage improvements.

The most common source of pollution from public sewers in the UK is that emanating from surface water sewer outfalls. Such polluting discharges are governed by the provisions of the various Rivers (Prevention of Pollution) Acts 1951 to 1961, the 1974 Control of Pollution Act as well as by various local Acts and Byelaws. The Water Authorities must also discharge their functions under the terms of the 1936 Public Health Act so as not to 'create a nuisance' and are additionally liable at common law if damage is caused as a result of a polluting discharge.

The detailed requirements for sewerage and drainage facilities resulting from urban development activity devolve to the local and county planning levels as laid down under various sections of the 1971 Town & Country Planning Act. Since the Water Authorities have limited powers of their own with respect to ensuring that proper drainage and source stormwater pollution control facilities are provided, they are reliant upon the cooperation of the planning authority to control any development through the attachment or refusal of consent conditions or though the direct imposition of statutory controls. The consultation procedures however are normally severely time-limited and do not always enable a proper or full identification of the detailed requirements and flow/quality problems to be assessed. The procedures and protocol for endorsement of design methods, subsequent adoption and maintenance can be anomolous, idiosyncratic and contradictory within and between the various controlling and statutory agencies. In the context of this organisational framework, the developer can be forgiven for often despairing at what he perceives as the unnecessary and time-wasting devolution and fragmentation of responsibilities.

This multiplicity and complexity of organisational, legal and financial arrangements in the UK necessarily means that management strategies and design practise for urban runoff control are often uncoordinated, ad-hoc, nonstandardised and certainly lack a clear framework of institutional responsibility at the individual site, local subcatchment and parent strategic catchment levels. A few Water Authorities are in the process of commissioning arterial drainage Management Plans as a means of achieving integrated river basin management and this process may well help to identify and formulate codes of practise for stormwater management.

Irrespective of these background problems and difficulties, the demand for cost-effectiveness in the provision of flow control infrastructures have increasingly identified source storage facilities as the most popular and least cost solution.

## HYDROLOGICAL DESIGN OF URBAN DETENTION PONDS

A recent survey (Hartsmere BC, 1987) commissioned by the Thames Water Authority (TWA) identified over 200 individual storage schemes in SE England, the large majority of which were associated with surface water control on new (58%) or existing (18%) developments. Although catchment sizes ranged from 1000 hectares, 82% were less than 100ha and 50% less than 50ha. The most common type of systems used are off-line (24%) and on-line (19%) surface storage, with below ground storage tanks and in-pipe storage making up a further 22% and 35% respectively. Siltation and oil pollution are invariably the cause of high inspection/maintenance costs which, although averaging £1.530 ($3.00) per $m^3$ storage per annum, reached some £10-15 ($18-25)/$m^3$/annum for 3 to 5% of reported storage schemes. Although the majority of surface storage systems within the TWA region are dry ponds, adjoining Water Authorities utilise a more even mix of wet and dry facilities. Figure 1 shows the range of design methods currently used to determine storage

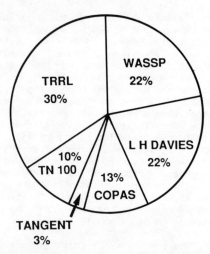

Figure 1. Detention Pond Design Methods.

and routing requirements with the TRRL Hydrograph, Lloyd Davies Rational and CIRIA TN 100 routing methods being most commonly adopted for surface storage schemes; the WASSP simulation method being frequently reserved for the design of on-line, below ground storage tanks. Whilst design return periods vary between 1:1 to 1:50, most surface storage schemes select a 1:10 year design return period which is usually considerably lower than the level of service or system performance the method will provide. The mixing of design approaches quite often produces oversized storage facilities that cannot be fully utilised as the sewer system, designed on a much higher return frequency, cannot convey the peak flows to the storage basin.

ENVIRONMENTAL QUALITY BENEFITS OF DETENTION PONDS

The dual or multifunctional use of urban detention ponds for flood control and mitigation of pollution runoff loads has received considerable attention in the United States in contrast to the UK where

# DETENTION POND DESIGN

storage ponds have traditionally been designed exclusively on the basis of flow criteria. The adoption of strategic and site storage provision is normally undertaken without reference to objectively made evaluations of the added value that may accrue from any water quality or ecological benefits.

The secondary amenity and recreational benefits offered by such basins have been recognised and often formally developed as exemplified in the case of many of the New Towns such as Telford, Bracknell and Milton Keynes where in the latter for example, some 30-40% of the costs can be attributed to providing amenity value and general environmental enhancement. The 16 storage ponds located in the Bracknell district have been formally adopted and developed by the New Town authority as Public Open Space to provide amenity, recreational as well as ecological attractions and are managed by a staff of six Countryside Rangers. The nature reserve, established by the Royal Society for the Protection of Birds, in the Sandwell Valley flood storage pond complex in the W Midlands also provides a good example of unique environmental opportunities for amenity and nature conservation in the heart of an urban connurbation. The complex attracts over 30,000 visitors a year to the landscaped pools with their islands and bays and some 170 species of birds have been recorded including jacksnipe and marsh harriers. A number of commercial developments such as the Aztec West Estate near Bristol have adopted extensive landscape planting in association with storage ponds to provide aesthetically attractive surroundings for companies located on the industrial park. The 25ha Lea Marston Lake on the River Tame in the Birmingham connurbation is the only example of a storage basin in the UK which has been designed exclusively to provide on-line purification control of urban and industrial effluent discharges. This Purification Lake affords a 2.8hr retention for the mean annual flood flow (9504Ml/d) and a 10.5hr retention period for the average daily flow (1320Ml/d) and provides overall reductions of some 40-47% for SS, 22-25% for BOD and 30-35% for toxic metals although persistent increases in ammoniacal nitrogen of 7-12% have been noted (Woods et al., 1984). Similar removal efficiencies have been noted for a 25,000 $m^3$ on-line wetpond located within a 60ha contributing residential catchment in NW London (Ellis, 1985).

Irrespective of these apparent advantages some Water Authorities regard the management problems, maintenance costs, as well as the safety and possible public health risks associated with surface storage ponds, as outweighing their perceived cost benefits and refuse to adopt them. This hostility is in large part influenced by the limited availability of clear and proven organisational and management structures and guidelines for their adoption, operation and maintenance. The negative attitude is further buttressed by a basic lack of information on the quality performance and efficiency of flood storage ponds as well as a deficiency in environmental/ecological evaluations that might be beneficially incorporated into design and performance criteria.

No simple methodologies exist to enable the engineer to estimate the expected operational quality effectiveness of a detention pond for any selected hydraulic design. There is little available case work or good practise guidance readily accessible to enable developers for example to evaluate the likely impact that toxic and bacterial contamination may have on the safe use of stormwater control wetlands for recreation and wildlife. Which features of stormwater storage facilities offer most value to wildlife and which types and forms of wetland habitat are appropriate for and compatible with different stormwater problems and site configurations? What habitats and species are of most value in surviving and tackling particular types of water quality problems? The ability to incorporate environmental and ecological criteria, as well as water quality considerations, as essential and recognised elements in detention pond design procedure is of vital importance in achieving integrated, best management practise for urban runoff control.

ENVIRONMENTAL DESIGN CONSIDERATIONS

Basin Mixing and Pollutant Routing

Poor mixing characteristics are undoubtedly primarily responsible for poor basin performance and for low sediment, pollutant and bacterial removal efficiencies. Pollutant inflows tend to take the form of zones of high concentration which are routed down the length of the pond following cut channels in the bottom sediment. Determination of hydraulic retention ($t_R$, based on the ratio of storage to flow volume (A/Q) or surface loading rates, can lead to gross underestimates; polluted outflow typically occurs within 4 to 6 hours of entering the pond (Fig. 2).

Bacterial Behaviour

Above rainfall depths of 3-5mm and intensities of 100mm/hr, the bacterial activation and supply is normally sufficient to maintain faecal coliform densities at levels above the EEC Bathing Water guidelines of 100MPN/100ml for the entire duration of the inflow event (Fig. 3).

Rainfall depths of above 10mm and 35mm are liable to violate the mandatory level of 2000 MPN/100ml for some 20% and 50% of the time respectively. The outlet bacterial quality does show considerable improvement however, in that the guideline limits are only likely to be exceeded some 40% of the time for rainfall depths below 10mm and are not exceeded at all for depths below 5/6mm with the mandatory limits probably being only rarely exceeded. On the basis of the US Federal Water Quality recommendations for recreational water use, the guideline criteria of 200MPN/100ml suggest that rainfall depths of 17mm would exceed this limit about half the time. A recent UK Institute of Environmental Health Officers review (IEHO, 1987) has drawn attention to the increasing proportion of inland recreational waters which are failing either the guidelines or mandatory EEC standards with over 80% failing the former level in the Midlands and 59% violating the latter criteria in the Greater London area. A considerable number of these waters include urban detention ponds which have been formally adopted for amenity and recreational purposes. The exceedance plots illustrated in Figure 3, which if taken with die-off rates, can provide

# DETENTION POND DESIGN

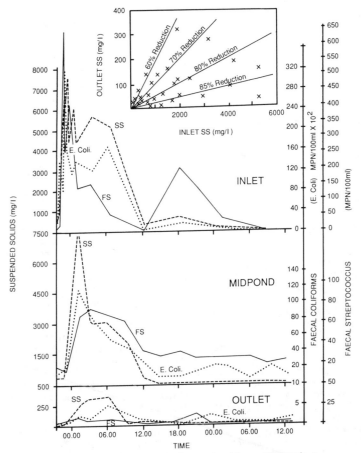

Figure 2. Pollutant Concentrations through a Flood Storage Pond.

useful general guidance for storage pond quality and recreational management purposes.

The observed faecal coliform $T_{90}$ values are of the order of 24-30 hours and as can just be seen from inspection of Figure 2, there is an apparent bacterial growth phase after some 10-15 hours. The initial bacterial levels fall in line with SS probably due to coliform attachment to and settlement with suspended particulate as well as to dilution and substrate depletion (Ellis, 1985). After the solids have settled in the critical 10-15 hour period, bacterial growths are no longer masked and a rise to maximum densities is commonly observed. Thus an optimum detention time of 24-30 hours will provide maximum protection of downstream water quality with a minimum design

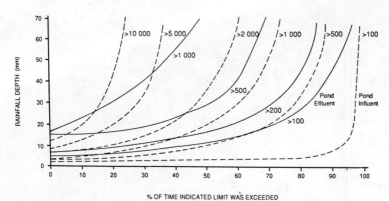

Fig.3 Exceedance of Faecal Bacterial Limits (MPN/100 ml)

recommendation of 12-16 hours. A twin level outlet configuration can provide an effective means of achieving the optimum retention time with a slow-acting minor outlet just above permanent pool level throttling discharge of the retained stormwater. Whilst this design concept has been generally advocated in the United States, it has not reached a wide or receptive audience in the UK.

Settlement Characteristics and Design

Efficient mixing is therefore prevented through a combination of flow turbulence (which retards settlement, as expressed by Vs the particle settling velocity, and also induces resuspension) and short-circuiting which maintains a plug flow routing. Thus in practice, the settlement characteristics of flood storage basins are somewhat removed from the classic settling theory developed by Hazen, (1904), who combined the physical effects of both turbulence and short-circuiting into a single trap efficiency parameter. Camp's three parameter solution for settling basin efficiency (Fig. 4) is widely used as a basis for estimating trap efficiency. The alternative absissa $Vs/v_*$ ratio given in Figure 4 (where $v_*$ is the shear velocity) can be regarded as a dimensionless indicator of the effect of the fluid turbulence on a given particle size. Thus for fixed flow depths and surface loading, flow-through must be reduced if the same scale of turbulence effects on settling is to be maintained when design particle size is reduced. The approach suggested by Vetters (1940) takes into account more accurately the effects of turbulence on settling and is virtually identical to the equation proposed in the 1971 US Bureau of Reclamation Sediment Investigation Technical Guide :

$$\text{trap efficiency (E)} = 1 - e^{-(VsA/Q)} \quad \quad \quad (1)$$

This is simply the 'best performance' solution of Hazen's equation (ie the curve of ($n = \infty$) and also corresponds to the turbulent side of

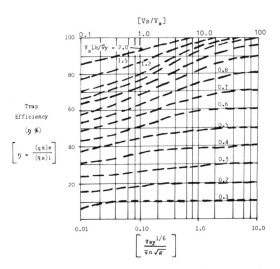

Figure 4. Detention Basin Trap Efficiency

the Camp solution (Fig. 4) and thus to implicit conditions of turbulence. Application of the Vetters solution generally reduces the trap efficiency values derived directly from the Camp method by some 20-30% and is probably theoretically sounder than attempting to apply adjustment or coefficient values to the particle settling velocities in the Camp equation (Hall, Hockin and Ellis, 1988).

Pollution Routing Algorithm

A simple mixing model for pollutant routing has been developed (Hall, Hockin and Ellis, 1988) for urban detention ponds which utilises the flow continuity equation :

$$Q - q = dS/dt \quad \quad (2)$$

where $Q$ is the inflow to and $q$ the outflow from the pond; $S$ is the pond storage and $t$ is time. The continuity equation for a specific pollutant parameter can be expressed as :

$$d(S.C.)/dt = Q.C_i - q.C_o - R \quad \quad (3)$$

where $C$, $C_i$ and $C_o$ are the pollutant concentrations in the pond, inflow and outflow respectively and where

$$R = S.C.V_s/H = A.C.V_s \quad \quad (4)$$

where $V_s$ is the particle or pollutant settling velocity, $H$ is the pond depth and $A$ the surface area. Hence :

$$d(S.C)/dt = S.dC/dt + C.dS/dt = Q.C_i - q.C_o - A.C.V_s \quad \quad (5)$$

# URBAN RUNOFF QUALITY CONTROLS

Substituting for $dS/dt$ from the flow continuity equation (2) and making the simplifying assumption that in a fully mixed pond $Co = C$ then :

$$S(dC/dt) = Q.(Ci - C) - A.C.Vs \quad \ldots\ldots(6)$$

This can be converted to a finite difference form and simplified to :

$$C_{n+1} K1 = Cn.K2 + K3 \quad \ldots\ldots(7)$$

where $Cn$ is the pollution concentration in the outflow at time step n of size $Dt$ and :

$$K1 = (S_{n+1} + Sn) + (Dt/2). (Q_{n+1} + Qn) + AVsDt \quad \ldots(8a)$$

$$K2 = (S_{n+1} + Sn) - (Dt/2). (Q_{n+1} + Qn) - AVsDt \quad \ldots(8b)$$

$$K3 = (Dt/2)(Q_{n+1} + Qn). (Ci_{n+1} + Ci_n) \quad \ldots\ldots(8c)$$

So that : $\quad C_{n+1} = Cn \ (K2/K1) + (K3/K1) \quad \ldots\ldots(9)$

Equation 7 is recursive in form with the concentration at time step n+1 being computed from that at time step n using the coefficients K1n+1, K2n+1, K3n+1 obtained from equations 8. The calculations may be conveniently set out in tabular form and both outflow chemographs and trap efficiencies determined for differing inflow concentration profiles or fixed EMC values and particle size associations as well as for selected design return events. The algorithm as currently developed also provides trap efficiency estimates for metal and hydrocarbon species based on simple partitioning ratios.

## Short-Circuiting and Design

The departure of the typical detention pond flow-through curve from that of ideal basin flow results from the instability caused by short-circuiting with consequent reductions in hydraulic and trap efficiency. The factors that have been identified in UK operational practise which are believed to be responsible for short-circuiting include :

- currents induced by poor inlet and outlet conditions as well as from basin shape

- dispersion in the horizontal plane due to turbulence

- density currents following in-cut channels

- wind induced surface currents

The first of these would appear to be by far the most significant with the others being increasingly important as flow-through momentum is reduced. The effects of wind on large ponds can be substantially reduced if the basin is aligned along the prevailing wind direction.

# DETENTION POND DESIGN

Having determined the surface area of the basin, the designer can improve the hydraulic behaviour (i.e. the shape of the flow-through curve) by having a minimum length to width ratio (L/W) of between 2 to 4. Basin shape can also be improved by subdivision with islands, which if longitudinal in shape can also act as divide walls. The strategic insertion of islands in the pond has been found to direct water flows more efficiently through the pond improving flow patterns and reducing 'dead' zones as well as increasing residence times and helping oxygenation of the pond. Additionally, the island can provide an aesthetic feature as well as acting as a natural refuge for wildlife and plants free from disturbance. If used in this way, the island must be graded to provide good drainage with appropriate cover established to prevent erosion and to provide bird nesting cover. A less aesthetically pleasing but equally efficient method of preventing short-circuiting is the insertion of a baffle wall extending across some one-third of the basin width and located at the upstream end of the detention facility.

## Inlet Design and Siltation

### Inlet Zone

A good flow distribution in the horizontal plane at the inlet zone will improve hydraulic efficiency and achieve an effective use of the settling zone. Therefore the need for a good inlet design cannot be overemphasised, for poor inlet design is probably one of the most important factors responsible for 'poor' basin performance in terms of the Hazen classification. The inlet design needs to distribute the influent uniformly over the cross-sectional area of the settling zone. Good flow distributions can be achieved through submerged weirs, bifurcating or gradually opening inlet expansion or by the use of inlet baffle walls which direct the inflow tangentially into the pond. Stepped inlets (and outlets) can also provide aeration to sustain minimum DO levels in both the pond and effluent waters. This can be extremely important where storm inflow occurs on the downward limb of the diurnal DO cycle causing prolonged and severe depression which is exacerbated by temporary high benthic oxygen demands due to turbulent disturbance.

### Siltation

One of the principal problems reported by many UK local authorities managing urban detention ponds has been that of rapid sedimentation adjacent to inlet zones. Silt traps at pond inlets should be recommended where constructional activity or redevelopment is likely to take place within the catchment or where commercial/industrial activity comprises a proportion of the urban land use. Traps should also be provided in basins where it is recognised that a regular maintenance commitment is unlikely to be adopted.

The inlet channel or trap design should give virtually constant flow velocity not exceeding 0.3m/sec at all discharges to ensure deposition. Oil interceptors, perhaps in combination with silt settlement facilities, should also be installed where there is a need to protect sensitive recreational, ecological or other amenity uses of the pond. By-pass interceptors capable of separating oil from flows of

1l/sec/ha of paved area have proved both efficient and economically acceptable in the UK and design recommendations appropriate for surface water drainage systems are available (HRS, 1982). The maintenance and removal costs of oil interceptors can be high and thus the insertion of these facilities should only be undertaken where there is a real likelihood of oil pollution or where any such occurrence will be ecologically unacceptable.

Some accumulation of sediments around pool inlets which form shoals, bars or spits can be of considerable value to wildlife particularly for wading birds and breeding waterfowl (Adams et al., 1985) and the ecological/amenity value of such sedimentation should not be ignored. The possible accumulation of potentially toxic sediments within a detention pond does however require a formal operational commitment to inspection and maintenance schedules and 'hardened' access should be provided for plant and transport. Sediment removal should be timed as to cause minimum disturbance for wildlife and ideally should be undertaken in stages over a few years to allow recolonisation from the undisturbed areas, and plants removed during maintenance should be transplanted whenever possible. Spoil should not be deposited close to the water's edge in order to prevent nutrients, heavy metals, hydrocarbons etc., leaching back into the water.

An agreed maintenance/inspection protocol should be drawn up with report forms completed and signed by personnel responsible for carrying out the work and this is becoming a more widespread practice in the UK. Information in respect of ponds should be centrally recorded on a database so that their long term performance can be checked and operating costs ascertained.

## OTHER ECOLOGICAL & AMENITY DESIGN CRITERIA

Whilst the primary impetus for constructing detention ponds in the UK is likely to remain the need to provide structural control for urban runoff flow with possible secondary quality considerations, it is becoming increasingly obvious that they can provide other tangible environmental cost benefits. The creation of new water-based resources in urban areas can also provide valuable opportunities for the development of wetland habitats which may help to offset some of the loss and despoilation of existing river corridor wetlands due to drainage, infilling or pollution. Urban detention ponds can be designed to maximise their value for both wildlife and other amenity uses providing that both the habitat requirements and the pollutional constraints are taken into account at an early stage of the design procedure and not added as an after thought. In the UK context, the water quality criteria and considerations are often being brought in via this 'back-door' pressure to enhance amenity or ecological needs.

### Basin Form and Geometry

The importance of island refuges and sediment shoals have already been mentioned, but the use of promontories and shallow embayments in the pond design will also provide shelter, seclusion and feeding grounds for wildlife as well as enabling spatial zonation for differing recreational uses. Access to conservation areas should be controlled through 'walkway' provision and by scrub areas landscaped into the design to deter the casual visitor as well as encourage those who wish access for quiet recreational and educational purposes.

A 2m wide level area should be provided about 0.5m above and abutting the edge of the permanent water level with side slopes between 8:1 and 12:1 to facilitate colonisation by marginal plants as well as providing safety. A side slope of between 10 to 15:1 is even more beneficial for wildlife and will also allow an additional margin of safety for children who may be attracted to the water's edge. The use of wire netting to contain shoreline rip-rap should be avoided as corroding ends and projections have been found to cause damage to wildlife and very young children.

Water depths should not exceed 0.5m for some 25-40% of the water surface area and an emergent vegetation/open water ratio of 50:50 should be maintained. This has been found to encourage a range of waterfowls and wetland birds such as mallards, herons, snipes, sandpipers, etc. At least 50-70% of the pond should have a depth of not less than 1.0-1.5m to encourage oxygenation as well as for roosting and feeding purposes. The deeper, permanent pools will also be more attractive to amphibians, reptiles and mammals whilst fish pools need to have a minimum depth of 2.5m. The fish species which have been found in UK experience to be least sensitive to urban stormwater pollution include tench, bream, carp, eel and perch although carrying capacities and restocking rates have presented continual and difficult management problems.

Planting and Landscape Design

Ecological design engineering is required to ensure that the type and rate of colonisation is appropriate in terms of any nature conservation or amenity objectives and that it reaches an early maturity. It is also widely recognised that reed ponds are effective in reducing solids, nutrients, bacteria and toxins as well as promoting DO recovery and in this respect the encouragement of a planting programme in the settling and marginal zones is desirable.

A framework of dominant marginal, floating - leafed and submerged macrophytic species, with emergent species located adjacent to inlets, will provide a basic management strategy. Of the latter species, Phragmites australis has been found to be amongst the most resistant, being able to withstand multiple oiling and heavy oil penetration as well as surviving polysaprobic conditions. Other wetland species such as reedmace, fennel pondweed and water forget-me-not also seem to grow well in substrates containing up to 50% weathered oil and are also well suited to fluctuating water levels.

Shoreline tree and shrub planting needs to be done with care as marginal and floating-leaf species can be shaded out and decaying organic mats will accumulate from fallen leaves, contributing to both nutrients and benthal oxygen demand. Nevertheless, such planting can provide useful shelter for waterfowl as well as protecting bird nesting locations. The open surrounds above the shoreline require landscaped grasslands which are of maximum erosion resistance and of minimum cutting/mowing requirements. A ryegrass-free mixture of slow growing cultivars (e.g. fescue, bent, meadow grass) can provide these requirements if the surface is kept sufficiently high above the water table. Surface grading and underdrainage, perhaps using a filter fabric covered drainage mat such as Enkazon, will render areas suitable for recreational use within an acceptable timescale following flooding.

## Maintenance

Macrophytic vegetation and associated filamentous algae may grow rapidly and present management and water quality problems given their propensity to remove nutrients and toxins. Regular weed harvesting is therefore essential by either mechanical or chemical means although the latter methods need very careful application and are not favoured in the UK. Experience would indicate that as much as 30-50kg/ha/yr of phosphorus and 400-500kg/ha/yr of nitrogen can be removed on a continuing basis by harvesting reed ponds.

## A DESIGN PROCEDURE FOR ENVIRONMENTAL QUALITY

(1) review estimated/expected pollutant inflow rates (including particle size distribution, concentrations and variability)

(2) decide on minimum particle size for selected removal rate(s). Estimate basin surface loading using Vetter's equation and calculate basin surface area. Subdivide influent into particle size bands and using routing algorithm estimate removal efficiency for each band for varying storm return periods and outlet control structures

(3) review local topographic, environmental, land use factors and receiving water quality objectives and make broad decisions on pond size, outflow rates and need for oil/silt traps

(4) review constraints on basin depth and flow-through velocity for preliminary pond layout from considerations of pond turbulence, practicality of construction and operation as well as likely ecological constraints

(5) review using Camp or Vetter's solution (with a conservative estimate of turbulence function) the initial estimate of basin area. Optimise, in terms of both primary and secondary uses, basin length, width and depth details and maximum flow-through velocity

(6) confirm final design deposition rates and sediment size distributions and outflow concentrations; confirm method/frequency of likely sediment removal

(7) review and decide on form detail of shoreline/islands and ecological planting profiles as well as general landscaping

(8) finalise hydraulic design detail of inlet and outlet zones

(9) draft management protocol for inspection and maintenance procedures together with reporting format.

## REFERENCES

Adams, L.W., Dove, L.E. and Franklin, T.M., "Mallard pair and brood use of urban stormwater control impoundments," <u>Wildlife Soc. Bull.</u>, 13, 46-51, 1985.

Ellis, J.B., "Structural control for urban stormwater quality, "In <u>Hydraulics of Floods and Flood Control</u>, Brit. Hydrodynamics Res. Assoc., Cranfield, Bedford, UK, 235-242, 1985.

Ellis, J.B., "Water and sediment microbiology of urban rivers and their public health implications, "<u>Public Hlth. Eng.</u>, 13, 95-107, 1985.

Hall, M.J., Hockin, D and Ellis, J.B., "<u>The Design of Flood Storage Ponds</u>, Butterworths, London, 1988.

Hartsmere Borough Council, "<u>Attenuation storage and flow control for urban catchments</u>, Interim Report, Thames Water Authority, Reading, UK, 1987.

Hazen, A., "On Sedimentation," <u>Trans. Amer. Soc. Civil Eng.</u>, 53, 53-61, 1904.

Hydraulics Research Ltd., "<u>Tests on compartment oil and petrol interceptors for surface water drainage systems</u>," Tech. Rpt., Wallingford, UK, 1982.

Vetters, C.P., "Technical aspects of the silt problem on the Colorodo River," <u>Civil Eng., 10, 698-701, 1940.</u>

Woods, D.R., Green, M.B. and Parish R.C., "Lea Marston Purification Lake: Operational and River Quality Aspects," <u>J Water Poll. Control</u>, 14, 226-242, 1984.

## Discussion of Dr. Ellis' Paper

Question:

Are there ponds designed with criteria you have derived for protection of the downstream water or for protection of the pond itself?

Answer:

A bit of both, really. Uses of the pond by people seeking secondary benefits such as habitat values, aesthetics, or recreational purposes mean that upstream controls--above the pond--can be required.

Question:

What is your data base with respect to the derivation of your ponds' design criteria? Is that base comparable to the U.S. NURP data base?

Answer:

Not at all. One wet pond's experience in 16 events is about all there is. There are no data at all on dry ponds.

# HABITAT AND WATER QUALITY CONSIDERATIONS IN RECEIVING WATERS

Lewis L. Osborne[1]
Edwin E. Herricks[2]

## Abstract

The effects of urbanization on stream water quality and physical habitat conditions have been well documented. Assessment of urban runoff impacts and subsequent design of runoff control technology have traditionally been criteria based. Although easily implemented and monitored, a criteria based management overlooks, or fails to incorporate temporal and spatial issues important to instream physical habitat and water quality condition. Temporal issues of importance include timing, magnitude, and return frequency of flow conditions. Temporal issues of importance include timing, magnitude, and return frequency of flow conditions as well as duration of exposure (concentration) and return frequency of specific exposure levels. Urbanization alters flow regime and produces unit-hydrograph changes which may effect stream biota by changing physical habitat conditions. Water quality is also altered by urban runoff, stream conditions may be dependent on both control technologies used and landuse activities present in the watershed. Spatial issues are important because biological conditions will vary with location in the watershed. Control technology, management approaches, and applicable criteria should reflect watershed location. This paper will view the streams receiving urban runoff in the context of the total watershed. Streams should not be viewed as conduits for the transfer of runoff. Suggested mitigation and control procedures incorporate both temporal and spatial dynamic aspects of the watershed.

## Introduction

When ecologists initiate a study of an area, the first step is establishment of boundaries or limits to the scope of investigation. One of the more convenient and arbitrary boundaries set by stream ecologists is the watershed. Studied as a unit, the watershed provides the

---
[1] Aquatic Biology Section, Illinois State Natural History Survey, 607 E. Peabody Drive, Champaign, IL 61820

[2] Department of Civil Engineering, University of Illinois, 205 N. Mathews St., Urbana, IL 61801

basis for defining both spatial and temporal limits. Watershed area is a convenient unit of measurement. The drainage net, a network of streams in a watershed, provides an interconnected system relating various locations in the watershed. Streams and rivers function as the primary conduit for the transfer of materials in the watershed. The stream can be viewed as an integrator of watershed activities upstream from the point of analysis.

In undisturbed watersheds a natural balance is developed between rainfall, runoff, and the materials which are removed from the watershed through stream channels. This balance, established over a long time period, allows the ecosystem, which includes organisms and the conditions providing their food and habitat, to reach a state of relative stability. Natural destabilizing events do occur in watersheds. Fire may destroy trees and floods may alter channel configuration and conditions, but the watershed responds by regrowth and channel adjustment to changed conditions resulting in a new stable state.

In addition to natural changes, human activities can alter watershed characteristics leading to changes in streams. Urbanization, the concentration of man's activities in small areas, fundamentally alters the watershed. The natural balance of the ecosystem will be changed leading to short-term and long-term instability in the flora and fauna. Urbanization does not always effect fundamental processes such as rainfall; however, derived characteristics such as runoff magnitude and frequency, may change. As we analyze the effects of urbanization and determine the severity of urbanization impacts, receiving waters come into focus as one of the more sensitive indicators of impact. Using stream ecosystem structure and function it is possible to assess urbanization effects and predict changes in stream dynamics. Streams serve to integrate watershed effects and allow spatial segregation of impacts.

## Effects of Urbanization on Receiving Waters

Urban runoff, by definition, occurs in areas changed from a natural state or condition. Land development physically alters the watershed. The magnitude of these changes varies with the percentage of a watershed urbanized. Typical effects of urbanization are modification of stream channels, increased erosion and sedimentation in the stream, modified hydrographs, and altered bankside or riparian vegetation which change stream temperature regime. Urbanization of a watershed also changes the release and delivery of chemicals naturally produced in the watershed, resulting in an increase in macro and micro nutrients for some time after disturbance (Vitousek, 1977). After development the stream usually receives a mix of chemical contaminants produced by man's activities. These contaminants will affect stream organisms and alter the character of the ecosystem. The result is a stream system which is fundamentally altered from its natural state. This fundamental alteration need not be considered an absolute loss. In fact, urban streams can and do provide important recreational opportunities and benefits and may be managed to sustain a healthy functioning ecosystem.

The modification of stream channels during urbanization is a consequence of this land use. The most severe modifications involve complete containment. In many cities, natural drainage ways have evolved into closed conduits to pass stormwater and sewage. Open stream channels in urban areas are often concrete lined to keep the channel from changing position, minimize roughness and drag to expedite storm flow passage, and maintained in a clear condition to assure free passage of water. Under less extreme conditions, the stream course may be modified to accommodate development or the stream may be channelized to provide for improved passage of water. Where natural channels remain in urbanized areas there are usually several reaches where the stream channel is constrained or efforts made to reduce erosion or channel location change by using revetments or rip-rap.

The direct modification of stream channels in urbanized areas is accompanied by channel alterations due to indirect effects of urbanization. For example, urbanization may increase erosion resulting in very high sediment loads to the stream. The effects of sedimentation on streams have been well documented (Gilbert, 1917; Meade, 1982; Kelsey, et al., 1987). Sediment begins a slow movement downward through the drainage net producing changes in stream morphology and channel characteristics. The short term urbanization process can create long term changes to the stream channel far downstream from the urbanized area.

A second set of indirect effects are changes in channel morphology produced by altered fluvial dynamics in urbanized areas. Urbanization has been shown to increase flood peaks and lengthened the flood season by increasing the danger from fall and spring floods (James, 1965). Wilson (1968) has observed that the mean annual flood in urbanized areas is 4.5 times similar rural basins and the 50-year flood is as much as three times that of a rural basin. Carter and Thomas (1968) have shown that basin morphometry will affect flooding from urbanization. For example, the mean annual flood in a steep basin may be eight times as large as that for a flat basin. The recognized result is that the main channel forming process is flood flow. If floods are larger, changes in the channel may be expected to be greater from one flood to the next. If flooding is more frequent, channel changes may be expected to be more frequent as well.

Urbanization will also affect the water quality of receiving waters. Heaney and Huber (1984) provide an assessment of urban runoff impact on receiving water quality. They observed that urban runoff dampen the diurnal fluctuation of dissolved oxygen and can reduce overall dissolved oxygen levels. They also observed that the expected water quality impacts from urban runoff are subtle and are often masked by the effects of sewage effluents or industrial waste. Recent attention has been directed to hydrocarbons in urban runoff. Fam, et al. (1987) found a relationship between commercial land use and aliphatic hydrocarbons on receiving waters. They also noted that n-alkanes > $C_{25}$ were common in undeveloped areas while motor oil and diesel fuel were the major source of hydrocarbons in urbanized areas. In another study Hoffman, et al. (1982) found petroleum hydrocarbons were likely produced by runoff from automotive crankcase oil. They observed elevated hydrocarbons in the first flush with indications that hydrocarbons would decrease in storms of high magnitude. Hydrocarbon was typically associated with particulate material in storm runoff.

It has been difficult to characterize a typical urban runoff. Common constituents include heavy metals, pesticides, putrescible organics, petroleum derivatives, ammonia, phosphorus, suspended solids, rubber, and a range of other contaminants which may be specific to the watershed and land-use. Bradford (1977) concluded that using the data then available, a relationship between land use and pollutants could not be shown.

Although it has been difficult to chemically characterize urban runoff, numerous studies have identified biological effects from urban runoff. Pratt and Coler (1976), Pratt, et al. (1981) and DiGiano (1975) studied the biological response of stream fish and macroinvertebrates to urban runoff. They observed a decrease in species diversity and generally observed that the composition of the fauna reflected the degree of pollutant loading. Of some importance is the statement by Medeiros, et al. (1983) that the variation in the composition, concentration, amplitude, rate of change and mass loading of toxicants in urban runoff prevents lotic communities from acclimating to urban runoff. The expected effect would be a community in constant flux and a ecosystem lacking essential stability. Medeiros et al. (1983) also observed that runoff toxicity is associated with the riverbed rather than the overlying water. They suggested a mechanism where urban runoff delivers toxicants to the stream with the effects delayed until dry weather flow.

In a recent study of urban runoff effects on benthic invertebrates Pedersen and Perkins (1986) observed that the urban stream was dominated by taxa which could tolerate the erosional/depositional character of the urban stream and utilize transient, low quality food sources. Of interest, as well, in Pedersen and Perkins findings was the conclusion that there was no long term loss of colonization potential in an urban stream. Other evidence (Gregory and Lamberti, unpublished data) indicates that a shorter retention time for organic matter is typical of channelized or urbanized stream reaches. This suggests stream communities may be affected by reduced food processing times in urban streams.

When viewed from an ecological perspective the results from biological studies suggest that urban runoff produces both habitat instability and chemical toxicity. Habitat instability can be expected to alter the composition of aquatic communities by favoring those species capable of withstanding continuous changes in habitat. Aquatic communities can also be expected to change composition and abundance reflecting the influence of urban runoff characteristics, not natural variability in environmental conditions. Chemical toxicity can be expected to differentially affect less tolerant species. The community present may be specific for the characteristic contaminants in urban runoff. The combined effect of habitat instability and chemical toxicity will be the absence of valued species.

In summary, urban runoff will produce short-term and long-term changes in receiving waters. When physical conditions are altered in a stream there will be corresponding alteration of habitat for stream biota. Urbanization will also alter water quality in receiving streams. The combination of changed physical habitat and altered water quality must be recognized as the major environmental consequences of urban runoff. We will now focus on specific physical habitat or water quality issues associated with urban runoff, explore techniques for assessing and predicting urbanization impacts on stream, and address strategies integrate receiving stream effects in the management of urban runoff.

General Considerations

Streams are capable of assimilating periodic, short term inputs of contaminants or pollutants without damage to stream flora and fauna. If urban stormwater runoff were the only major source of contaminants acting upon a stream the periodic or short term effects may be mitigated by stream assimilative capacity. Unfortunately, streams, particularly urban streams, are stressed from persistent as well as periodic sources of pollution within the watershed. These include domestic sewage and industrial wastes generated in the urban area and non-point pollutants generated throughout the watershed. If successful management and effective engineering control practices are to be implemented for urban runoff, it is imperative that instream impacts be addressed from the perspective of the stream reach, while considered within the framework of the watershed. All factors causing impact in the watershed must be addressed when considering reach-specific management or control. We advocate the use of a watershed perspective to assess state or condition and ground predictions of the instream effects of urban runoff. Our approach is discussed below.

Assessing impacts from pollutants in a watershed requires development of spatial perspective in the analysis process. A useful focus which allows integration of spatial issues is habitat analysis. We can view habitat in the context of an individual species, a guild, or a community developing location specific requirements for management or control. For example, species specific habitat assessment may require microhabitat analysis (Osborne, et al. 1987) while habitat assessments for guilds of organisms or stream communities require macrohabitat assessment, usually on a stream reach (Herricks, 1985; Herricks and Braga, 1987). Although the focus on habitat analysis typically emphasizes physical and chemical factors, biological factors also contribute to identification of habitat specifications.

In the absence of pre-impact habitat conditions, a watershed perspective is useful for assessing the general conditions of biotic communities at a given location (reach) within a river network. A family of empirical relationships can be developed which relate expected species richness (diversity), abundance, and biotic indices (Karr, 1981, Armitage, et al., 1983, Tamm, 1988) to location in a watershed. Such relationships between geomorphology and biology are helpful in developing stream classification procedures (Osborne et al., 1987) and can be used in impact assessments. For instance, Figure 1 shows the empirical relationships between stream order, watershed area, and link number with fish abundance in the Sangamon River Basin, Illinois. As can be seen from these data, different fish abundance patterns can be expected depending upon the reach location within a drainage network. Herricks and Himelick (1981) have developed similar relationships for seven watershed area categories for all of the river basins in Illinois. Osborne et. al. (1987) also demonstrated that drainage area provides better statistical relationships for predictive and management purposes than did stream order or link number. Keeping in mind confidence limits and the general limitations associated with any regression application, such geomor-

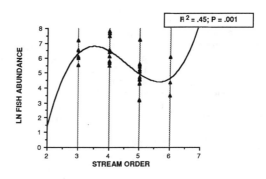
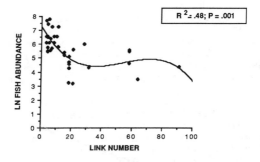
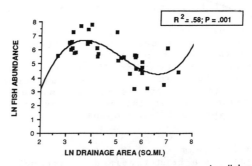

Figure 1. The relationship between fish abundance and stream order, link number, and drainage area in the Sangamon River, Illinois (data obtained from the Illinois Stream Information System, IDOC).

phological/biological relationships can provide an indication of impacts on biological systems. This is particularly useful when comparing streams within the stream geographic region or streams in one drainage net. Knowledge of species specific habitat requirements can be used to specify instream habitat requirements or compare expected and actual fauna in other stream reaches of similar size and drainage location within a given region. This provides a powerful tool in the management of urban streams. Existing fauna can be assessed based on fauna expected for a location with similar drainage area. This is a basic assumption in the formulation of biotic indices, including the Index of Biotic Integrity (IBI) (Karr 1981). Habitat and water quality specifications can be generated to support management of urban streams.

Using spatial location within a watershed to formulate criteria can be useful. Location can help determine the extent of expected impact and facilitate the choice of models, criteria, or metrics for specific stream or watershed characteristics. For example, a lake or reservoir receiving stream flow may influence management strategy. We know that the addition of nutrients accelerates the eutrophication process. Since nutrient levels can be high in urban stormwater runoff (e.g., Rimer et al., 1978) streams which eventually flow into a lake must be managed differently than streams that don't. Regardless of eventual flow destination, impact analysis can not ignore the effects of urban runoff on the ecosystem of the receiving stream.

The practical use of concentration/loading management deserves some review. The determination of loading begins with an analysis of concentration. Mass discharges are calculated from analyzed concentrations and measured discharge flow. Stream loading is estimated based on dilution capacity. The watershed location may affect whether concentration or loading values will be used for management. A small load may produce an unacceptable concentration in a small stream where dilution capacity is limited. Thus in upper reaches of a watershed management should focus on concentration. Because even a large load may produce an instream concentration with limited biological or ecological consequence in a receiving stream with greater dilution capacity, the management focus in the lower watershed can emphasize contaminant loading rather than concentration limitation. Managing loading is particularly important when downstream lakes or reservoirs are present.

Schlosser and Karr (1981) suggest that both loading and concentration be determined. Although loading is considered in the NPDES permitting process, some issues surrounding the use of loading in management can be mentioned. Accurate measures of discharge volume and contaminant concentration must be available. There are many opportunities for introduction of error in the analysis of contaminant concentration (Herricks, et al., 1985) which can influence loading calculations. Care should be taken that loading estimates reflect reality rather than calculation expediency. Receiving stream discharge must be available. The highest quality discharge information will include sufficient historical data to calculate flow frequency with some accuracy. This will support the use of minimum expected recurrence flows for accurate dilution modeling. When historical discharge information is not available, management may rely on conservative estimates of dilution flows or develop mechanisms of controlling discharge based on observed stream flow/dilution data.

We suggest that an emphasis on concentration or loading in management should vary depending on the water resource problem being addressed. The classical approach in environmental planning has been to apply deductive models (e.g. Streeter-Phelps) and view land use effect on water quality from the perspective of loading. Using this approach a loading limit need not be based on the health or "integrity" of a stream and its associated communities because increased dilution during high discharge periods would allow high loading. If the assessment objective is to determine impacts on the stream, rather than downstream reservoirs, concentration measures may provide a more meaningful indication of impact than loading. This contrasts with some opinions (e.g., Huber,1986) that it is seldom necessary to consider more than just total pollutant loads from storms for receiving water analysis. These apparent conflict in views may simply reflect differences between an engineering design perspective and a biological perspective concerned with maintenance of instream integrity (see Herricks, 1987).

Within any stream reach, concentrations of nutrients and toxicants are largely dependent upon upstream landuse/cover activities. An analysis by Rimer et. al. (1978) indicated that the level of nonpoint source pollution within a watershed generally increases with increasing impervious area. Using multiple regression models consisting of the proportion of urban, forested, agricultural, and total drainage basin areas within specified distances from the stream channel (i.e., buffer strips), Osborne and Wiley (1988) have shown that instream nutrient concentrations generally increase with increasing urban areas in low gradient streams. These authors suggested the use of multiple regression models for assessing impacts on stream nutrient concentrations associated with urbanization. While such models are capable of predicting average seasonal concentrations of nutrients within a stream reach, they are not appropriate for addressing impacts associated with single storm events in their present form. Other "continuous simulation" modeling efforts (see Huber, 1986 for discussion) may therefore be more appropriate for assessing eventual stormwater inputs from individual storm events.

The principal contaminants and physical alterations associated with urban stormwater runoff were discussed earlier. Many investigators (e.g., Polls et al., 1980) have questioned the magnitude of long-term impacts associated with the short-term inputs of urban stormwater runoff. In fact, the spatial extent and degree of impact attributable to stormwater runoff within any stream reach will be by both the characteristics of the urban runoff and the quantity of exogenous materials introduced from points upstream. Further, impacts associated with some stormwater runoff contaminants will be largely related to prevailing instream biological processes. For instance, short-term exposure to high concentrations of nutrients may not, in and of themselves, pose a toxic risk to stream fauna. Short-term nutrient introduction can, however, indirectly impact downstream communities, by increasing stream productivity and therefore night-time oxygen demand. Similarly, short-term storm events can alter instream physical habitat which will exist until altered by another storm of equal or greater magnitude. In large areas of the United States, and particularly in the midwest, oxygen depletion and habitat degradation are major factors which impair stream ecosystem quality. For these reasons we address methods of dissolved oxygen and physical impact below.

## Dissolved Oxygen Impacts

Dissolved oxygen is one of the most important water quality parameters utilized to evaluate stream condition. Rimer et al. (1978) reported that storm related events depress DO in some North Carolina receiving streams by an average slightly greater than 1.0 mg/l and suggested that only moderate, if any impacts would result. Their study was conducted in a watershed where autochthonous production was apparently low or stream gradient was relatively high. Low-gradient, warmwater streams, typical of much of the midwest, are likely to be much more susceptible to nutrient and BOD loading from urbanization. They generally have little shading, warm temperatures, and high background nutrient levels due to agricultural landuse in the watershed. This combination of factors requires a basic understanding of instream primary production and respiration demands if instream impacts are to be predicted. Prediction of dissolved oxygen impacts can be made using diel oxygen sag models which can be improved by incorporating realistic measures of instream oxygen demands.

# RECEIVING WATERS CONSIDERATIONS

In downstream oxygen sag analyses, variations in time are often ignored (i.e., $\partial O/\partial t = 0$) and the following equation used to describe the change in oxygen content of a unit volume as it flows downstream:

$$d_O/d_x = 1/v \; (-[K_c L_c + K_n L_n] + K_a D + P_t - S_b) \quad (1)$$

where: $(d_O/d_x)$ is the change in oxygen with distance from the source;

$[K_c L_c + K_n L_n]$ is the change due to waterborne BOD ($L_c$ and $L_n$ are carbonaceous and nitrogenous demand, $K_c$ and $K_n$ are deoxygenation coefficients)

$K_a D$ is diffusion

$P_t$ is photosynthesis which is itself time dependent

$S_b$ is benthic respiration (incorporating both algal and microbial respiration demand)

This assessment of a potential DO-sag has been the foundation of engineering analysis since the 1920's. This analysis allows modelling the effects of BOD inputs without requiring computationally intensive numerical methods for a solution.

If time-dependent dynamics of oxygen at a particular location (x) in the channel are of interest, $d_O/d_x$ can be set to zero and $d_O/d_t$ solved using the equation:

$$d_O/d_t = -[K_c L_c + K_n L_n] + K_a D + P_t - S_b \quad (2)$$

Wiley et al. (1987b) have used this type of oxygen model to predict the effects of changes in stream discharge on instream oxygen dynamics.

The most common approach to predicting the effect of discharge of effluents with high BOD uses a modified Streeter-Phelps analysis of the oxygen profile of the receiving stream under some specific discharge (e.g., Q7,10) condition (Wiley and Osborne, 1987). Unfortunately, in many areas the modified equation used is derived from equation 1 by eliminating the terms for photosynthesis and community respiration (i.e., $P_t$ and $S_b$) and setting $L_c = L_{ac} \exp(-K_c t)$ and $L_n = L_{an} \exp(-K_n t - t_d)$ (incorporating first-order reaction kinetics for the BOD terms). The basic differential equation is then converted into terms of oxygen deficit (D = saturation concentration – oxygen concentration) and solved analytically. In essence, photosynthesis and respiration demands other than that associated with the effluent are assumed to be zero.

Intensive studies in the Vermilion River watershed of east-central Illinois (Wiley et al., 1987a; Wiley et al., 1987b) demonstrated that during the summer growing season the respiratory demands for waterborne BOD (2-5 mg/l per day at 25 C for BOD's of 5-10 ppm) are typically dwarfed by the influences of photosynthesis and non-BOD respiration of the benthic micro-flora. Daily photosynthetic rates measured during the summers of 1984 and 1985 ranged from 5 to over 120 mg/l per day and over all seasons averaged 14 mg/l per day. Total community respiration ranged from 5 to over 100 mg/l per day and had an average annual value of 24 mg/l per day (Wiley and Osborne, 1987). The effects of such instream demands are illustrated below for a typical site in the Vermilion River with a 200 square mile drainage area (Q = 17 cfs, exceedence frequency = 0.7, $K_a$ = 9.49). A modified Streeter-Phlelps (disregarding photosynthesis and benthic respiration) would predict that given a carbonaceous BOD of 10 ppm, the maximum deficit occurring downstream would be 0.45 ppm. When photosynthesis and respiration terms are given reasonable values for the Vermilion River (P = 38.7 mg/l per day; R = 24.6 mg/l per day; based upon Illinois Natural History Survey (INHS) regression models), the maximum deficit for a BOD of 10 ppm becomes 2.94 ppm, a 653% difference. The bias which results in this underestimation of the deficit is an inescapable consequence of assuming that benthic demand is zero.

Rectifying this omission is not difficult (Thoman 1972). In the case of urban stormwater runoff benthic respiration can be considered a constant, and included in the analytical solution of equation 1. With this further modification the Streeter-Phelps model takes the following form:

$$D = (K_c L_{ac}/K_a - K_c)(\exp(-K_c t) - \exp(-K_a t)) + (K_n L_{an}/K_a - K_n) \\ (\exp(-K_n t - t_0) - \exp(-K_a t - t_0)) + D_a \exp(-K_a t) + \\ (1 - \exp(-K_a t))(S_b/K_a) \qquad (3)$$

where $D_a$ is the initial deficit, $L_{ac}$ is the ultimate carbonaceous demand, $L_{an}$ is the ultimate nitrogenous demand, $S_b$ is benthic demand in mg/l per day, and other terms are as given previously. This equation will accurately estimate the minimum daily concentration in the stream unless the reaeration coefficient is so low that the night-time concentration of oxygen does not approach it equilibrium value.

The major difficulty in implementing equation 3 is obtaining an estimate of $S_b$ (benthic demand) for a reach in question. Two approaches are possible. $S_b$ may be estimated from a diel oxygen curve, based on monitoring oxygen concentrations over 24 hours during low flow and summer conditions. This approach would not only provide a reach-specific estimate of Sb which could be used in equation 3, but also has the added benefit of providing an empirical estimate of the reaeration coefficient and provides data on the prevailing range in oxygen concentrations. Alternately, estimates of Sb could be based on empirical models relating respiration rates to more easily measured parameters like temperature or depth. For example, Wiley et al., (in submission) have used a regression model based on their Vermilion river studies to estimate total respiration and photosynthetic rates from depth, turbidity, riparian shading, and temperature:

$$P = 1.59 \ T^{0.709} \ NTU^{-0.41} \ H^{-0.82} \ S^{-0.229}$$

$$R = 5.03 \ e^{0.109P} \ T^{0.32}$$

where P is photosynthesis (mg/l per day); T is temperature in degrees C; H is mean depth in meters; S is maximum proportion of the channel surface shaded (0-1); and R is total reach respiration (mg/l per day).

The above regression estimators were based on a series of 191 diel oxygen measurements made over a 2 year period in the Vermilion River watershed. Similar models could be constructed for watershed throughout the country. Either of these approaches would provide a realistic basis for predicting minimum oxygen concentrations due to urban runoff. The major advantage of the former method is that it would provide the most accurate assessment of impact due to stormwater loading. The later method is inherently less accurate, but requires no site-specific data collection once a watershed or regional relationship between stream characteristics and watershed (riparian) conditions have been developed. Most importantly, the above discussion points to the fact that instream impacts from urban runoff cannot be accurately assessed by assuming that instream metabolic processes are negligible.

### Physical Habitat Assessment

It has been suggested (Porcella and Sorenson, 1980; Perkins, 1982) that physical changes such as altered fluvial processes associated with urban stormwater runoff may be the fundamental determinant in the functioning of urban stream ecosystems. Stream fauna may be directly effected through specific habitat modifications by the pulses of urban runoff during storms which influence erosional/ depositional characteristics and may favor development of a fauna adapted to bed instability (Pendersen and Perkins, 1986). Richey (1982) has also hypothesized that the same altered flow regime may influence food resource base by increasing through-put of materials, reducing residence and instream-processing time of organic matter. The use of nutrients and the processing of organic matter in streams has been described in terms of a spiral (Newbold, et al., 1981). If urban runoff alters physical habitat a functional indicator would be a change in the distance required for nutrient and energy spiralling. Although inadequately tested as an assessment procedure measurement of spiraling distance may be useful in assessing the effects of physical habitat alteration on ecosystem function. If spiralling length in unaffected stream reaches are used for comparison, it will be possible to assess nutrient and energy dynamics as affected by urban runoff or other contaminants.

In addition to measurement of functional alterations of ecosystems it is possible to assess organism specific effects associated with physical habitat alteration. The most common of these are habitat suitability analyses as incorporated in habitat evaluation procedures (HEP) (USFWS, 1980) and instream flow needs (IFN) analyses. IFN analysis approaches have a direct application to urban runoff analysis. Several instream flow assessment methods have been developed over the past 15-20 years (see review in Loar and Sale, 1981; Morhardt, 1986) which cover a broad spectrum of sophistication and technical requirements. Many of these are designed to assess the impacts of flow modification on fish populations. Three general categories of instream flow methods are commonly recognized: discharge methods; single transect methods; and multiple transect-habitat rating methods. Among the most widely advocated procedures in this latter category is the U.S. Fish and Wildlife Service's Instream Flow Incremental Methodology (IFIM; Bovee, 1982). The major component of the IFIM is the Physical Habitat Simulation (PHABSIM). PHABSIM combines output from one of two hydraulic simulation models (IFG-4 or Water Surface Profile Model) with habitat frequency of use information for selected target species to estimate relative amounts of fish habitat for a given reach of stream under varying discharges (see Milhous et al., 1984). The habitat use information is based on suitability curves for depth, velocity and substrate.

Although based on procedures widely used in habitat assessment, serious questions have recently arisen regarding the adequacy of the IFIM as a management tool (Mathur, et al., 1985; Morhardt, 1986; Scott and Shirvell, 1987; Conder and Annear, 1987; Osborne et al., in press). Further, PHABSIM does not presently incorporate a routine to account for changing substrate conditions (i.e., substrate parameters are assumed to remain constant). As previously discussed substrate variability caused by high sediment loading and instream processing of sediment added by urbanization is important in the analysis of urban runoff. Also, the methodology has not been widely employed to examine periodic, short-term increases in stream discharge. For these reasons, use of the IFIM model is not a preferred method for assessing urban stormwater runoff although methods based on a habitat suitability analysis should be considered. For example, regression-based habitat assessment procedures have been proposed (e.g., Morhardt, 1986; Osborne et al., 1988) for use in habitat evaluation. Although these procedures have not been adequately validated for widespread use they do hold promise as simple empirical methods which can be used in combination with water quality models (e.g. DO-sag, nutrient spiralling models, etc.).

Because high magnitude, low frequency storm events shape physical habitat and pulsed flow events, characteristic of urban stormwater runoff modify some habitat characteristics, what is needed is a method which predicts the effects of pulse changes in the intensity and duration of a flood event on instream habitat conditions. Frequency of use curves for velocity, depth, and substrate conditions have been generated for many species and geographic regions. Future research should be directed to the prediction of substrate, velocity, and depth alterations in a channel following pulse runoff events.

## Summary

Assessing the effects of urban runoff on stream habitat and water quality requires separation of the effects of natural and anthropogenic factors affecting the receivng system. For this reason we advocate a watershed perspective for both the prediction and assessment of instream impacts. The structure and function of ecosystems, and expected water quality all can be related to location of a reach in the watershed. Making comparisons between reaches with similar characteristics drainage net permits a general assessment of expected conditions within a reach. Spatial location can be used to select models, criteria, or metrics for specific stream or watershed management strategies. A useful integrator for the design of control technology or management programs is instream habitat. We advocate integration of physical and chemical habitat analyses. For example, analysis of instream physical conditions coupled with analysis of contaminant concentration effects on ecosystems provides the basis for preserving stream integrity. We suggest that an emphasis on concentration or loading in management vary depending upon the water resource problem being addressed. We also advocate the incorporation of accurate measures of instream photosynthesis and respiration terms for calculating dissolved oxygen deficits. Finally additional research is needed to predict periodic, short-term changes in flows on instream habitat modifications, particularly substrate and channel configuration.

## References

Bovee, K. 1982. A Guide to Stream Habitat Analysis Using the Instream Flow Incremental Methodology. U.S. Fish Wildlife Service, Instream Flow Paper 12. FWS/OBS-82/26, Washington, D.C.

Conder, A.L. and T.C. Annear. 1987. Test of weighted usable area estimates derived from a PHABSIM model for instream flow studies on trout streams. North Amer. Journ. of Fish Managemt. 7:339-350.

Carter, R. W. and D. M. Thomas. 1968. Flood frequency in metropolitan areas. Proc. of the 4th Am. Water Resources Assoc. Conf. Urbana, IL, pp 56-67.

DiGiano, F. A., R. A. Coler, R. C. Dahiya, and B. B. Berger. 1975. A projection of pollutional effects of urban runoff in the Green River, MA. Proc. Am Water Res. Assoc. 20:28-37.

Gilbert, G. K., 1917. Hydraulic-mining debris in the Sierra Nevada. U.S. Geol. Surv. Prof. Pap. 105, 154 pp.

James, L. Douglas, 1965. Using a digital computer to estimate the effects of urban development on flood peaks. Water Resources Res. 1(2).

Fam, S., M. K. Stenstrom, and G. Silverman. Hydrocarbons in urban runoff. J. Env. Eng. 113(5):1032-10456.

Herricks, E. E., and D. E. Himelick. 1981. Illinois Water Quality Management Information System Biological Component - Fisheries Users Manual. Civil Engineering Studies, Environmental Engineering Studies #64, UILU-ENG-81/2009.

Herricks, E. E., D. J. Schaeffer, J. C. Kapsner. 1985. Assessing water quality management success - When is a violation a violation. J. Water Poll. Cont. Fed. 57(2):109-115.

Herricks, E.E. 1986. Disciplinary integration: The solution. In: Modeling Runoff Quality. H.C. Torno, J. Masalek and M. Desbordes (eds.). Springer-Verlag, Heidelberg.

Herricks, E. E. and M. I. Braga. 1987. Habitat elements in river basin management and planning. Wat. Sci. Tech. 19(9):19-29.

Hoffman, E. J., J. S. Latimer, G. L. Mills, and J. G. Quinn. Petroleum hydrocarbons in urban runoff from a commercial land use area. J. Water Poll. Cont. Fed. 54(11):1517-1525.

Huber, W.C. 1986 Modeling urban runoff quality: State-of-the-art. In: Modeling Runoff Quality. H.C. Torno, J. Masalek and M. Desbordes (eds.). Springer-Verlag, Heidelberg.

Kelsey, H. M., R. Lamberson, M. A. Made. 1987. Stochastic model for the long-term transport of stored sediment in a river channel. Water Resources. Res. 23(9):1738-1750.

Loar, J.M. and M.J. Sale. 1981. Analysis of environmental issues related to small-scale hydroelectric development V. Instream flow needs for fishery resources. Oak Ridge National Lab., Environmental Science Div. Public. 1829, 123 pp.

Mathur, D., W. Bason, E. Purdy, and C.A. Silver. 1985. A critique of the Instream Flow Incremental Methodology. Can. J. Fish. Aquat. Sci. 42:825-831.

Meade, R. H. 1982. Sources, sinks, and storage of river sediment in the Atlantic drainage of the United States. J. Geol. 90:235-252.

Medeiros, C. R. LeBlanc, and R. A. Coler. 1983. An in situ assessment of the acute toxicity of urban runoff to benthic macroinvertebrates. Env. Toxicology and Chem. 2:119-126.

Milhous, R.T., D. Wegner, and T.Waddle. 1984. User's Guide to the Physical Habitat Simulation System. Instream Flow Information Paper No. 11. Revision FWS/OBS81-42. U.S. Fish and Wildlife Service, Ft. Collins, Co

Morhardt, J.E. 1986. Instream flow methodologies. Electric Power Research Institute Report. EPRI EA-4819. Palo Alto, CA.

Newbold, J.D., J.W. Elwood, R.V. O'Neill, and W. Van Winkle. 1981. Measuring nutrient spiralling in streams. Can. J. Fish. Aquat. Sci. 38:860-863.

Osborne, L.L., R.W. Larimore, and M.J. Wiley. 1987. Stream classification: Stream order vs. watershed are as metrics. Paper Presented at 25th Anniversary Meeting of Illinois Chapter of the American Fisheries Society. March 3-5, 1987. Marion, IL.

Osborne, L.L. and M.J. Wiley. 1988. Empirical relationships between land use/cover and stream water quality in a midwestern agricultural watershed and their application to environmental planning. Jour. Environ. Managemt. 26:9-27.

Osborne, L.L., M.J. Wiley, and R.W. Larimore. 1988. Evaluation of the PHABSIM instream flow model: Accuracy of predicted fish habitat conditions in low-gradient, warmwater streams. Regulated Rivers Journal (in press).

Pedersen, E.R. and M.A. Perkins. 1986. The use of benthic invertebrate data for evaluating impacts of urban runoff. Hydrobiol. 139: A3-22.

Perkins, M.A. 1982. An evaluation of instream ecological effects associated with urban runoff in a lowland stream in Western Washington. Final Report. U.S.E.P.A., EPA R 80637, 26 pp.

Polls, I., C. Lue-Hing, D.R. Zenz, and S.J. Sedita. 1980. Effects of urban runoff and treated municipal wasterwater on a man-made channel in northeastern Illinois. Wat. Res. 14(3):207-215.

Porcella, D.B. and D.L. Sorenson. 1980. Characteristics of nonpoint source urban runoff and its effect on stream ecosystems. EPA-600/3-80-032, U.S.E.P.A., Corvallis Environ. Res. Lab., Corvallis, OR 110 pp.

Pratt, J. M. and R. A Coler. 1976. A procedure for routine biological evaluation of urban runoff in small rivers. Water Res. 10:1014-1025.

Pratt, J. M., R. A. Coler, and P. J. Godfrey. 1981. Ecological effects of urban stormwater runoff on benthic invertebrates inhabiting the Green River, MA. Hydrobiologia 83:29-42.

Rimer, A.E., J.A. Nissen, and D.E. Reynolds. 1978. Characterization and impact of stormwwater runoff from various land cover types. J. Water Pollution Control 50(2): 252-264.

Schlosser, I.J. and J.R. Karr. 1981. Riparian vegetation and channel morphology impact on spatial patterns of water quality in agricultural watersheds. Environmental Management 5:233-243.

Scott, D. and C.S. Shirvell. 1987. A critique of the instream flow incremental methodology and observations on flow determination in New Zealand. In: Regulated Streams-Advances in Ecology. J.F. Craig and J.B. Kemper (eds.). Plenum Press, New York

Thoman, R.V. 1972. Systems Analysis and Water Quality Management. McGraw-Hill Book Co., New York. 286 pp.

Vitousek, P. M. 1977. The regulation of element concentrations in mountain streams in the northeastern United States. Ecol. Mono. 47:65-87.

Wiley, L.L. and L.L. Osborne. 1987. Potential environmental impacts of lagoon exemption proposal R86-17. Illinois State Natural History Survey Position Paper. Champaign, IL 17 pp.

Wiley, M.J., L.L. Osborne, R.W. Larimore and T.J. Kwak. 1987a. Augmenting concepts and techniques for examining critical flow requirements of Illinois stream fishes. Aquatic Biology Section Technical Report 87/5. Illinois State Natural History Survey. Final Report F-43-R. Champaign, IL.

Wiley, M.J., L.L. Osborne, D. Glosser, and S.T. Sobaski. 1987b. Large-scale Ecology of the Vermilion River Basin. Aquatic Biology Section Technical Report 87/6. Illinois State Natural History Survey. Final Report ENR Project EH24. Champaign, IL.

Wilson, K. V. 1968. A preliminary study of the effect of urbanization on floods in Jackson, Mississippi. U. S. Geol. Survey. Prof. Pap. 575D, pp 259-261.

## Discussion of Dr. Osborne's Paper

Question:

Were the data you discussed measured dynamically?

Answer:

Yes--two years, around the clock.

Question:

Were the major impacts you noticed attributable to point or nonpoint sources?

Answer:

In some cases, even in agricultural areas, the biggest phosphorous impact is caused by point sources from urban areas, exacerbated by urban nonpoint sources.

Question:

Diurnal variations at the point of interest you define for yourself appear to be insufficient characterizations. Spatial variations along the stream are important as well. Determinations of K-rates and the effects of resuspension and sediment oxygen demand appear to have been left out of the relationships you described.

Answer:

The terms you mention do not appear to have been so big a problem. The respiration term I discussed tends to include the net effects of all those things. The more complex unresolved matters appear to be in the hydraulic context--particle flows versus water flows and the like.

## State Perspectives on Water Quality Criteria

### Eric H. Livingston[1]

### ABSTRACT

The Federal Clean Water Act framework for water quality regulation is reviewed. The role of water quality standards in the regulation of stormwater discharges and the protection of beneficial uses is discussed. The relationship between the implementation of best management practices (BMPs) and water quality criteria is explored with emphasis on how Florida's stormwater regulatory program was established. Florida's BMP design and performance standards for stormwater discharges from new developments are summarized. A conceptual program to address the modification of existing stormwater discharges to reduce their pollution loading to receiving waters is proposed.

### INTRODUCTION

The Federal Clean Water Act establishes the framework for our nation's water quality management regulations and programs by setting forth a national objective to "restore and maintain the chemical, physical and biological integrity of the nation's waters". Furthermore, the Act sets forth several goals to help achieve this objective including:
o    Wherever attainable, water quality shall provide for the protection and propagation of fish, shellfish and wildlife and provide for recreation in and on the water.
o    Programs to control nonpoint sources of pollution shall be developed and implemented expeditiously.

To carry out the purposes of the Act, the Environmental Protection Agency and the states have adopted water quality criteria and standards. The word "criterion" should not be used interchangeably with the word "standard" (EPA, 1976). Criterion represent a constituent concentration or level associated with a degree of environmental effect upon which scientific judgement may be based. Within the context of the water environment, criterion has come to mean a designated concentration of a

---

[1]Environmental Administrator, Nonpoint Source Management Section, Florida Department of Environmental Regulation, 2600 Blair Stone Road, Tallahassee, Florida 32399-2400

constituent that, when not exceeded, will protect an organism, or organism community or a prescribed water use or quality with an adequate degree of safety. Water quality criteria, therefore, reflect a knowledge of the capacity for environmental accumulation, persistence or effects of specific pollutants in specific aquatic systems. In some cases, a criterion may be a narrative statement instead of a constituent concentration.

A water quality standard includes designated beneficial uses (e.g., potable waters, fishable/swimmable waters), the water quality criteria to protect those uses as well as an antidegradation policy. A standard may use a water quality criterion as a basis for regulation or enforcement, but the standard may differ from a criterion because of prevailing local natural conditions or because of the importance of a particular waterway, economic considerations or the degree of safety to a particular ecosystem that may be desired.

Water quality criteria traditionally have been based on acute and chronic toxicity bioassay tests that were derived for continuous pollutant sources. Unfortunately, urban runoff is an intermittent pollutant source which represents a shock loading of relatively short duration with a longer time between exposures. In addition, urban stormwater impacts on receiving waters are highly variable depending on numerous factors such as rainfall and runoff characteristics, water body type, land use, soils, geology, chemistry and topography. Therefore, the validity of applying traditional water quality criteria to urban stormwater pollution sources is highly debatable.

Stormwater also exerts long term impacts on receiving waters thereby raising further questions about the applicability of current water quality criteria. Stormwater carries relatively high levels of suspended solids which alone can cause severe impacts on receiving water biota. Suspended solids also carry many other contaminants (e.g., metals, phosphorus) which accumulate as bottom sediments (a pollutant sink). Exposure to these sediment bound toxicants can pose threats to the aquatic biota and environment and can also impose a sediment oxygen demand. If the pH falls or bottom conditions become anaerobic, these contaminants can also release from the sediments and enter the water (Harper, 1985).

From the foregoing discussion it seems evident that evaluating stormwater impacts on receiving waters, and consequently the design of control programs, cannot be based solely on traditional water quality criteria. Mancini (1983) proposes a methodology for the development of wet weather criteria while Mancini and Plummer (1986) further discuss this concept. While wet weather criteria may be applicable, their development and implementation would be difficult and impractical. Especially since there is so little information on the toxicity of urban runoff thereby precluding the development of whole effluent information that would be needed for permitting stormwater discharges (EPA, 1983a). Sediment criteria is also essential with different criteria for fresh and marine environments (FDER, 1988). Finally, the basic foundation of any stormwater program should be the protection of beneficial uses.

## REGULATORY FRAMEWORK OF FLORIDA'S STORMWATER PROGRAM

Section 208 of the Federal Clean Water Act required the development of areawide water quality management plans to control point and nonpoint sources of pollution. As part of Florida's program conducted during the late 1970's and early 1980's, many investigations were undertaken to assess the impacts of stormwater and the effectiveness of various best management practices (Livingston, 1984). These studies demonstrated that stormwater, whether from agriculture, forestry or urban lands, was the primary source of pollutant loading to Florida's receiving waters. Subsequently, it was concluded that the ability to meet the Clean Water Act objective of fishable and swimmable waters would require the implementation of stormwater programs to reduce the delivery of pollutants from stormwater discharges.

Recognition of this problem, along with the availability of Federal funds, led Florida to draft regulations to control stormwater in the late 1970's. The first official State regulation specifically addressing stormwater was adopted in 1979 as part of Chapter 17-4, Florida Administrative Code (F.A.C.). Chapter 17-4.248 was the first attempt to regulate this source of pollution that, at the time, was not very well understood. Under Chapter 17-4.248 the Department based its decision to order a permit on a determination of the "insignificance" or "significance" of the stormwater discharge. This determination seems reasonable in concept; however, in practice, such a decision can be as variable as the personalities involved. What may appear insignificant to the owner of a shopping center may actually be a significant pollutant load into an already overloaded stream.

In adopting Chapter 17-4.248, the Department intended that the rule would be revised when more detailed information on nonpoint source management became available. About one year after adoption, the Department began reviewing the results of research being conducted under the 208 program. The Department also established a stormwater task force with membership from all segments of the regulated and environmental communities. A revised stormwater rule, Chapter 17-25, F.A.C., was developed after two years, more than 100 meetings between department staff and the regulatory interests, and the dissemination of 29 official rule drafts for review and comment. The adopted rule required a stormwater permit for all new stormwater discharges and for modifications to existing discharges that were modified to increase flow or pollutant loading.

The Stormwater Rule had to be implemented within the framework of the federal Clean Water Act. The Act establishes two types of regulatory requirements to control pollutant discharges--technology-based effluent limitations which reflect the best controls available considering the technical and economic achievability of those controls; and water quality-based effluent limitations which reflect the water quality standards and allowable pollutant loadings set up by state permit (EPA, 1983). The latter approach can be developed and implemented through a biomonitoring approach based on whole effluent toxicity making it very applicable to stormwater. However, Florida's tremendous growth and the accompanying creation of tens of thousands of new stormwater discharges together with our lack of data on stormwater loading toxicity made this approach unimplementable.

Guidance on the development of stormwater regulatory programs and the role of water quality criteria has been issued by the Environmental Protection Agency (1987). The guidance recognizes that BMPs are the primary mechanism to enable the achievement of water quality standards. For the purposes of this paper, a BMP is a control technique that is used for a given set of site conditions to achieve stormwater quality and quantity enhancement at a minimum cost. Further, the guidance recommends that state programs should include the following steps:

1. design of BMPs based on site specific conditions; technical, institutional and economic feasibility; and the water quality standards of the receiving waters.

2. monitoring to ensure that practices are correctly designed and applied.

3. monitoring to determine
   a) the effectiveness of BMPs in meeting water quality standards
   b) the appropriateness of water quality criteria in reasonably assuring protection of beneficial uses.

4. adjustment of BMPs when it is found that water quality standards are not being protected to a designed level and/or evaluation and possible adjustment of water quality standards.

## WATER QUALITY CRITERIA

It is intended that proper installation and operation of state approved BMPs will achieve water quality standards. While water quality standards are to be used to measure the effectiveness of BMPs, EPA recognizes that there should be flexibility in water quality standards to address the impact of time and space components of stormwater as well as naturally occurring events. If water quality standards are not met, then the BMPs should be modified, or the discharge cease or, in some cases, reassessment of the water quality standards be undertaken.

### RATIONALE FOR STORMWATER RULE STANDARDS

The overriding standards of the Stormwater Rule are the water quality standards and appropriate regulations established in other Department rules. Therefore, an applicant for a stormwater discharge permit must provide reasonable assurance that stormwater discharges will not violate state water quality standards. Because of the potential number of discharge facilities and the difficulties of determining the impact of any facility on a waterbody or the latter's assimilative capacity, the Department decided that the Stormwater Rule should be based on design and performance standards.

The performance standards established a technology-based effluent limitation against which an applicant can measure the proposed treatment system. If an applicant can demonstrate treatment equivalent to the description in the performance standard systems, then the applicant should be able to meet applicable water quality standards. The actual design and performance standards are based on a number of factors which will subsequently be discussed.

Stormwater Management Goals - Stormwater management has multiple objectives including water quality protection, flood protection (volume, peak discharge rate), erosion and sediment control, water conservation and reuse, aesthetics and recreation. The basic goal for new development is to assure that the post-development peak discharge rate, volume, timing and pollutant load does not exceed predevelopment levels. However, BMPs are not 100% effective in removing stormwater pollutants while site variations can also make this goal unachievable at times. Therefore, for the purposes of stormwater regulatory programs, the Department (water quality) and the Water Management Districts (flood control) have established performance standards based on risk analysis and implementation feasibility.

Rainfall Characteristics - An analysis of long term rainfall records was undertaken to determine statistical distribution of various rainfall characteristics such as storm intensity and duration, precipitation volume, time between storms, etc. It was found that nearly 90% of a year's storm events occurring anywhere in Florida produce a total of 2.54 cm (1 inch) of rainfall or less (Anderson, 1982). Also, 75% of the total annual volume of rain falls in storms of 2.54 cm or less.

Runoff Pollutant Loads - The first flush of pollutants refers to the higher concentrations of stormwater pollutants that characteristically occur during the early part of the storm with concentrations decaying as the runoff continues. Concentration peaks and decay functions vary from site to site depending on land use, the pollutants of interest, and the characteristics of the drainage basin. Florida studies (Wanielista and Shannon, 1977; Miller, 1985) indicated that for a variety of land uses the first 1.27 cm (.5 inch) of runoff contained 80-95 percent of the total annual loading of most stormwater pollutants. However, first flush effects generally diminish as the size of the drainage basin increases and the percent impervious area decreases because of the unequal distribution of rainfall over the watershed and the additive phasing of inflows from numerous small drainages in the larger watershed. In fact, as the drainage area increases in size above 40 ha (100 ac) the annual pollutant load carried in the first flush drops below 80% because of the diminishing first flush effect.

BMP Efficiency and Cost Data - Numerous studies conducted in Florida during the Section 208 program generated information about the pollutant removal effectiveness of various BMPs and the costs of BMP construction and operation. Analysis of this information revealed that the cost of treatment increased exponentially after "secondary treatment" (removal of 80% of the annual load) (Wanielista, et al., 1982).

Selection of Minimum Treatment Levels - After review and analysis of the above information, and after extensive public participation, the Department set a stormwater treatment objective of removing at least 80% of the average annual pollutant load for stormwater discharges to Class III (fishable/swimmable) waters. A 95% removal level was set for stormwater discharges to sensitive waters such as potable supply waters (Class I), shellfish harvesting waters (Class II) and Outstanding Florida Waters. The Department believed that these treatment levels would protect beneficial uses and thereby establish a relationship between the rule's BMP performance standards and water quality standards. The actual stormwater treatment volumes for various BMPs are set forth in Table 1.

## ADMINISTRATION OF THE STORMWATER RULE

Under the Florida Water Resources Act of 1972, the

between storms, etc. It was found that nearly 90% of a year's storm events occurring anywhere in Florida produce a total of 2.54 cm (1 inch) of rainfall or less (Anderson, 1982). Also, 75% of the total annual volume of rain falls in storms of 2.54 cm or less.

Runoff Pollutant Loads - The first flush of pollutants refers to the higher concentrations of stormwater pollutants that characteristically occur during the early part of the storm with concentrations decaying as the runoff continues. Concentration peaks and decay functions vary from site to site depending on land use, the pollutants of interest, and the characteristics of the drainage basin. Florida studies (Wanielista and Shannon, 1977; Miller, 1985) indicated that for a variety of land uses the first 1.27 cm (.5 inch) of runoff contained 80-95 percent of the total annual loading of most stormwater pollutants. However, first flush effects generally diminish as the size of the drainage basin increases and the percent impervious area decreases because of the unequal distribution of rainfall over the watershed and the additive phasing of inflows from numerous small drainages in the larger watershed. In fact, as the drainage area increases in size above 40 ha (100 ac) the annual pollutant load carried in the first flush drops below 80% because of the diminishing first flush effect.

BMP Efficiency and Cost Data - Numerous studies conducted in Florida during the Section 208 program generated information about the pollutant removal effectiveness of various BMPs and the costs of BMP construction and operation. Analysis of this information revealed that the cost of treatment increased exponentially after "secondary treatment" (removal of 80% of the annual load) (Wanielista, et al., 1982).

Selection of Minimum Treatment Levels - After review and analysis of the above information, and after extensive public participation, the Department set a stormwater treatment objective of removing at least 80% of the average annual pollutant load for stormwater discharges to Class III (fishable/swimmable) waters. A 95% removal level was set for stormwater discharges to sensitive waters such as potable supply waters (Class I), shellfish harvesting waters (Class II) and Outstanding Florida Waters. The Department believed that these treatment levels would protect beneficial uses and thereby establish a relationship between the rule's BMP performance standards and water quality standards. The actual stormwater treatment volumes for various BMPs are set forth in Table 1.

## ADMINISTRATION OF THE STORMWATER RULE

Under the Florida Water Resources Act of 1972, the

Department of Environmental Regulation, a water quality agency, serves as the umbrella administering agency delegating authority to five regional water management districts whose primary functions historically have been related to water quantity management. Therefore, a second objective in developing the Stormwater Rule was to coordinate the water quality considerations of the Department's stormwater permits with the water quantity aspects of the Districts' surface water management permits.

---

Table 1. BMP Treatment Volumes for Stormwater Discharges to Class III Waters

SWALES – Infiltrate 80% of the runoff generated by a 3 yr-1 hr storm (typically about 2 inches of runoff)

RETENTION – Off-line infiltration of the first 0.5 inch runoff (DA <100 acres) or the runoff from the first 1 inch of rainfall, whichever is greater (DA >100 acres)

DETENTION WITH FILTRATION – Filtration of retention volume

WET DETENTION – Detention of the first 1 inch of runoff or volume calculated by 2.5 times % impervious, whichever is greater
 – Commercial, industrial land uses must pretreat up to 0.5 inch runoff by infiltration BMPs

WETLANDS – Detention and assimilation of retention volume

NOTES:
1. Discharges to sensitive waters must treat 50% more stormwater volume and may require infiltration pretreatment
2. Discharges to sinkhole watersheds must treat first two inches of runoff (SRWMD criterion)

---

In addition, the delegation of the stormwater permitting program allows for minor adjustments to the Stormwater Rule design and performance standards to better reflect local conditions. Florida is a very diverse state with major differences throughout in soils, geology, topography, rainfall, etc., which can directly affect the usability and treatment effectiveness of a BMP. Such problems can be minimized by adoption by the Districts of slightly different design and performance standards which are approved by the Department prior to implementation. Both the Department's and Districts' stormwater rules essentially require a new development to include a comprehensive stormwater management system. The system should be viewed as a "BMP treatment train" in which a number of different BMPs are integrated into a

# WATER QUALITY CRITERIA

comprehensive system that provides aesthetic and recreational amenities in addition to the traditional stormwater management objectives.

## BMP GUIDELINES

In addition to the stormwater treatment volumes, other design and performance standards have been set to assure that BMPs function optimally to attain the stormwater treatment goal and other management objectives (Livingston, et al., 1988). These guidelines will be discussed for each of the BMPs currently being used extensively in Florida.

Swales - are defined by Chapter 403, Florida Statutes, as manmade trenches that:
- have a top width-to-depth ratio of the cross section equal to or greater than 6:1, or side slopes equal to or greater than 3 feet horizontal to 1 foot vertical.
- contains contiguous areas of standing or flowing water only following a rainfall event.
- is planted with or has stabilized vegetation suitable for soil stabilization, stormwater treatment and nutrient uptake.
- is designed to take into account the soil erodibility, soil percolation, slope, slope length and drainage area so as to prevent erosion and reduce pollutant concentration of any discharge.

Swale treatment of stormwater is accomplished primarily by infiltration of runoff and secondarily by adsorption and vegetative filtration and uptake (Yousef, et al., 1985). Recent investigations have concluded that Florida soil, slope and water table conditions essentially preclude the use of swales as the sole BMP to treat stormwater (Wanielista, et al., 1985). Therefore, the greatest utility of swales is as a pretreatment BMP within a BMP treatment train stormwater system. Infiltration pretreatment can be easily accomplished by using raised storm sewer inlets within the swale, or by elevating driveway culverts or using swale blocks to create mini-retention areas.

Retention - Off-line retention areas which receive the first flush volume only while the later runoff is diverted to a flood control BMP are the most effective stormwater treatment practice. Treatment is achieved through diversion and infiltration of the first flush thereby providing total pollutant removal for all stormwater that is retained on-site. To reduce operation needs, increase aesthetics and reduce the land area needed for stormwater treatment, retention areas should be incorporated into a site's landscaping and open space areas. Effectiveness of retention areas can be increased and groundwater impacts decreased by:

1. Infiltrating the stormwater treatment volume within 72 hours or within 24 hours if the retention area is grassed.
2. Grassing the retention area bottom and side slopes-- reduces maintenance and maintains soil infiltration properties.
3. Maintaining at least three feet between the bottom of the retention area and seasonal high water tables or limerock.
4. In Karst sensitive areas, using several small, shallow infiltration areas to prevent formation of solution pipe sinkholes within the system.

Exfiltration trenches typically are used in highly urbanized areas where land is unavailable for retention basins. They consist of a rock filled trench, surrounded by filter fabric, in which a perforated pipe is placed. The stormwater treatment volume is stored within the pipe and exfiltrates out of the perforations into the gravel envelope and into the surrounding soil. Pretreatment with catch basins to remove sediments and other debris is essential to prevent clogging.

<u>Detention with Filtration</u> - systems were proposed as an alternative to retention for those areas of Florida where local conditions, especially flat topography and high water tables, prevent infiltrating the stormwater treatment volume. The filters must consist of two feet of natural soil or other suitable fine textured granular media which meets certain specifications including:
- Filters must have pore spaces large enough to provide sufficient flow capacity so that the filter permeability is equal to or greater than the permeability of the surrounding soil.
- The design shall assure that particles within the filter do not move.
- When sand or other fine textured material other than natural soil are used for filtration, the filter material will meet the following criteria:
   a) Be washed (less than 1 percent silt, clay or organic matter) unless filter cloth is used to retain such materials within the filter
   b) Have a uniformity coefficient between 1.5 and 4.0
   c) Have an effective gain size of 0.20 to 0.55 mm in diameter.
- Be designed with a safety factor of at least two.
- Will recover the treatment volume (bleed down) within 72 hours.

Filters are placed in the bottom or sides of detention areas where the filtered stormwater is collected in an underdrain pipe and then discharged. Experience has shown that these filters are very difficult to design and construct. Operation is also difficult because of low hydraulic head and maintenance is nearly impossible. It

## WATER QUALITY CRITERIA

is not a question of if a filter will clog, only when it will clog. In addition, filters are designed to remove particulate pollutants and do not remove dissolved pollutants such as phosphorus or zinc. Therefore, filtration systems are no longer recommended for use except under very special conditions and where a full time maintenance entity such as a local government will assume such responsibilities.

Wet Detention - systems consist of a permanent water pool, an overlying zone in which the stormwater treatment volume temporarily increases the depth while it is stored and slowly released and a shallow littoral zone (biological filter). In addition to their high pollutant removal efficiencies (EPA, 1983b), wet detention systems can also provide aesthetic and recreational amenities, a source of fill for the developer and even "lake front" property which brings a premium price.

Wet detention criteria are listed in Table 2. These have been developed to take full advantage of the biological, physical and chemical assimilation processes occurring within the wet detention system. If the system is designed as a development amenity, the use of pretreatment BMPs integrated into the overall stormwater management system is highly recommended to prevent algal blooms or other perturbations that would reduce the aesthetic value. Placing raised storm sewers in grassed areas such as parking lot landscape islands, using swale conveyances, or a perimeter swale/berm system along the detention lake shoreline are techniques that have been used frequently.

---

### Table 2. Wet Detention Guidelines

- Treatment volume as per Table 1.
- Treatment volume slowly recovered in no less than 120 hours with no more than half of the volume discharged within the first 60 hours following the storm.
- Volume in the permanent pool should provide a residence time of at least 14 days.
- At least 30% of the surface area shall consist of littoral area with slopes of 6:1 or flatter that is established with appropriate native aquatic plants selected to maximize pollutant uptake and aesthetic value.
- Littoral zone plants shall have a minimum 80% survival rate and coverage after two years. Cattails and other undesirable plants shall be removed.
- Littoral zone is concentrated near the outfall or in a series of shallow benches ending at the outfall.
- Side slopes no steeper than 4:1 out to a depth of two feet below the level of the permanent pool.
- Maximum depth of 8-10 feet below the invert of the discharge structure is recommended. Maximum depth

shall not create aerobic conditions in bottom sediments and waters.
- Maximize flow length possible between inlets and outlet; a length to width ratio of at least 3:1 is recommended. Diversion barriers such as baffles, islands or a peninsula should be used if necessary to increase flow length and length to width ratio.
- An oil and grease skimmer shall be designed into the outlet structure.
- If the system is planned as a "real estate lake", pretreatment by infiltration is recommended.
- Inlet areas should include a sediment sump.

---

Wetland Treatment - was authorized by the 1984 Henderson Wetlands Protection Act which allows stormwater treatment in wetlands that are connected by an intermittent water course which flows in direct response to rainfall thereby causing the water table to rise above ground surface. Not only does this take advantage of natural treatment mechanisms but it gives another economic value to wetlands, an incentive to the developer to use not destroy the wetland, and it revitalizes ditched and drained wetlands by providing water.

Wetlands may be viewed as nature's kidneys--they store stormwater, dampen floodwaters, and also transform pollutants and even retain them thereby providing natural stormwater treatment (Richardson, 1988). However, care must be taken to protect the numerous assimilation mechanisms within the wetland plants and sediments. In addition, the wetland hydroperiod--the duration that water stays at various levels--must be protected or restored since it determines the form, function and nature of the wetland. Therefore, pretreatment practices to attenuate stormwater volume and peak rate and to reduce oil, grease and especially sediment are essential. Normally, a pretreatment lake is constructed adjacent to the wetland.

The following guidelines are presented for incorporating wetlands into a stormwater management system:
- Treatment volume as per Table 1.
- Treatment volume slowly recovered in no less than 120 hours with no more than half of the volume discharged within the first 60 hours following the storm.
- Stormwater must sheet flow evenly through the wetland to maximize contact with the wetland plants, sediments and microorganisms. Spreader swales, distribution systems or a level spreader between the pretreatment lake and the wetland have been used extensively.
- Swales should be used for stormwater conveyance throughout the development.
- The hydroperiod must be protected or restored. Treatment capacity of the wetland is determined by the storage volume available between the normal low and

high elevations. These elevations are determined by site specific indicators such as lichen and moss lines, water stain lines, adventitious root formation, plant community zonation, hydric soils distribution and rack/debris lines.
○ Erosion and sediment control during construction is essential since only a few inches of sediment deposited in the wetland will destroy the wetland filter.
○ Inflow/outflow monitoring, sediment metal levels and vegetative transect monitoring is required to help evaluate the effectiveness of these systems and the impacts of stormwater additions to wetlands.

## THE CHALLENGE AHEAD

Probably the biggest stormwater management problem facing Florida is how to reduce pollutant loadings discharged by older systems, especially local government master systems, constructed before the Stormwater Rule. These systems were designed solely for flood protection and rapidly deliver untreated stormwater directly to rivers, lakes, estuaries and sinkholes. Retrofitting urban stormwater discharges will become a national priority because of the recently enacted Section 402 of the 1987 Federal Clean Water Act. This section requires the establishment of a NPDES permitting program for stormwater discharges as described elsewhere in these proceedings (Gallop, et al.). The Act requires that these stormwater pollutant discharges be reduced to "the maximum extent practicable". Interpretation, implementation and enforcement of this standard could prove extremely difficult similar to the problems Florida encountered in implementing its first stormwater regulations (17-4.248)

Establishing a stormwater program to retrofit existing systems presents many technical, institutional and financial dilemmas. The unavailability and cost of land in urbanized areas make the use of conventional BMPs infeasible in most instances. Current state laws and institutional arrangements promote piecemeal, crises solving approaches aimed at managing stormwater within political boundaries yet stormwater follows watershed boundaries. Land use planning and management must be fully integrated into the stormwater management scheme. Retrofitting is also prohibitively expensive and many local governments are already short of funds. Therefore, solving our existing urban stormwater problems will require imaginative, innovative approaches.

## STORMWATER LEGISLATION

During the next year, stormwater legislation will be proposed in Florida to enhance the state's current program and initiate a long term effort to reduce pollutant loadings from older stormwater systems. Following is a brief discussion of some of the essential elements of such a program.

Watershed Management - A watershed approach which integrates land use planning with the development of stormwater infrastructure is essential. After all, it is the intensification of land use and the increase in impervious surfaces within a watershed that creates the stormwater and water resources management problems. Consequently, a "watershed management team" effort involving state and local governments together with the private sector is necessary. In fact, local governments are the primary team member since they determine zoning and land use, they issue building permits and inspect projects, and they have code enforcement powers that can help to assure that stormwater systems are properly operated and maintained.

Local governments need to identify and map the existing natural stormwater system--the creeks, wetlands, flood plains, drainageways and natural depressional areas. Once mapped, these areas need to be zoned for conservation or low intensity uses compatible with the functions provided by the natural system. The existing manmade stormwater system must also be mapped and essential characteristics such as pipe size, drainage areas, invert elevations, etc., be determined. This information should then be fully integrated with the existing and future land use plan for the watershed and a master stormwater management plan developed and implemented. The Growth Management Act of 1985, which requires all local governments to adopt comprehensive plans addressing current and future land use with infrastructure needs, establishes a base structure that could promote a watershed management approach.

Treatment Requirements for Older Systems - Numerous problems inherent in a highly urbanized area prevent the application of new development stormwater treatment standards from being imposed on older systems. Instead a "watershed loading" concept is proposed which considers the beneficial uses of the receiving waters, the total stormwater load that can be assimilated by the receiving waters, a hierarchy of treatment levels based on BMP feasibility and local public perceptions about receiving water quality. For example, treatment standards would progressively increase as one moves from the downtown central business district to the suburbs to rural areas. However, the actual treatment level would depend on the watershed's total allowable loading which is based on citizen desires for certain beneficial uses of the receiving water.

Selective Targeting - The extremely high cost of retrofitting older urban stormwater systems also implies a need for careful evaluation of pollutant reduction goals. A long term (25 years) plan based on prioritization of watersheds such that existing systems are selectively targeted for modification is needed to assure that citizens receive the greatest benefit (pollutant load reduction, flood protection) for the dollar. The upgrading of older systems must also be coordinated with other already planned infrastructure improvements such as road widenings. An excellent example of this approach is the Orlando Streetscape Project. While downtown streets were torn up for this downtown renovation, the existing stormwater system was modified by the addition of off-line exfiltration systems to reduce pollution loads to downtown lakes.

Education - Education programs for the general public and for professionals involved in stormwater management are vital. Citizens must understand how their everyday activities contribute to stormwater pollution. For example, citizens should not discard leaves, grass clippings, used motor oil or other material into swales or storm sewers. Even more importantly, comprehensive training and certification programs are needed for those in the private and public sectors who design, construct, inspect, operate or maintain stormwater management systems.

Funding - The cost of providing needed stormwater infrastructure improvements to address current and future flooding and water quality problems is gigantic. Yet local governments are already struggling financially and traditional revenue sources such as property taxes cannot be relied upon to pay for stormwater management. Instead a dedicated source of revenue based on contributions to the stormwater problem is needed. The stormwater utility can provide this. The City of Tallahassee implemented Florida's first stormwater utility in October 1986 and many other local governments have or are following this example.

Innovative BMPs - The infeasibility of using traditional BMPs to reduce stormwater pollutant loads in highly urbanized areas means that creative and innovative BMPs will be needed such as those discussed in this section.

Alum injection within storm sewers was used in Tallahassee to reduce stormwater loadings to Lake Ella (Harper, et al., 1986). A sonic flow meter measures storm sewer flow causing a flow proportional dose of aluminum sulfate to be injected and mix with the polluted stormwater. As the alum mixes with the stormwater, a small floc is produced which attracts suspended and dissolved pollutants by adsorption and enmeshment into and onto the floc particles. The floc then settles to the lakes bottom sediments, gradually blanketing and incorporating into the sediments thereby reducing internal recycling of nutrients and metals. Other advantages of alum injection include excellent pollutant reduction (>85%) and relatively low construction and operations costs, especially for the highly urbanized areas.

**Pervious concrete** consists of specially formulated mixtures of Portland cement, uniform open graded coarse aggregate and water. When properly mixed and installed pervious concrete surfaces have a high percentage of void space which allows rapid percolation of rainfall and runoff. Pervious concrete is being used widely in Florida, especially for parking lots, and could be an important BMP to reduce stormwater loadings in highly urbanized areas. Recent field investigations of pervious concrete parking areas that have been in place for up to 12 years, revealed that the infiltration capacity of the concrete has not decreased significantly, a major concern. Further information about the use, design and construction of pervious concrete surfaces is available (FCPA, 1988).

**Improved street sweepers** that will pick up the small particles (<60 microns) that contain high concentrations of metals and other pollutants could also prove valuable in reducing stormwater loadings, especially from downtown business districts where other BMPs will usually be infeasible. Initial evaluation of an innovative sweeper design indicates that this sweeper can remove these small particles (Stidger, 1988). However, extensive watershed tests to evaluate the machine's cost-effectiveness, pollution removal effectiveness and other advantages or disadvantages will not be undertaken until later this year.

**Regional stormwater systems** which manage stormwater from several developments or an entire drainage basin offer many advantages over the piecemeal approach that relies upon small, individual on-site systems. They provide economies of scale in construction, operation and maintenance. Regional systems can also help manage stormwater from existing and future land uses and will be a central part of any retrofitting program. Another reason for a watershed management approach that fully integrates land use and stormwater management.

<u>The Southeast Lakes Program--A Model</u> - Many of the above elements of a watershed-wide master stormwater planning approach are being implemented by the City of Orlando. The city has adopted an excellent local stormwater ordinance, developed a fine community education program and a prioritized urban lake management program (Zeno and Palmer, 1986). One of the most innovative programs is the Southeast Lakes Project which is designed to correct flooding problems and to reduce stormwater pollutant loads to 15 urban lakes and 58 drainage wells which currently convey untreated stormwater to an aquifer. A corrective

watershed management plan was cooperatively developed by the city, its consultants, the Department of Environmental Regulation and the St. Johns River Water Management District. The project was initiated not because of enforcement of water quality standards but because of a loss of beneficial uses and local citizen desires and perceptions. Modifications to the existing stormwater systems will be made over a ten-year period with treatment requirements based on "net environmental improvement" and total watershed load.

One of the most important aspects of the project is the use of innovative BMP designs which promote multiple objectives and take advantage of city-owned properties. At Al Coith Park a spreader swale will be built on the park's perimeter. When it rains, runoff will enter and fill the swale, overtopping the sidewalk berm and sheet flow across the grassed parkland where it will percolate into the ground. At Lake Greenwood, the surrounding city-owned land is being converted into an urban wetland and expanded lake. The wetland and lake is a complex treatment train which incorporates many BMPs into a very aesthetically pleasing stormwater system and park. In addition to improved stormwater management, the citizens are receiving added benefits of recreation and open space. In addition, the retrofitting project has stimulated redevelopment and renovation of existing properties thereby providing citizens with economic benefits as property values rise.

**REFERENCES**

Anderson, D.E. (1982). "Evaluation of Swale Design", M.S. Thesis, University of Central Florida, College of Engineering, Orlando.
E.P.A. (1976). Quality Criteria for Water, EPA 440/9-76-023.
E.P.A. (1983a). Water Quality Standards Handbook.
E.P.A. (1983b). "Results of the Nationwide Urban Runoff Program--Final Report".
E.P.A. (1987). "Nonpoint Source Controls and Water Quality Standards", Chapter 2, pp 2-25, Water Quality Standards Handbook.
Florida Concrete and Products Association (1988). Pervious Pavement Manual.
Florida Dept. of Environmental Regulation (1988). "A Guide to the Interpretation of Metal Concentrations in Estuarine Sediments", Report prepared by Coastal Zone Management Section.
Gallop, J., et al. (1988). "Current Status of EPA Stormwater Rule Making", these proceedings.
Harper, H.H. (1985). "Fate of Heavy Metals from Highway Runoff In Stormwater Management Systems", Ph.D. dissertation, University of Central Florida, College of Engineering, Orlando.

Harper, H.H., M.P. Murphy and E.H. Livingston (1986). "Inactivation and Precipitation of Urban Runoff Entering Lake Ella by Alum Injection in Stormsewers", Proceedings North American Lake Management Society International Symposium, Portland, Oregon, November 5-8.

Livingston, E.H. (1984). "A Summary of Activities Conducted Under the Florida Section 208 Water Quality Management Planning Program, February 1978-September 1984", Final report submitted to U.S.E.P.A.

Livingston, E.H, et al. (1988). The Florida Development Manual: A guide to Sound Land and Water Management.

Mancini, J.L. (1983). "A Method for Calculating Effects on Aquatic Organisms of Varying Concentrations", Water Resources 17(10):1355-1362.

Mancini, J.L. and A.J. Plummer (1986). "Urban Runoff and Water Quality Criteria", Urban Runoff Quality--Impact and Quality Enhancement Technology, Engineering Foundation Conference, Henniker, New Hampshire, June 23-27.

Miller, R.A. (1985). "Percentage Entrainment of Constituent Loads in Urban Runoff, South Florida", USGS WRI Report 84-4329.

Richardson, C.J. (1988). "Freshwater Wetlands: transformers, filters or sinks?" FOREM 11(2):3-9, Duke University School of Forestry and Environmental Studies.

Stidger, R.W. (1988). "Can a Curb and Gutter Unit Clean Up Pollutants", Better Roads 58(2):18-23.

Wanielista, M.P. and E.E. Shannon (1977). "Stormwater Management Practices Evaluations", Report Submitted to the East Central Florida Regional Planning Council.

Wanielista, M.P., et al. (1982). "Stormwater Management Manual", Prepared for Florida Dept. of Environmental Regulation

Wanielista, M.P., et al. (1985). "Enhanced Erosion and Sediment Control Using Swale Blocks", Report FL-ER-35-87 submitted to Florida Dept. of Transportation (FDOT).

Yousef, Y.A., et al (1985). "Removal of Highway Contaminants by Roadside Swales", Report FL-ER-30-85 submitted to FDOT.

Zeno, D.W. and C.N. Palmer (1986). "Stormwater Management in Orlando, Florida", Urban Runoff Quality--Impact and Quality Enhancement Technology, Engineering Foundation Conference, Henniker, New Hampshire, June 23-27.

## Discussion of Mr. Livingston's Paper

Question:

Have you considered ground water impacts?

Answer:

Yes, in two studies. But I admit there isn't much information for criteria developed yet.

Question:

Do you have a data base for developing sediment-related criteria?

Answer:

Some, in salt water. We're getting an excellent data base on heavy metals on sediments.

Technological, Hydrological and BMP Basis
for Water Quality Criteria and Goals

Wayne C. Huber[*], M. ASCE

ABSTRACT

Water quality criteria and goals may be addressed through at least four means: institution of 1) technology-based controls, 2) best management practice (BMP) based controls, 3) water quality standards-based controls, and 4) hydrology-based controls. This paper considers the relative characteristics, merits and linkages of the four options in the context of control of stormwater runoff quality.

INTRODUCTION

The specter of a permit requirement for all for stormwater discharges under Section 405 of the Clean Water Act Amendments of 1987 lies ahead for the entire United States. The time-table and implications of this act for local governments are nicely summarized by Tucker (1987). Among other requirements, the act stipulates that any permits for discharges from municipal separate storm sewers shall "require controls to reduce the discharge of pollutants to the maximum extent practicable, including management practices, control techniques and system [management], design and engineering methods and such other provisions as the administrator or state determines appropriate for the control of such pollutants." EPA is currently formulating the regulations and permit requirements for the act; the agency has until February 4, 1989 to prepare this material for municipalities with separate storm sewer systems serving populations of 250,000 or more, with a delayed schedule for lesser populations.

Section 405 of the Clean Water Act and its predecessors are the general motivation behind these conference proceedings. This paper discusses generic methodologies that might be applied for the purpose of permitting of storm water discharges. These are based on:

1) available technology,
2) best management practices (BMPs),
3) water quality criteria and standards, and/or
4) hydrology.

The four methods do not constitute the only available taxonomy for analysis of stormwater runoff quality, but cover most available methods. For example, the related paper by Heaney (1988) discusses risk-based methods. Also please note that water quality criteria and various BMPs are discussed extensively in other papers in these proceedings.

-----
[*] Professor of Environmental Engineering Sciences, University of Florida, Gainesville, FL 32611.

The methods are obviously interrelated. For instance, it should be possible to categorize any control measure as an "available technology" or BMP, but the separation is not clear since both terms imply devices or measures for pollutant reduction. In addition, historically, a water quality standards-based approach for publicly owned treatment work (POTW) design meant simulation of a receiving water under assumed worst-case conditions (usually the 7-day, 10-year low flow or 7Q10). The discharge of treated waste water was readily characterized and changed relatively little with time. Thus, simpler steady-state modeling tools could be used. For storm water, none of these simplifications hold: the effluent is highly variable and is a direct function of the catchment hydrology, and 7Q10 may not be the worst case condition at all. Hence, for storm water control, the water quality standards-based approach must be coupled with the hydrology-based approach.

The following discussion points out such ambiguities and does not attempt to provide a keen definition for each method. Rather, the methods are compared on the basis of their relative ability to address the needs for compliance with control of stormwater discharges.

TECHNOLOGY-BASED CONTROL

Technology-based refers to implementation of treatment processes that remove a pollutant at a specified level (e.g., 90 % removal) or to a specified concentration (e.g., $\leq$ 30 mg/l) using available, standard technology. The usual example is the requirement for a minimum of secondary treatment at all POTWs in the U.S. -- the technology is well understood and "standard," although any number of treatment configurations could be used in an individual plant design. The requirement for treatment up to a specified technological level is independent of the need for or effect of the treatment. That is, the resultant effect on receiving waters is not considered explicitly, and the treatment may or may not be enough to improve receiving water quality to desired levels. In fact, it may have no discernible effect on receiving waters at all (because of large dilution).

The cost of a technology-based control is usually well defined. However, the benefits are not defined at all since receiving water effects are not evaluated. Hence, there is no optimization performed other than in minimizing the cost of the treatment facility. The method is appropriate for POTWs because of universally poor (and uniform) quality of untreated sewage and because of the public demand for such treatment (EPA, 1979). Stormwater, however, has entirely opposite characteristics, and it is hard to imagine a universal requirement for, say, secondary treatment of every stormwater outfall.

Other difficulties with technology-based control of storm water are the lack of pollutant removal or control standards and the lack of a standard technology in general. For example, secondary treatment at a POTW typically consists at least of BOD and suspended solids control to a maximum of 30 mg/l for each in the effluent (standards will vary with different states). No such consensus exists (or may even be possible) for storm water. Because of its intermittent nature, high variability, and generally low concentrations of pollutants (compared to raw sewage), a standard or "customary" technology (analogous to activated sludge at a

POTW) does not exist for storm water. Thus, treatment at an outfall will require a highly site-specific design, thus increasing the cost of implementation.

If standards were developed, it would be possible in theory to require treatment at every outfall to meet these standards, ignoring the obvious costs, not to mention impossibility, of treating every one of millions of stormwater discharges across the country. A size criterion (e.g., minimum flow, minimum pipe diameter, minimum area drained) might be used to set a threshold level for treatment, but this would still ignore the fundamental need for some evaluation of receiving water effects. Technology-based control of combined sewer overflows (CSOs) is difficult to implement for all of the reasons discussed above (EPA, 1979), and it appears to be even less suited for storm water control.

BEST MANAGEMENT PRACTICE-BASED CONTROL

Best management practice (BMP) options for control of storm water are many. Retention/detention (i.e., treatment associated with storage) is probably the most widely used method, with use of wetlands, grassed swales, porous pavement, street cleaning, skimmers for floatables and oil and grease, earth filtration, etc. often considered as options for improvement of the quality of storm water. Several such BMPs are discussed in detail in related papers in these proceedings. "Good housekeeping" and maintenance are vital to reduction of combined sewer overflows (CSOs) and the concept is readily extended to storm sewers as well. Removal of illicit connections to storm sewers is a requirement of the Clean Water Act Amendments and can result in an enormous improvement to stormwater quality (e.g., Schmidt and Spencer, 1986). Since construction costs are usually small for good maintenance practices (e.g., removal of solids in the collection system, repair of hydraulic structures, control of infiltration), they tend to be very cost-effective as well (Lager et al., 1977).

The principal difference between a BMP and a technology-based control is that there is no "guarantee" of performance (pollutant reduction) from a BMP. The removal effectiveness of a detention pond, for example, depends upon the nature of the influent, treatability of the pollutants (especially important is the degree of particulate matter and settleability), detention time, possibility of short-circuiting, etc. The inherent variability of storm water means that the same control measure is likely to behave differently in different settings, unlike the consistency with which raw sewage may be treated. Moreover, the performance of many BMPs is still largely undocumented, with clear design parameters missing. In general, particulates and solids can be reduced by many means, but dissolved constituents are much more difficult to reduce by most BMPs.

Notwithstanding the lack of certainty about performance, BMPs can be consistently effective for some problems, notably erosion control at construction sites. Erosion control BMP implementation can be easily monitored, and receiving water (and land surface) benefits are visible to the public, without need for quality monitoring.

Ease of regulation is perhaps the primary advantage of BMPs. A developer can be required to provide a certain amount of storage of runoff; it is a simple matter to ensure compliance. This approach has been used in Florida, for example, in which Florida Department of Environmental Regulation Chapter 17-25 states in part that stormwater discharge facilities are exempt from regulation if they "provide retention, or detention with filtration, of the runoff from the first one inch of rainfall; or, as an option, for projects or project sub-units with drainage areas less than 100 acres, provide retention, or detention with filtration, of the first one half inch of runoff." (Several other options are also included in the Stormwater Rule.)

The primary disadvantage of purely a BMP approach is the same as for technology-based control: there is no evaluation of receiving water effects. This is partly due to lack of knowledge of the actual treatment effectiveness, although in some instances post-project monitoring may be required to establish whether or not receiving water quality goals are being met.

Some forms of BMP implementation also suffer from maintenance problems. For example, storage on private property may be required, perhaps through zoning regulations, on roof tops, parking lots, or other areas on which the owner may be less than eager to have standing water. There is thus little incentive for maintenance that may result in a perceived economic disadvantage, and it is difficult for an agency to regulate such a distributed system of controls. Publicly managed facilities do not have these problems -- if adequate funds are budgeted for maintenance.

WATER QUALITY STANDARDS-BASED CONTROL

Water quality standards-based refers to implementation of treatment sufficient to meet specified receiving water quality standards. The determination of the required level of control is usually made on the basis of receiving water quality modeling. The receiving water standard can be based on any number of criteria, e.g., aquatic life (toxics, DO), human health risks (bacteria, toxics), lake eutrophication (nutrients), or shellfish (bacteria). Both acute (short-term) and chronic (long-term) effects can be considered.

Concentration and/or loading levels for DO, bacteria and nutrients are fairly well defined for control purposes, but a problem with this approach is that concentration standards for effects of most toxics are not well defined at either the acute or chronic levels. The EPA Nationwide Urban Runoff Program (NURP) provides an extensive discussion of receiving water quality criteria for storm water, including the effect of the intermittent nature of stormwater discharges (EPA, 1983). A comparison of these criteria for heavy metals is shown in Table 1. The NURP team compared the various levels shown in Table 1 to the frequency distribution (alternatively expressed in terms of return period) of instream concentrations during storm events (see Figure 1 for an example). A judgment must then be made as to the relative benefits and costs of reducing a criterion violation from, say, once a year to once every five years. (The process of determining the concentration-frequency relationships lends itself nicely to the hydrology-based approach discussed below.)

Table 1. Summary of Receiving Water Target Concentrations (ug/l) Used in Screening Analysis for Target Substances (EPA, 1983).

| Contaminant | Water Hardness mg/l (as Ca CO$_3$) | Freshwater Aquatic Life | | Saltwater Aquatic Life | | Human Ingestion | Estimated Effect Level For Intermittent Exposure | |
|---|---|---|---|---|---|---|---|---|
| | | 24 Hour | Max | 24 Hour | Max | (1) | Threshold | Significant Mortality |
| Copper | 50 | 5.6 | 12 | 4.0 | 23 | NP | 20 | 50 - 90 |
| | 100 | 5.6 | 22 | 4.0 | 23 | | 35 | 90 - 150 |
| | 200 | 5.6 | 42 | 4.0 | 23 | | 80 | 120 - 350 |
| | 300 | 5.6 | 62 | 4.0 | 23 | | 115 | 265 - 500 |
| Zinc | 50 | 47 | 180 | 58 | 170 | NP | 380 | 870 - 3,200 |
| | 100 | 47 | 321 | 58 | 170 | | 680 | 1,550 - 4,500 |
| | 200 | 47 | 520 | 58 | 170 | | 1,200 | 2,750 - 8,000 |
| | 300 | 47 | 800 | 58 | 170 | | 1,700 | 3,850 - 11,000 |
| Lead | 50 | 0.75 | 74 | | | 50.0 | 150 | 350 - 3,200 |
| | 100 | 3.8 | 172 | (25) | (670) | | 360 | 820 - 7,500 |
| | 200 | 12.5 | 400 | | | | 850 | 1,950 - 17,850 |
| | 300 | 50.0 | 660 | (C) | (A) | | 1,400 | 3,100 - 29,000 |
| Chrome (+3) | 50 | | 2,200 | | | 170.00 | | |
| | 100 | (44) | 4,700 | N.P. | (10,300) | | 8,650 | |
| | 300 | (C) | 15,000 | | (A) | | | |
| Chrome (+6) | - | 0.29 | 21.0 | 18 | 1260 | 50.0 | | |
| Cadmium | 50 | 0.01 | 1.5 | | | 10 | 3 | 7 - 160 |
| | 100 | 0.02 | 3.0 | 4.5 | 59.0 | | 6.6 | 15 - 350 |
| | 300 | 0.08 | 9.6 | | | | 20 | 45 - 1,070 |
| Nickel | 50 | 56 | 1,090 | | | 13.4 | | |
| | 100 | 96 | 1,800 | 7.1 | 140.0 | | | |
| | 300 | 220 | 4,250 | | | | | |

NOTES:
- NP = No criteria proposed.
- Some toxic criteria are related to Total Hardness of receiving water. Where this applies, several values are shown. Other values may be calculated from equations presented in EPA's Criteria Document (Federal Register, 45,231, November 28, 1980). Where a single value is shown, water hardness does not influence toxic criteria.
- Concentration values shown within parentheses ( ) are not formal criteria values. They reflect either chronic (C) or acute (A) toxicity concentrations which the EPA toxic criteria document indicated have been observed. Values of this type were reported where the data base was insufficient (according to the formally adopted guidelines which were used in developing the criteria) for EPA to develop 24 hour and Max values.
- Note (1): The "Human Ingestion" criteria developed by the EPA Toxic Criteria documents are indicated to relate to ambient receiving water quality. The Drinking Water Criteria relate to finished water quality at the point of delivery for consumption.
- Estimated Effects levels reflect estimates of the concentration levels which would impair beneficial uses under the kind of exposure conditions which would be produced by Urban Runoff. They are an estimate of the relationship between continuous exposure and intermittent, short duration exposures (several hours once every several days). Threshold concentrations are those estimated to cause mortality of the most sensitive individual of the most sensitive species.
  Significant Mortality concentrations are shown as a range which reflects 50 percent of the most sensitive species and mortality of the most sensitive individual of the 25th percentile species sensitivity.

A conservative, non-degradation policy (e.g., zero discharge or zero risk) is sometimes applied to prevent any deterioration of receiving water quality due to a discharge. This can result in an effluent that is "cleaner" than the receiving water; in others instances it may prevent a discharge altogether if the technology is insufficient or too costly to remove a pollutant to a desired level.

The costs associated with a given level of control of storm water can be evaluated, although, as discussed earlier, the technology is not standardized. In principle, the benefits accrued due to maintenance of receiving water quality standards can also be evaluated (e.g., through damages attached to beach closings, fish kills, user-days of recreation, etc.). In practice, benefit evaluation can be very difficult; see the related paper by Heaney (1988) for a better discussion.

In principle, establishment of stormwater control levels based on receiving water impacts is intellectually pleasing and can address the actual issues of beneficial use of receiving waters in a scientifically defensible manner. In practice, however, establishment of receiving

# WATER CRITERIA AND GOALS

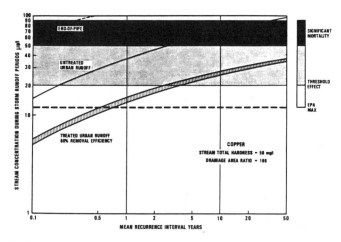

Figure 1. Hypothetical relationship between recurrence interval and pollutant concentration, from NURP screening methodology (EPA, 1983).

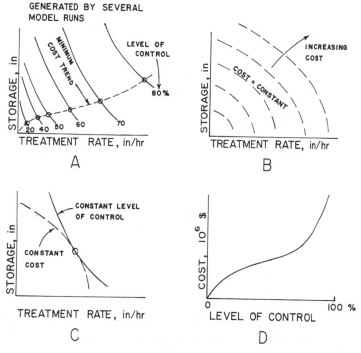

Figure 2. Conceptual cost optimization.

water standards or criteria under wet-weather conditions is difficult, as is the determination of dollar benefits due to pollutant reduction. The overall analysis required for this approach is also probably the most difficult of any considered.

## HYDROLOGY-BASED CONTROL

Hydrology-based refers to interception, storage and/or treatment of the T-year event (or more likely for quality control, the T-month event), where the return period is based on peak flow or (for quality control) total volume of runoff. (Return period and other probability measures such as cumulative frequency may be used interchangeably in the following discussion.) Depending upon the simulation model used, return period could also be based directly on predicted pollutant load, maximum load rate (e.g., g/sec), maximum concentration, event mean concentration, etc. (e.g., Robinson and James, 1984). The output of the continuous simulation is the time series for the parameters of interest, some options for which have just been mentioned. If receiving water quality is also simulated, the time series of concentrations in the receiving waters can also be predicted. The time series may then be analyzed in various ways, most often in the form of a frequency distribution (e.g., Figure 1), as in the NURP studies (EPA, 1983).

The usual procedure is thus to perform a continuous simulation of the rainfall-runoff-quality processes (an alternative is discussed below) and possibly the receiving water processes as well. The continuous simulation approach avoids the use of an arbitrary design storm as the forcing function for the stormwater runoff processes, and easily incorporates the effect of antecedent conditions on both quantity and quality processes. A decision must then be made as to what is the desired level of control, e.g., what return period event should be controlled? One way is to base the decision on predicted receiving water quality -- in essence, water quality standards-based control, with "level of control" optionally tied to return period of an event. This method accounts for receiving water effects, but costs may not be directly considered. Alternatively, 1) optimal (minimal) control costs may be determined for a given level of control (Figure 2, using storage and treatment as alternative control options), and 2) the costs plotted against level of control (dollars vs. return period) from which the "knee of the curve" might used as a logical (?) cut-off point for control (Figure 2D). The very high costs associated with capture of rare events cause the cost curve to rise dramatically near 100 percent control. A decision made on this basis ignores receiving water benefits, but avoids high marginal costs of control.

A third option is to perform a cost-benefit analysis if the benefits attributed to a given level of control can be determined. A hypothetical example is presented by Donigian and Linsley (1979). If the benefits are based on receiving water quality this method is probably the best of any, but also the most difficult and subject to errors related directly to the accuracy of the runoff and receiving water simulation. Unfortunately, the errors inherent in urban runoff water quality simulation are quite large (Huber, 1985).

Continuous simulation is not the only method for generation of the frequency distribution of hydrologic or water quality parameters; the derived distribution method is also useful. First applied for frequency analysis of flood peaks (Eagleson, 1972), it was later applied to problems of urban runoff by Howard (1976), Chan and Bras (1979), Zukovs (1983), and Loganathan et al. (1985). These analyses dealt primarily with runoff quantity. Hydroscience (1979) and Di Toro (1984) developed derived distribution methods that resulted in predictions for the in-stream frequency distribution of a runoff quality parameter. Their methods were refined and applied to the EPA NURP assessment methodology (EPA, 1983), alluded to earlier. For example, an in-stream analysis of urban runoff effects on copper concentrations is shown in Figure 1. The analysis could be carried further by assuming recurrence interval is a function of level of control (e.g., removal of copper from the urban runoff), and costs are a function of level of control. Thus, it is possible to construct a plot of cost vs. level of control, or cost vs. return period of copper concentration at a given concentration level, from which control decisions can be made. Since control effectiveness and cost are site specific, this was not done as part of the NURP study.

CASE STUDIES

San Francisco Bay

Details of the master plan for control of combined sewer overflows into San Francisco Bay are presented by McPherson (1974). Even though this relates primarily to CSOs, it is important for stormwater control because the San Francisco study led to the development of the STORM model (Roesner et al., 1974, HEC, 1977), the first continuous simulation model applied to urban runoff problems. STORM computes runoff quantity and quality in a simple manner using an hourly time step, followed by storage and/or treatment at the downstream end. STORM's methodology does not account for the fact that storage and treatment are basically inseparable; in particular, no treatment (removal) at all is simulated during storage. As an approximation, however, various combinations of storage and treatment can be applied (and traded off) at the downstream end to reduce the number of overflows (or, say, BOD loading), as indicated in Figure 3. When costs are assigned to levels of storage and treatment, the optimal combination of storage and treatment can be found, and the cost plotted vs. level of control, as conceptualized in Figure 2.

Receiving water quality modeling was not used to settle upon a design criterion for San Francisco Bay. Rather, data similar to those of Figure 3 were coupled with cost evaluations to determine a goal for reduction of CSO events to the Bay from an estimated 82 per year to 1 overflow every 2 years. This was later relaxed to 2 overflows per year because the costs for the original goal were too high. Even two overflows per year is an ambitious goal. As shown in Figure 4, in general, control of low return period events (e.g., weeks to months) is enough to remove most of the pollutant loads from urban runoff (Heaney et al., 1979).

Des Moines River

Another city with CSOs, Des Moines, Iowa was studied as part of a

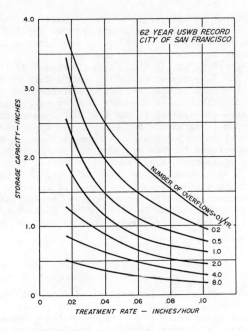

Figure 3. STORM-generated relationship between number of CSOs and level of storage and treatment (Roesner et al., 1974).

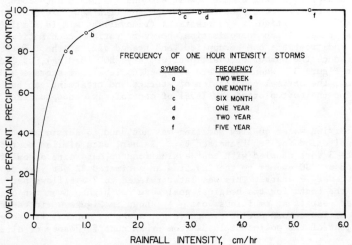

Figure 4. Overall percent rainfall control (capture) vs. hourly rainfall intensity, for Atlanta, GA, 1948-1972 (Heaney et al., 1979). Rainfall can be used as surrogate for runoff. Capture of small storms leads to high level of overall control.

demonstration of the use of continuous simulation for evaluation of receiving water impacts (Medina et al., 1981). STORM was used to generate pollutant loads to the river and Level III Receiving (Medina, 1979) was used to compute exceedance frequencies for in-stream DO. Control costs are shown in Figure 5 as a function of percent of storm events that create a violation of a 4 mg/l standard -- in essence a water quality standards based approach with the recognition that it is infeasible to meet this standard 100 percent of the time. As can be seen, incremental costs are high for reduction of the percent violations to low levels.

Little Juniata River

A very similar approach to the Des Moines study was applied to evaluation of the impact of CSOs on the Little Juniata River at Altoona, Pennsylvania (Warwick and Edgmon, 1988). STORM was again used to drive a simplified version of the Level III Receiving model (all input lumped, no dispersion). DO rate coefficients were adjusted to calibrate the model to event-based DO measurements. A cumulative frequency analysis of river DO from a 3-yr simulation indicated that DO was less than or equal to 4 mg/l 42 percent of the time (for 208 CSO events). The process was highly seasonal, as shown in Figure 6. A river scour model was also used to simulate generation of BOD by sludge resuspension, and parameter uncertainty was assessed. A cost analysis is not given in the paper, but the frequency basis is clear.

Halifax River

As part of the EPA 208 area-wide waste water management plan for Volusia County, Florida, the effect of storm water and waste water discharges into the Halifax River, a 40-km tidal estuary near Daytona Beach (part of the Intracoastal Waterway), was studied using continuous simulation for a typical three-month summer wet season (Scholl et al., 1980). The Receive model was used to simulate dynamic variations of DO, total nitrogen (TN) and total phosphorus (TP) under various combinations of estimated storm water runoff with and without control and estimated year 2000 POTW input with and without control. In general, POTW loadings of ultimate oxygen demand, TN and TP were much greater than for storm water. Hence, the conclusions of the study (POTW control is most important) were not surprising.

Results were presented in the form of cumulative duration curves for concentration (Figure 7). It is apparent from the study that POTWs had the dominant effect on TN and TP since reduction (hypothetical elimination) of POTW effluent led to the greatest concentration reductions in the receiving water. The effect on Halifax River DO of either POTW effluent or storm water was small due to the high background oxygen demand. For the Halifax River at least, control of POTW effluent (for estimated year 2000 loadings) should precede stormwater control. One implication of this study is that it is possible to assess the relative contribution of storm water to receiving water quality on a regional basis, with results presented on a frequency basis. A similar procedure is described by Donigian and Linsley (1979), including a hypothetical damage assessment due to DO reduction.

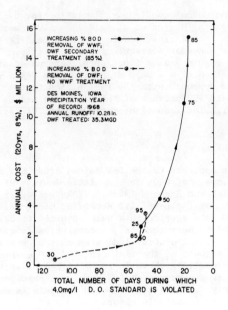

Figure 5. Cost-effectiveness of control alternatives for Des Moines River, one-year simulation (Medina et al., 1979).

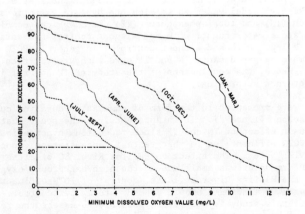

Figure 6. Minimum DO-frequency curves for seasonal storm groupings, Juniata River (Warwick and Edgmon, 1988).

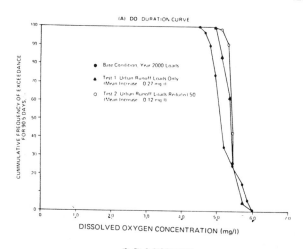

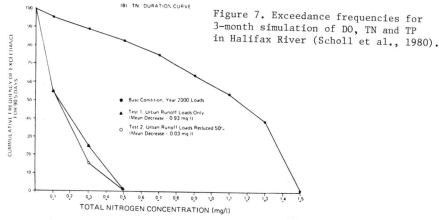

Figure 7. Exceedance frequencies for 3-month simulation of DO, TN and TP in Halifax River (Scholl et al., 1980).

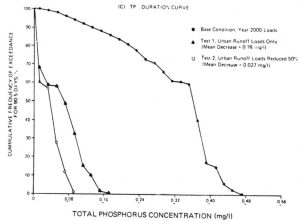

## Megginnis Arm

The Megginnis Arm tributary to Lake Jackson, Florida is an urbanized catchment of approximately 2230 acres in northern Tallahassee (Figure 8). As part of the EPA Clean Lake Program, a combination of BMPs -- detention pond, earthen filter and artificial marsh -- was installed to alleviate degradation of Lake Jackson due to rapid urbanization of its watershed, with sedimentation and phosphorus being of primary concern (Esry and Bowman, 1984). The filter and downstream marsh are designed to pass approximately 25 cfs, with overflow over the spillway expected on the order of once per year. Detailed design information is provided by Esry and Bowman (1984). The project was designed as part of a research and development effort without benefit (or need) of continuous simulation or receiving water quality modeling to demonstrate the effectiveness of the various BMPs. Preliminary monitoring indicates that the filter is doing a good job of removing clay particles, with relatively little effect on phosphorus. While effluent from the marsh is visually cleaner than the influent, the long-term phosphorus removal capability of the marsh is yet to be determined. Megginnis Arm has also been used to demonstrate the difficulties of synthetic design storms when compared to use of historic storms chosen from continuous simulation results using the SWMM model (Huber et al., 1986).

SUMMARY AND CONCLUSIONS

Several options exist for analysis of control of stormwater discharges. Water quality standards-based methods (that must incorporate hydrology-based techniques for stormwater analysis) are the most sophisticated and probably provide the most meaningful information for analysis of stormwater impacts on receiving waters in terms of impacts upon beneficial use and the potential for economic optimization. "Sophistication" carries a price, however, in terms of level of effort of the required analysis and considerable uncertainty in the results of water quality simulation (unless extensive data are available for calibration and verification). Most examples of this type of analysis for storm water are for large city-wide or basin-wide simulations. The detail of individual storm sewer outfalls is seldom addressed. Such a large-scale view of the problem is probably justified, however, in an effort to determine whether or not storm water constitutes a receiving water problem in the first place. The analysis is quite feasible, as shown in CSO and other assessment analyses, but is most likely to be performed by a public agency for determination of policy rather than by a consultant or developer at the micro-scale level.

Another approach is to provide some control measure or BMP regardless of the need and cost and regardless of whether or not the measure will be effective enough. This practice is relatively easy to monitor and implement in regulations, and its value will increase as additional research provides new data on BMP effectiveness and suitability under various situations. However, this practice does not directly deal with effects on receiving water, at least not on a case by case basis. Hence, there will always be room for doubt as to whether a law or rule based only on BMPs is adequate.

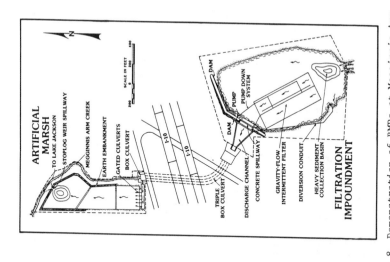

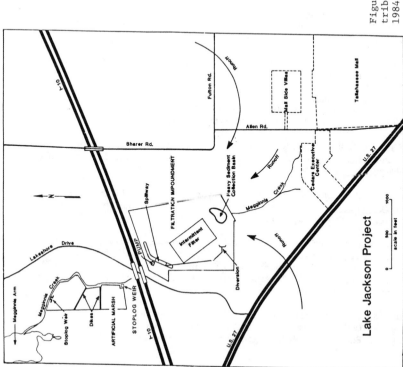

Figure 8. Demonstration of BMPs at Megginnis Arm tributary, Tallahassee, FL (Esry and Bowman, 1984).

## REFERENCES

Chan, S. and R.L. Bras, "Urban Storm Water Management: Distribution of Flood Volumes," Water Resources Research, Vol. 15, No. 2, pp. 371-382, April 1979.

Di Toro, D.M., "Probability Model of Stream Quality Due to Runoff," J. Environmental Engineering, ASCE, Vol. 110, No. 3, pp. 607-628, June 1984.

Donigian, A.S. and R.K. Linsley, "Continuous Simulation for Water Quality Planning," Water Resources Bulletin, Vol. 15, No. 1, pp. 1-16, February 1979.

Eagleson, P.S., "Dynamics of Flood Frequency," Water Resources Research, Vol. 8, No. 4, pp. 878-898, August 1972.

Environmental Protection Agency, "Benefit Analysis for Combined Sewer Overflow Control," Seminar Publication, EPA-625/4-79-013, Environmental Protection Agency, Cincinnati, OH, April 1979.

Environmental Protection Agency, "Results of the Nationwide Urban Runoff Program, Volume I, Final Report," NTIS PB84-185552, Environmental Protection Agency, Washington, DC, December 1983.

Esry, D.H. and J.E. Bowman, "Final Construction Report, Lake Jackson Clean Lakes Restoration Project," Northwest Florida Water Management District, Havana, FL, June 1984.

Heaney, J.P., "Cost Effectiveness and Urban Stormwater Quality Criteria," These Proceedings, July 1988.

Heaney, J.P., Huber, W.C., Field, R.I. and R.H. Sullivan, "Nationwide Cost of Wet-Weather Pollution Control," J. Water Pollution Control Federation, Vol. 51, No. 8, pp. 2043-2053, August 1979.

Howard, C.D.D., "Theory of Storage and treatment Plant Overflows," J. Environmental Engineering Division, Proc. ASCE, Vol. 102, No. EE4, pp. 709-722, August 1976.

Huber, W.C., "Deterministic Modeling of Urban Runoff Quality," in Urban Runoff Pollution, H.C. Torno, J. Marsalek and M. Desbordes, eds., NATO ASI Series, Vol. G10, Springer-Verlag, New York, pp. 167-242, 1985.

Huber, W.C., Cunningham, B.A. and K.A. Cavender, "Use of Continuous SWMM for Selection of Historic Design Events in Tallahassee," Proc. Stormwater and Water Quality Model Users Group meeting, Orlando, FL, EPA/600/9-86/023 (NTIS PB87-117438/AS), Environmental Protection Agency, Athens, GA, pp. 295-321, March 1986.

Hydrologic Engineering Center, "Storage, Treatment, Overflow, Runoff Model, STORM, User's Manual," Generalized Computer Program 723-S8-L7520, Corps of Engineers, Hydrologic Engineering Center, Davis, CA, August 1977.

Hydroscience, Inc., "A Statistical Method for Assessment of Urban Stormwater Loads -- Impacts -- Controls," EPA-440/3-79-023, Environmental Protection Agency, Washington, DC, January 1979.

Lager, J.A., Smith, W.G., Lynard, W.G., Finn, R.M. and E.J. Finnemore, "Urban Stormwater Management and Technology: Update and User's Guide," EPA-600/8-77-014 (NTIS PB-275654), Environmental Protection Agency, Cincinnati, OH, September 1977.

Loganathan, V.G., Delleur, J.W. and R.I. Segarra, "Planning Detention Storage for Stormwater Management," J. Water Resources Planning and Management, ASCE, Vol. 111, No. 4, pp. 382-398, October 1985.

McPherson, M.B., "Innovation: A Case Study," Tech. Memorandum No. 21 (NTIS PB-232166), ASCE Urban Water Resources Research Program, New York, February 1974.

Medina, M.A., "Level III: Receiving Water Quality Modeling for Urban Stormwater Management," EPA-600/2-79-100 (NTIS PB80-134406), Environmental Protection Agency, Cincinnati, OH, August 1979.

Medina, M.A., Huber, W.C., Heaney, J.P. and R.I. Field, "River Quality Model for Urban Stormwater Impacts," J. Water Resources Planning and Management Div., Proc. ASCE, Vol. 107,* No. WR1, pp. 263-280, March 1981.

Robinson, M.A. and W. James, "Chedoke Creek Flood Storage Computed by Continuous SWMM and Dynamic Storms," Computational Hydraulics Inc., Report R124 to the Hamilton-Wentworth Regional Engineering Department, Hamilton, Ontario, December 1984.

Roesner, L.A., Nichandros, H.M., Shubinski, R.P., Feldman, A.D., Abbott, J.W. and A.O. Friedland, "A Model for Evaluating Runoff-Quality in Metropolitan Master Planning," Tech. Memorandum No. 23 (NTIS PB-234312), ASCE Urban Water Resources Research Program, New York, April 1974.

Schmidt, S.D. and D.R. Spencer, "The Magnitude of Improper Waste Discharges in an Urban Stormwater System," J. Water Pollution Control Federation, Vol. 58, No. 7, pp. 744-748, July 1986.

Scholl, J.E., Heaney, J.P. and W.C. Huber, "Water Quality Analysis of the Halifax River, Florida," Water Resources Bulletin, Vol. 16, No. 2, pp. 285-293, April 1980.

Tucker, L.S., "The Haunting Specter of the Clean Water Act Amendments," Flood Hazard News, Vol. 17, No. 1, Urban Drainage and Flood Control District, Denver, CO, p. 3, December 1987.

Warwick, J.J. and J.D. Edgmon, "Wet Weather Water Quality Modeling," J. Water Resources Planning and Management, ASCE, Vol. 114, No. 3, pp. 313-325, May 1988.

Zukovs, G., "Development of a Probabilistic Runoff Storage/Treatment Model," Master of Engineering Thesis, University of Toronto, Dept. of Civil Engineering, Toronto, Ontario, 1983.

Cost-Effectiveness and Urban Storm-Water Quality Criteria

James P. Heaney, M., ASCE*

Abstract

This paper reviews alternative methods of evaluating the effectiveness of investments in urban stormwater quality improvement. Experience in the related area of control of dry-weather flows is presented. Results indicate that no single measure of performance or effectiveness has emerged. Primary reliance is still placed on specifying a required level of treatment. Even with proven treatment technology and relatively uniform influent, the effluent variability is similar to urban runoff. Next, methods of risk assessment are presented from the perspective of their potential applicability to evaluating urban runoff control programs. Interest in benefit-cost analysis, cost-effectiveness, and cost-risk analysis has been rekindled in the 1980's due to concern over the enormous cost of control programs. Urban runoff control in particular is an example of an area where policy makers could not be convinced of the wisdom of spending 100's of millions of dollars to control a source of contamination which had only ill-defined impacts on receiving waters. This paper summarizes methods for doing a cost-benefit assessment of urban stormwater problems. Experience in assessing the value of dry-weather sewage treatment facilities is used as a guideline in recommending feasible approaches for examining the wet-weather problem.

Introduction

Methods for evaluating water quality management programs with particular emphasis on urban stormwater are reviewed. The first part of the paper summarizes several recent attempts to evaluate the cost-effectiveness of the federal investment of $37 billion in wastewater treatment from 1972 to 1985. These assessment efforts include analysis of water quality trends in rivers, surveys of changes in fisheries, comparisons of water quality before and after added wastewater treatment was installed, comparisons of standards violations, and tabulations of general perceptions of water quality changes. Also, EPA has sponsored numerous research efforts directed at evaluating recreational benefits associated with water quality improvement. The steady state dry-weather problem is relatively simple as compared to the transient wet-weather problem so it is instructive to see what progress has been made in solving this easier problem.

Also, in the past several years, research has continued in developing improved analytical techniques including risk analysis methods to evaluate a variety of water resources problems. These techniques will be reviewed within the context of their potential relevance to urban

---
*Director, Florida Water Resources Research Center, and Prof., Dept. of Environmental Engineering Sciences, Univ. of Florida, Gainesville.

stormwater quality problems. Lastly, the specific problem of evaluating the urban stormwater quality problem is examined based on experience gained from evaluating dry-weather problems and past efforts to define the cost-effectiveness of investments in wet-weather quality control.

Cost-Effectiveness of Dry-Weather Treatment

From 1972 to 1985, the Federal government invested $37 billion in sewage treatment plant construction. EPA Needs surveys indicate an additional $109 billion will be needed between 1985 and the year 2000 for dry-weather facilities alone (US Government Accounting Office, 1986). The Government Accounting Office (GAO) reviewed five studies done in the early 1980's to evaluate the cost-effectiveness of the Federal investment over the period from 1972 to 1985. A summary of these five studies, shown in Table 1, compares how each of them defined water quality, what indicators were used, the method of data collection and analysis, and whether any generalizations were made (GAO, 1986).

The Inventory report is based on the 1982 water quality inventory by EPA (US EPA, 1984). It summarizes the 1982 reports from the states on progress in achieving a variety of water quality goals. However, no consistent methodology was used in developing these summaries. Rather, this Inventory report is a combination of aggregate statistics on control programs, and anecdotal information regarding water quality improvements.

The STEP report was prepared by the Association of State and Interstate Water Pollution Control Administrators (1984) under an EPA contract. Using a survey form, the states evaluated about 700,000 of the 1.8 million stream miles in the United States. The results were tabulated in terms of whether these streams supported, partially supported, or did not support the designated use based on examination of chemical and biological indicators and direct observation and professional judgment.

The 1982 National Fisheries Survey used a survey of biologists to evaluate the condition of the receiving waters of the United States (U.S. Environmental Protection Agency and U.S. Fish and Wildlife Survey, 1984). The biologists' responses were based on a mixture of objective biological and chemical data and subjective judgments. The results are summarized in Table 2.

The biologists responding to the survey ranked the most important factors affecting the nation's fisheries. The results, shown in Table 3, indicate that nearly 70 percent of the impact is associated with turbidity, high water temperature, and nutrient surplus. Toxic substances and dissolved oxygen deficiency account for 17 percent of the impact. This result suggests that the most important indicators for fisheries impacts are not the same ones used in most of our receiving water impact assessments in urban areas. Also, some of the most important factors are not significantly affected by wastewater treatment plants, e.g., temperature.

Table 1. Methodological Features of the Five Reports Reviewed by 1986 GAO Study.

| Report | Definition of water quality | Indicator | Data collection | Analysis | Generalization |
|---|---|---|---|---|---|
| Inventory | Various (effluent load, in-stream characteristics, designated use) | Various (standards violations, fish kills) | Various (monitoring stations, intensive surveys, perceptions) | Raw data, stream-by-stream description, summary tables | No national summary, reports generalize to the state level |
| STEP | Capability for designated use | Various (state officials' perceptions of chemical and biological measures) | Instruments designated by state water-quality officials (who often relied on data similar to Inventory data) | Summation of responses | Yes |
| Fisheries survey | Capability for designated use, fishability | Biologists' estimates of presence of fish | Sample survey; questionnaires to experts | Summary tables and graphs | Yes |
| Before-and-after case studies | In-stream characteristics | Dissolved oxygen, dissolved oxygen deficit, ammonia, biological measures | Intensive surveys, computerized data base | Stream modeling and statistical techniques | No |
| Geological Survey | In-stream characteristics | Chemical measures | Chemical data from monitoring stations | Statistical analysis with flow adjustment | No |

Table 2. Results of Fisheries Survey for Determining a River's Support of Fish Life.

| Scale Value | Attribute | % in Category |
|---|---|---|
| 5 | Maximum ability to support sport fish, species of special concern, or both | 4 |
| 4 | No description provided | 16 |
| 3 | No description provided | 25 |
| 2 | Minimum ability to support sport fish, species of special concern, or both | 22 |
| 1 | Ability to support nonsport fish only | 10 |
| 0 | No ability to support any fish population | 23 |
| | Total | 100 |

Table 3. Ranking of Factors Adversely Affecting the Nation's Fisheries in 1982.

| Rank | Factor | Percent | Cumulative Percent |
|---|---|---|---|
| 1 | Turbidity | 29 | 29 |
| 2 | High water temperature | 19.1 | 48.1 |
| 3 | Nutrient surplus | 11.3 | 59.4 |
| 4 | Toxic substances | 9.1 | 68.5 |
| 5 | Dissolved oxygen deficiency | 7.9 | 76.4 |
| 6 | Nutrient deficiency | 3.9 | 80.3 |
| 7 | Low water temperature | 2.9 | 83.2 |
| 8 | pH too acidic | 2.5 | 85.7 |
| 9 | Salinity | 1.5 | 87.2 |
| 10 | Sedimentation | 1.5 | 88.7 |
| 11 | Siltation | 0.8 | 89.5 |
| 12 | Low flow | 0.7 | 90.2 |
| 13 | Gas supersaturation | 0.6 | 90.8 |
| 14 | Herbicides and pesticides | 0.5 | 91.3 |
| 15 | pH too basic | 0.3 | 91.6 |
| 16 | Channelization | 8.2 | 91.8 |
| 17 | Other | 8.2 | 100 |
| | Total | 100 | |

Source: EPA and U.S. Fish and Wildlife Service, 1984.

Hydroqual (1984) performed a detailed before and after evaluation of 13 recipients of EPA Construction Grants funds. The principal water quality indicators were changes in dissolved oxygen, biochemical oxygen demand, ammonia, and biological measures. The results indicate improvements for 12 of the 13 bodies of water. The database for this study was two periods of time and not a complete time series that would show trends more clearly. This type of study is probably the most relevant for the type of assessments to be made for specific areas.

Lastly, the 1983 national water summary published by the U.S. Geological Survey (1984) analyzed trends in water quality for 22 indicators at approximately 300 monitoring stations located throughout the United States. Results for suspended sediment, fecal coliform, and dissolved oxygen are shown in Table 4. For these three important parameters, no change was observed 71 to 83 percent of the time with improvements in dissolved oxygen 11.2 percent of the time and a decrease in fecal coliform 12.6 percent of the time. These results show no significant impact of the improved treatment that was installed during the 1970's. These results might also suggest the need to more carefully select the sampling stations since the impacts of improved wastewater treatment would not be that apparent in larger receiving waters located significant distances from the treatment plant or plants.

Table 4. Water Quality Trends From 1974-81 Based on USGS Survey (1984).

| Indicator | No. of Stations | Percentages Increase | Decrease | No change | Total |
|---|---|---|---|---|---|
| Suspended sediment | 289 | 15.2 | 14.2 | 70.6 | 100 |
| Fecal coliform | 269 | 7.1 | 12.6 | 80.3 | 100 |
| Dissolved oxygen | 276 | 11.2 | 6.2 | 82.6 | 100 |

Performance of Dry-Weather Wastewater Treatment Plants

It is instructive to evaluate the effectiveness of dry-weather treatment plants before suggesting reasonable guidelines for the more complicated wet-weather systems. The results of a 1980 GAO study of dry weather plants indicate the following:

"The EPA's statistical reports on plant performance indicate that at any given point in time 50 to 75 percent of the plants are in violation of their National Pollutant Discharge Elimination Permit. GAO's random sample of 242 plants in 10 States shows an even more alarming picture-- 87 percent of the plants were in violation of their permit; 31 percent were, in GAO's opinion, in serious violation".

A summary of the frequency of the violations for 242 plants in three EPA regions, shown in Table 5, indicates that only 31 of the 242 plants had no violations. Of the 211 plants with violations, these violations lasted from one to twelve months. Violations were measured in terms of total suspended solids, biochemical oxygen demand, and fecal coliforms.

Table 5. Effluent Violations That Occurred in GAO Sample During the 12 Month Period During 1978-1979* (GAO, 1980).

| Region | Sample Number | No Violations | Facilities in violation Number of months | | | |
|---|---|---|---|---|---|---|
| | | | 1-3 | 4-6 | 7-9 | 10-12 |
| Boston | 100 | 6 | 13 | 20 | 28 | 33 |
| Chicago | 92 | 18 | 23 | 15 | 13 | 23 |
| San Francisco | 50 | 7 | 17 | 4 | 16 | 6 |
| Total | 242 | 31 | 53 | 39 | 57 | 62 |

* All plants were classified as being physically capable of providing secondary or better levels of treatment and were generally processing at least 1 mgd.

The results of special studies of a sample of the plants in violation were done in order to determine the reasons. The reasons, shown in Table 6, range from design to operating problems.

Table 6. Reasons for Plant Noncompliance in 1980 GAO Study.

| Major category | Number of plants |
|---|---|
| Design deficiencies | 10 |
| Equipment deficiencies | 2 |
| Infiltration/inflow problems | 3 |
| Industrial waste overloads | 5 |
| O&M deficiencies | 9 |
| Total* | 29 |

* Total number of plants studied equaled 15. Most plants have more than one problem.

The findings of the GAO study strongly suggest chronic deficiencies in the average performance of dry weather treatment plants. Haugh et al. (1980) evaluated the variability of performance of 43 activated sludge plants in the United States (see Table 7). The daily variability of activated activated sludge plants is compared with the event variability in urban runoff based on the results of the Nationwide Urban Runoff Program (1983). Interestingly, the variability of the dry-weather effluent BOD is similar to the variability of BOD in urban runoff. One would have expected a much higher variability in the urban runoff.

Table 7. Statistical Characteristics of Wastewater Effluent and
Stormwater Runoff (Haugh et al., 1981; US EPA, 1983).

| Pollutant | Source-all values in mg/L | | | | | |
|---|---|---|---|---|---|---|
| | Daily wastewater effluent(1) | | Urban runoff event medians | | | |
| | | | Residential | | Commercial | |
| | Median | C. of Var. | Median | C. of Var. | Median | C. of Var |
| BOD | 11.21 | 0.39 | 10 | 0.41 | 9.3 | 0.31 |
| Suspended solids | 14.42 | 0.55 | 101 | 0.96 | 57 | 0.39 |

1. Percent of variability due to measurement error is 20 for BOD and 33 for suspended solids.

Given the above results of the performance of dry-weather plants, attempts to develop refined criteria for receiving waters impacted by wet-weather flows seem premature. For example, the criterion for control of combined sewer overflows used in the 1978 EPA Needs Survey was as follows:

> The minimum receiving water dissolved oxygen concentration shall not average less than 2.0 mg/l for more than 4 consecutive hours; nor shall the minimum receiving water dissolved oxygen concentration average less than 3.0 mg/l for more than 72 consecutive hours(3 days). In addition, the annual average receiving water dissolved oxygen concentration shall be greater than 5.0 mg/l for all waters which will support warm water species and shall be greater than 6.0 mg/l for all waters which will support cold water species."

Use of these more refined criteria is a natural extension of the use of mathematical models to estimate receiving water quality. While it is relatively easy to write a computer program to tabulate the statistics for the above criterion, it is an onerous effort to develop the actual field and laboratory information to support such a recommendation, and it is also difficult to enforce such a regulation.

Risk Analysis

Interest in risk analysis in environmental and water resources engineering has been rekindled during the past decade due to the projected large costs of environmental controls, and the occurrence of large-scale disasters. A large number of papers and books dealing with risk analysis in environmental and water resources engineering are available. Many of the applications to environmental and water resources problems are summarized in Ricci (1985) and Haimes and Stakhiv (1986). Methods for evaluating risk can be divided into four categories: risk aversion, cost-risk analysis, cost-effectiveness analysis, and benefit-cost analysis (Rowe, 1977). Each of these methods is discussed below in the context of its present and potential relevance to the urban runoff quality problem.

### Risk Aversion

The Clean Water Act, as amended, sought to achieve, by mid-1983, wherever attainable, water quality that provides for the protection and propagation of fish, shellfish, and wildlife; and recreation in and on the water. These conditions are commonly referred to as the fishable/swimmable goal. This approach is an example of a risk averse policy option in that the goal is to be achieved independent of the benefits and/or costs of such a program. This anti-degradation philosophy was popular in the early and mid-1970's. Interest waned when cost estimates indicated enormous amounts of money would be required, e.g., $266.1 billion for urban runoff control estimated by the 1974 EPA Needs Survey (U.S. EPA, 1975). By comparison, from 1972 to 1985, the Federal government had invested $37 billion in the construction grants program. Thus, it is easy to see why an estimated "need" of $266.1 for urban runoff alone caused quite a stir.

### Cost-Risk Analysis

Cost-risk analysis estimates the costs required to reduce risks to some "acceptable" level as judged by finding the area where the added cost of further reductions in risk becomes relatively high. A result of this approach is to define "reasonable" levels of required treatment such as "best practicable" or "best available" technology. The use of best practicable or best available technology was popular in the 1970's as part of the EPA program. However, the agency did not usually specify a measure of risk to be reduced. A problem with this method is to define "risk" in a meaningful manner. The seriousness of a water quality problem depends on the magnitude, duration, season of the year, and other factors. Thus, it is usually difficult to agree on a single measure of risk such as the frequency of "violations" for a single constituent where the violation is measured as exceeding a prescribed value.

### Cost-Effectiveness Analysis

Cost-effectiveness analysis is similar to cost-risk analysis in that total costs are compared to some measure of performance. Heaney et al. (1979) performed a nationwide assessment of urban runoff control costs using % BOD removal as the measure of effectiveness. The results indicate that it is not cost effective to go to a very high level of control. The suggested reasonable level of control is the so-called knee of the curve, located at a removal rate of about 70 %. The resultant control program costs about 10 % of the estimate of the earlier Needs Survey. This cost can be reduced substantially by taking advantage of multi-purpose utilization of stormwater control facilities. The primary reason why the earlier estimates of stormwater control costs were so high is that a traditional design storm for drainage was used, e.g., 2 year, 1 hour storm. However, a high level of capture can be achieved by designing for more frequent events such as the one month storm. Such a control system captures all of the smaller storms and part of the larger storms. Medina et al. (1981) performed a continuous simulation of stormwater quality in Des Moines, Iowa in order to derive a cost-effectiveness relationship wherein effectiveness is defined in terms of the total number of days during which a 4.0 mg/l dissolved oxygen standard is violated. The percentage of storm events violating the dissolved oxygen standard is used as an alternative criterion. Huber (1988) discusses other examples of cost-effectiveness analysis as applied to urban runoff and CSO problems.

Benefit-Cost Analysis

Benefit-cost analysis is the preferred approach for comparing alternatives. The objective is to maximize benefits -costs. Freeman (1979) classifies environmental improvement benefits into four categories: property values, public health, recreation, and productivity. As Bockstael, Hanneman, and King (1987) point out, most of the water quality benefits are for recreational uses. Renewed interest in the early 1980's by EPA led to support for research on methods of assessing water quality benefits. Smith and Desvouges (1986) summarize this research as it relates to recreation. Freeman (1979) can be consulted for an overview of the other benefit categories. While significant strides have been made in the past several years in developing improved methods of benefit assessment, very little information is available on the relationship between water quality and benefits. For example, recreational value assessments are done using survey techniques which ask people to choose among the levels of water quality shown in Figure 1. Also, very simple relationships between water quality and recreational uses are assumed, e.g., Table 8. Thus, it is apparent from reviewing the work of economists assessing recreational benefits associated with water quality improvements that public perceptions of water quality fall into relatively few categories, and they are not able to differentiate among relatively small changes in specific water quality constituents. Rather, they respond to major shifts in water quality only as it affects obvious beneficial uses such as boating and swimming. On the positive side, examination of recreation and other benefits does permit us to have an integrative measure of the impact of improvement of water quality in its many dimensions.

The introductory section on comparing methods of selecting alternatives presented some of the measures of effectiveness that have been used to evaluate stormwater quality programs. This section presents a more detailed review of these options and suggests some of the more productive directions of future inquiry.

Effectiveness Measures for Urban Runoff Control

Receiving water measures

Frequency approach

Di Toro (1984) has developed a frequency approach for estimating the probability distribution of stream water quality due to storm runoff. This work was done as part of EPA efforts to summarize the results of studies in 30 cities as part of the EPA Nationwide Urban Runoff Program (1983). Log-normally distributed runoff and streamflows and concentrations are assumed. The result of this analysis is a relatively easy way to obtain a preliminary estimate of the probability of pollutant concentrations during storm runoff periods. This method is quite

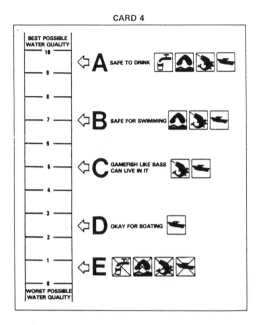

Figure 1. Water Quality Choices Presented in Assessing Benefits (Smith and Desvouges, 1985).

Table 8. Dissolved Oxygen and Recreation Activities (Smith and Desvouges, 1986).

| Dissolved oxygen required, %(1) | Feasible recreation activity |
|---|---|
| 45 | Boatable conditions |
| 64 | Fishable conditions |
| 83 | Swimmable conditions |

(1) Percent of saturation.

useful for statewide, regional, or national assessments. However, as Roesner and Dendrou (1985) point out, this method is not accurate for analysis of specific stormwater problems due to its simplifications in terms of the assumed distributions, and its lack of cause-effect relationships.

### Continuous Simulation

As discussed earlier, Medina et al. (1981) use a continuous simulation model to obtain the probability distribution of dissolved oxygen over a period of interest, e.g., the summer months. Warwick and Edgmon (1988) used this approach to evaluate a specific combined sewer overflow problem in Altoona, Pennsylvania. The expected frequency of violation of dissolved oxygen standards was evaluated using a three year historical record from 1975-1977, containing 208 storm events. The model was calibrated using detailed measurements from six storm events. The results of their analysis indicated that 42 % of all storms analyzed caused in-stream dissolved oxygen concentrations to fall below 4.0 mg/L. On a seasonal basis, 77 % of the summer storms caused a violation in the 4.0 mg/L standard. One of the most important,and yet difficult to quantify, impacts is that due to resuspension of benthal deposits. Warwick and Edgmon (1988) were able to document this phenomena for a large event with stream velocities of .689m/s (2.26 ft/sec). Overall, their study presents an excellent example of how receiving water impact assessments need to be done to develop sound policies for wet-weather control.

### Stream Reconnaissance

Heaney and Huber (1984) summarize the results of a nationwide assessment of the impact of urban runoff on receiving water quality. They concluded that documented case studies of these impacts are rare for several reasons. First, under the anti-degradation philosophy espoused by PL 92-500 in 1972, receiving water impact studies were generally not required. Next, many of the receiving waters are already polluted with other waste sources so it is difficult to isolate the individual impact of urban runoff. Lastly, the impacts of storms are transient. They concluded that a problem solving framework was necessary to adequately quantify impacts and their origins.

Schmidt and Spencer (1986) studied the Allen Creek storm drain system which was suspected of causing water quality problems. A water quality survey was performed in four phases: comprehensive drain survey, intensive bacteriological study, business survey and dye testing, and chemical sampling. Their results showed that, for this catchment, urban runoff pollution is not strictly limited to road runoff and contains wastes from permitted and illegal discharges as shown in Table 9.

### Effluent Quality Measures

The earlier discussion on dry-weather treatment plants indicated that the quality of the effluent varies considerably for activated sludge plants treating an influent of relatively uniform quantity and quality. Thus, even for this relatively uncomplicated case of treating

Table 9. Summary of Storm Drain Connections by Business Type for
Allen Creek Drainage Basin, Ann Arbor, Michigan
(Schmidt and Spencer, 1984).

| Business type | % connected to storm drain |
|---|---|
| Auto repair shops/tire stores | 65 |
| Service stations | 63 |
| Printers/copiers | 9 |
| Manufacturers | 56 |
| Dry cleaners/laundries | 0 |
| Government facilities | 80 |
| Auto parts stores | 40 |
| Auto body shops | 75 |
| University facilities | 75 |
| Muffler/transmission shops | 50 |
| Car washes | 50 |
| Auto dealerships | 100 |
| Auto rental agencies | 33 |
| Photo processors | 33 |
| Utilities | 33 |
| Plating shops | 100 |
| Private homes | 100 |

a uniform waste source with a proven technology, it is still common for the effluent quality to exhibit considerable variation. Furthermore, as the GAO (1980) study showed, many of these dry-weather plants are not meeting their average requirements for periods ranging from one month to a year or longer. Thus, given this situation for dry-weather, it seems prudent to develop relatively simple performance measures for wet-weather systems wherein the influent varies widely and the performance of the control units is uncertain.

An even more fundamental question is to what extent are the existing effluent standards meaningful? The results of Haugh et al. (1980) clearly indicate the need for explicit recognition of the need to specify performance criteria in terms of the expected variability of the output. Standard procedures exist for specifying such performance measures (see Stephanopoulos, 1984 for example) using concepts from control theory. A difficult problem with waste treatment systems is that the control processes are removing numerous pollutants, all of which are of concern in protecting beneficial uses in the receiving water. However, these pollutants are not necessarily discharged in a uniform manner, e.g., they may result from a specific episode or rainfall event. Thus, the control unit serves to reduce the threat of dangerous levels of many of these contaminants simultaneously. Correspondingly, multiple measures of effectiveness are needed with criteria of concentrations and durations for each potential pollutant. Because of the resultant complexity, past efforts have selected a single proxy of performance such as percent removal of BOD or percent of days that the D.O. standard was violated and have used it as the effectiveness measure.

A vexing question in cost effectiveness analysis of wastewater treatment plants has been to estimate the cost of removing a specific pollutant in a plant that removes numerous pollutants. Payne (1988) has shown how this problem can be solved using the simplified Shapley Value from cooperative game theory as the cost allocation device. The cost analysis is done based on assuming that the plant is built to remove one, two, three, etc. of the pollutants to the extent required for each subset of all of the pollutants of interest. Then, it is possible to prorate the cost among all of the pollutants.

Summary and Conclusions

This purpose of this paper is to review alternative ways to evaluate the cost-effectiveness of investments in urban runoff control. First, the results of similar efforts to evaluate the effectiveness of major Federal investments in dry-weather treatment facilities are presented. Even for this relatively simple case of continuous flows of known quality, no agreed upon measures of performance have emerged either for the effluent or the receiving water. Rather, numerous measures of performance have been used. These results should signal a caution to people attempting to devise similar measures for wet-weather problems.

Risk analysis methods are reviewed within the context of their relevance to urban stormwater problems. These methods provide a useful conceptual framework. However, they do not provide a solution to the problem since they still rely upon selecting a single measure of effectiveness based on the risk of violating a single water quality criterion.

The review of various measures of performance for stormwater systems reveals the same problem of trying to select a single measure of performance in terms of a standard and some required level of reliability. This exercise is difficult and may not be meaningful since numerous measures of water quality are of concern. Methods for prorating the cost of the control system are available. However, the preferred situation is to have a direct measure of benefits such as enhanced recreational opportunities. Within this context, a sound water quality management plan must provide a control system that can reduce the levels of a wide variety of pollutants that may be chronically or occasionally present in urban runoff. Then, the public will have assurances that this water is relatively safe. Secondly, priority needs to be given to implementing urban runoff controls as part of an overall water quality management plan for an area that will provide demonstrable improvements in benefits, e.g., open closed beaches or shellfish harvesting areas. The current Federal approach to this problem should provide an effective combination of requiring treatment and monitoring outfalls to reduce or eliminate illicit discharges. This analysis suggests that it might be productive to conduct some comprehensive water quality benefit assessments to show the value of the proposed major investments.

References

Association of State and Interstate Water Pollution Control Administrators. 1984. America's Clean Water: The State's Evaluation of Progress 1972-1982, Washington, D.C.

Bockstael, N.E., W. Hanemann, and C. Kling. 1987. Estimating the Value of Water Quality Improvements in a Recreational Demand Model, Water Resources Research, 23: 951-960.

Di Toro, D.M. 1984. Probability Model of Stream Quality Due to Runoff, Jour. of Environmental Engineering, ASCE, 110,3, pp. 607-628.

Freeman, A.M. III. 1979. The Benefits of Environmental Improvement: Theory and Practice, Johns Hopkins Press for Resources for the Future, Baltimore, Md.

General Accounting Office. 1980. Costly Wastewater Treatment Plants Fail to Perform as Expected, Report by the Comptroller General of the United States, Washington, D.C., 66 pp.

Haimes, Y.Y. and E.Z. Stakhiv (Eds.). 1986. Risk-Based Decision Making in Water Resources, American Society of Civil Engineers, New York, 333 pp.

Haugh, R., S. Niku, E.D. Schroeder, and G. Tchobanoglous. 1981. Performance of Trickling Filter Plants: Reliability, Stability, and Variability, EPA-600/S2-81-228, Cincinnati, Oh., p. 6.

Heaney, J.P., W.C. Huber, R.Field, and R.H. Sullivan. 1979. Nationwide Cost of Wet-Weather Pollution Control, Jour. Water Pollution Control Federation, Vol. 51, No. 8, pp. 2043-2053.

Heaney, J.P., and W.C. Huber. 1984. Nationwide Assessment of Urban Runoff Impact on Receiving Water Quality, Water Resources Bulletin, Vol. 20, no. 2, pp. 35-42.

Huber, W.C. 1988. Technological, Hydrological and BMP Basis for Water Quality Criteria and Goals, Proc. ASCE Conference on Current Practice and Design Criteria for Urban Runoff Water Quality Control, New York.

Hydroqual. 1984. Before and After Case Studies: Comparisons of Water Quality Following Municipal Treatment Plant Improvements, U.S. Environmental Protection Agency, Washington, D.C.

Medina, Jr., M.A., W.C. Huber, J.P. Heaney, and R. Field. 1981. River Quality Model for Urban Stormwater Impacts, Jour. of the Water Resources Planning and Management Division, ASCE, 107, WR1, pp. 263-280.

Payne, S.N. 1988. Efficiency/Equity Analysis of Water Resource Problems, M.E. Thesis, Department of Environmental Engineering Sciences, U. of Florida, Gainesville, 100 pp.

Ricci, F.R (Ed.). 1985. Principles of Health Risk Assessment, Prentice-Hall, Inc., 417 pp.

Roesner, L.A. and S.A. Dendrou. 1985. Discussion of Di Toro (1984), Probability Model of Stream Quality Due to Runoff, Jour. of Environmental Engineering, ASCE, 110,3, Jour. of Environmental Engineering, 111,5, pp. 738-740.

Rowe, W. 1977. An Anatomy of Risk, John Wiley and Sons, New York.

Schmidt, S.D. and D.R. Spencer. 1986. The Magnitude of Improper Waste Discharges in an Urban Stormwater System, Jour. of Water Pollution Control Federation, 58,7, pp.744-748.

Smith, V.K. and W.H. Desvouges. Measuring Water Quality Benefits, Kluwer.Nijhoff Publishing, Boston, 327 pp.

Stephanopoulos, G. 1984. Chemical Process Control, Prentice-Hall, Englewood Cliffs, N.J., 696 pp.

U.S. Environmental Protection Agency. 1975. Cost Estimates for Construction of Publicly Owned Wastewater Treatment Facilities-1974 Needs Survey, Washington, D.C.

U.S. Environmental Protection Agency. 1983. Final Report of the Nationwide Urban Runoff Program, Water Planning Division, Washington, D.C.

US Environmental Protection Agency. 1984. National Water Quality Inventory: 1982 Report to the Congress, Washington, D.C.

U.S. Environmental Protection Agency and U.S. Fish and Wildlife Survey. 1984. 1982 National Fisheries Survey, 3 vols.,Washington,D.C.

U.S. Geological Survey. 1984. National Water Summary 1983-Hydrologic Events and Issues, Washington, D.C.

Warwick, J.J. and J.D. Edgmon. 1988. Wet Weather Quality Modeling, Jour. of Water Resources Planning and Management, Vol. 114, No. 3, pp. 313-325.

## Discussion of Dr. Heaney's Paper

Question:

In your cost-effectiveness or benefit-cost surveying, have you asked: Who's going to pay?

Answer:

Most surveys have included the question of "willingness to pay": How much would **you** be willing to pay for your perceived water usage values?

Question:

Is that perception usually based on recreational usage of receiving waters, as opposed to any or all other uses?

Answer:

Yes, essentially.

Question:

Do we ever ask people whether simple "protectin of the natural resources" is a value or a benefit sufficient to justify treatment?

Answer:

Some economics researchers have identified and attempted to quantify values that the population placed on unique water resources, such as Lake Tahoe or the Grand Canyon or Okeefenokee Swamp--even if the people doing the valuing never visited or "used" those waters themselves. These benefits are known as "vicarious benefits," and, yes, that question is asked this way in some surveys.

# FEDERAL REQUIREMENTS FOR STORM WATER MANAGEMENT PROGRAMS

James D. Gallup [1]
Kevin Weiss [2]

## Introduction

In 1972, the Federal Water Pollution Control Act (referred to as the "Clean Water Act" or CWA), was amended to provide that the discharge of any pollutant to surface waters without a National Pollutant Discharge Elimination System (NPDES) permit is unlawful. Since that time, efforts to improve water quality under the NPDES program have focused primarily on reducing pollutants in discharges of industrial process wastewater and municipal sewage. As pollution control measures were initially developed for these discharges, it became evident that more diffuse sources (occurring over a wide area) of water pollution, such as non-point sources (NPS) and urban storm water runoff were also major causes of water quality problems. Controls for diffuse sources including storm water have been slower to develop than controls for other point sources such as publicly owned treatment works (POTWs).

In 1973, EPA promulgated its first storm water regulations exempting from permit requirements those conveyances carrying storm water runoff uncontaminated by industrial or commercial activity unless the particular storm water discharges had been identified by the NPDES Director as a significant contributor of pollution (38 FR 13530 (May 22, 1973). The Agency maintained that, while these sources fell within the definition of a point source, they were nonetheless ill-suited to the traditional end-of-pipe, technology-based controls that are the basis of the NPDES program. Because of the intermittent, variable, and unpredictable nature of a storm water, EPA reasoned that the problems caused by storm water discharges were better managed at the local level through nonpoint source controls such as the imposition of specific management practices to prevent the pollutants from entering the runoff. The Agency also justified its decision by noting that issuing individual NPDES permits for the hundreds of thousands of storm water point sources in the United States would create an overwhelming administrative burden and would divert resources away from control of industrial process wastewater and municipal sewage, which at the time, were more pressing and identifiable environmental problems.

---

[1] Chief, Technical Support Branch, Office of Water Enforcement and Permits, Office of Water, U.S. Environmental Protection Agency, Washington, DC, 20460.

[2] Chemical Engineer, Technical Support Branch, Office of Water Enforcement and Permits, Office of Water, U.S. Environmental Protection Agency, Washington, DC, 20460.

In what was to become the first in a series of challenges to the storm water regulations, the Natural Resources Defense Council (NRDC) brought suit in the U.S. District Court for the District of Columbia, challenging the Agency's authority to selectively exempt categories of point sources from permit requirements, NRDC v. Train, 396 F. Supp. 1393 (D.D.C. 1975), aff'd, NRDC v. Costle, 568 F. 2nd 1369 (D.C. Cir. 1977). The District Court held that EPA could not exempt discharges identified as point sources from regulations under the NPDES permit program.

In response to the District Court's decision in NRDC v. Train, EPA issued a rule on March 18, 1976, (41 FR 11307) establishing a permitting program for all storm water discharges except for rural runoff uncontaminated by industrial or commercial activity. On June 7, 1979 and May 19, 1980, EPA published comprehensive revisions to the NPDES regulations (44 FR 32854 (June 7, 1979); 45 FR 33290 (May 19, 1980). These rules required the same application information for storm water point sources as that required of all industrial and commercial process wastewater discharges. These requirements included testing under certain circumstances for pollutants identified in the 1977 amendments to the Clean Water Act (CWA) which stressed the control of toxic pollutants.

This regulation brought suits in several Courts of Appeals and District Courts by a number of major trade associations, several of their member companies, NRDC and Citizens for a Better Environment (NRDC v. EPA, 673 F. 2d. 392 (D.C. Cir. 1980). After two years of settlement negotiations, the Agency and industry petitioners signed the NPDES Settlement Agreement on July 7, 1982, which addressed a number of issues relating to the NPDES program, including storm water. Under the terms of the Agreement, EPA proposed changes to the storm water regulations (47 FR 52073 (November 18, 1982) and published final storm water regulations on September 26, 1984 (49 FR 37998). These storm water regulations generated considerable controversy and, once again, suits were filed. Several changes to the application requirements for storm water discharges were proposed in notices issued on March 7, 1985 and August 12, 1985.

## Section 405 and the WQA

At the same time that EPA was evaluating the appropriate means to regulate storm water discharges, the Congress was examining the storm water issue in the course of the reauthorization of the Clean Water Act, and on February 4, 1987, Congress passed the Water Quality Act (WQA).

The WQA contains three provisions which specifically address storm water discharges. The central provision governing storm water discharges is section 405 which adds section 402(p) to the CWA. Section 402(p)(1) provides that EPA or NPDES States cannot require a permit for certain storm water discharges until October 1, 1992 except for storm water discharges exempted under section 402(p)(2). Section 402(p)(2) lists five types of storm water discharges which are required to obtain a permit prior to October 1, 1992:

    A. A discharge with respect to which a permit has been issued prior to February 4, 1987;

B. A discharge associated with industrial activity;

C. A discharge from municipal separate storm sewer system serving a population of 250,000 or more;

D. A discharge from a municipal separate storm sewer system serving a population of 100,000 or more, but less than 250,000; or

E. A discharge for which the EPA or a State with an approved NPDES program determines that the storm water discharge contributes to a violation of a water quality standard or is a significant contributor of pollutants to the waters of the United States.

Section 402(p)(4) requires EPA to promulgate final regulations governing storm water permit application requirements for storm water discharges associated with industrial activity and discharges from large municipal separate storm sewer systems (systems serving a population of 250,000 or more) by "no later than two years" after the date of enactment (i.e. no later than February 4, 1989). The WQA also requires EPA to promulgate final regulations governing storm water permit application requirements for discharges from medium municipal separate storm sewer systems (systems serving a population of 100,000 or more, but less than 250,000) by "no later than four years" after enactment (i.e. no later than February 4, 1991).

In addition, Section 402(p)(4) provides that permit applications for storm water discharges associated with industrial activity and large municipal separate storm sewer systems "shall be filed no later than three years" after the date of enactment of the WQA (i.e. no later than February 4, 1990). Permit applications for discharges from medium municipal systems must be filed "no later than five years" after enactment (i.e. no later than February 4, 1992). The Conference Report accompanying the WQA provides that after October 1, 1992, the permit requirements of the CWA are restored for municipal separate storm sewer systems serving a population of fewer than 100,000.

The WQA clarified and amended the requirements for permits for storm water discharges in the new CWA section 402(p)(3). The Act clarified that permits for storm water discharges associated with industrial activity must meet all of the applicable provisions of section 402 and section 301 including technology and water quality based standards. However, the WQA made significant changes to the permit standards for discharges from municipal separate storm sewers. Section 402(p)(3) (B) provides that NPDES permits for such discharges:

o may be issued on a system- or jurisdiction-wide basis;

o shall include a requirement to effectively prohibit non-storm water discharges into the storm sewers; and

o shall require controls to reduce the discharge of pollutants to the maximum extent practicable, including management practices, control techniques and system, design and engineering methods, and such other provisions as the Director determines appropriate for the control of such pollutants.

The EPA, in consultation with the States, is required to conduct two studies on storm water discharges that are in the class of discharges for which EPA and NPDES States cannot require permits prior to October 1, 1992. The first study will identify those storm water discharges or classes of storm water discharges for which permits are not required prior to October 1, 1992 and determine, to the maximum extent practicable, the nature and extent of pollutants in such discharges. The second study is for the purpose of establishing procedures and methods to control storm water discharges to the extent necessary to mitigate impacts on water quality. Based on the two studies, EPA is required to issue regulations by no later than October 1, 1992 which designate additional storm water discharges to be regulated to protect water quality and establish a comprehensive program to regulate such designated sources. The program must, at a minimum

(a) establish priorities,
(b) establish requirements, and
(c) establish expeditious deadlines.

The program may include performance standards, guidelines, guidance, and management practices and treatment requirements, as appropriate.

Section 401 of the WQA also amends section 402(1)(2) of the CWA to provide that the EPA shall not require a permit for discharges of storm water runoff from mining operations or oil and gas exploration, production, processing, or treatment operations or transmission facilities if the storm water discharge is not contaminated by contact with, or does not come into contact with, any overburden, raw material, intermediate product, finished product, byproduct, or waste product located on the site of such operations.

Section 503 of the WQA amends section 502(14) of the CWA to exclude agricultural storm water discharges from the definition of point source. This provision exempts agricultural storm water discharges from regulation under the section 402 program.

Current Rulemaking.

EPA is currently preparing a notice of proposed rulemaking to establish NPDEs permit application requirements for storm water discharges associated with industrial activity and discharges from large and medium municipal separate storm sewer systems. The public will be invited to submit comments during the comment period following the publication of the proposed rule in the Federal Register. EPA will then evaluate the comments in developing a final regulation.

The rulemaking concerning permit applications for storm water discharges, will, within the framework of the WQA, balance the need for addressing the environmental risk associated with storm water discharges with the administrative burden associated with processing permits and permit applications for the large number of storm water discharges. Some of the major issues associated with the rulemaking are discussed below.

## Storm Water Discharges Associated with Industrial Activity.

**Individual Permit Applications.** Permit application requirements for storm water discharges must provide permit writers sufficient information to develop appropriate permit conditions for the discharge. Traditionally, applications for NPDES permits have focused primarily on effluent sampling data which is used for developing concentration based effluent limitations for pollutants present in the discharge as permit conditions. However, because of the variability associated with pollutant concentrations in storm water discharges, obtaining truly representative sampling data for storm water discharges can be extremely difficult.

EPA will consider if it is appropriate to shift the emphasis of permit application requirements for storm water discharges away from the collection of extensive representative data towards the collection of additional qualitative information needed for evaluation of the nature of the storm water discharge. Sampling requirements could be designed to provide limited screening information as opposed to detailed representative data.

**Group Permit Applications.** Procedures for submitting group applications can provide industrial facilities with an option to submitting individual applications. The concept of group applications involves industrial facilities that are part of the same effluent guideline subcategory or that are otherwise sufficiently similar to submit a single application which covers all facilities in the group. The burden of submitting permit applications can be reduced because only a limited number of facilities participating in the group application would be required to submit representative storm water sampling data. Group applications could then be used for issuing general permits or, for NPDES States without authority to issue general permits, for developing a model for issuing individual permits.

**Discharges to Municipal Separate Storm Sewer System.** Municipal separate storm sewers receive storm water drainage from a wide variety of land use activities, including residential, commercial, and industrial lands. Although the WQA clearly holds municipalities responsible for storm water discharges from non-industrial lands, EPA must evaluate whether storm water discharges associated with industrial activity which discharge into a municipal separate storm sewer have to be covered by an individual permit or can be covered by a permit issued to the operator of the municipal separate storm sewer. Under the latter approach, EPA must evaluate if industrial facilities which discharge storm water to municipal separate storm sewers should be required to notify the municipality of the discharge. The notification requirement would aid municipalities in developing and implementing local storm water programs that would address pollutants in their municipal storm water discharges.

## NPDES Permit Applications for Discharges from Municipal Separate Storm Sewer Systems.

**System-Wide Approach.** EPA's permit application requirements have traditionally required an individual permit application for each outfall. However, it is often more appropriate to address discharges from municipal storm sewers with a jurisdiction-wide program approach that allows system-wide planning and implementation and appropriate targeting of controls based on an evaluation of priorities. Under a system-wide approach, one permit application may be submitted for all discharges from large or medium municipal separate storm sewer systems. System-wide permit applications can in turn be used to issue system-wide permits which could cover all discharges in the system.

**Municipal Storm Water Management Programs.** Permit requirements for municipal separate storm sewer system should be developed in a flexible manner to allow site-specific permit conditions to reflect the wide-range of impacts that can be associated with these discharges. Permits for different municipalities should place different emphasis on controlling various components of discharges from municipal storm sewers.

The development of site-specific permit conditions can be facilitated by requiring permit applicants to submit, along with other information, a description of existing structural and non-structural measures to reduce pollutants in discharges from municipal separate storm sewers, and the applicant's proposal for a storm water management program with an implementation schedule which will be implemented during the term of the permit. The process of identifying components of a comprehensive control plan should begin early in the permitting process and municipal applicants should be given the opportunity to identify and propose the components of the plan that they believe are appropriate for reducing discharges of pollutants. Proposed plans could be used by the permitting authority to develop permit conditions to reduce pollutants in the discharges from municipal separate storm sewer systems to the maximum extent practicable. This overall scheme recognizes that local government entities have a critical responsibility for evaluating the nature and sources of pollutant discharges from municipal separate storm sewer systems and for devising appropriate methods of control.

The WQA requires that permits for discharges from municipal separate storm sewers shall include a requirement to effectively prohibit non-storm water discharges into the storm sewers. EPA is considering emphasizing the detection of illicit connections in permit applications for discharges from large and medium municipal separate storm sewer systems. Under one approach that is being considered, a screening analysis for illicit connections could be initiated by visual observations of selected outfalls during dry weather periods. If dry weather flow is observed, inexpensive field testing could be conducted by color, odor, turbidity, the presence of an oil sheen or surface scum, pH, chlorine, detergents, and certain metals. If the preliminary field test indicates the potential presence of illicit discharges, follow-up sampling could be performed before beginning extensive sewer investigations.

Summary

The WQA has provided EPA with the framework for developing a comprehensive storm water management program. The initial phases of the program include the development of application requirements and deadlines for storm water discharges associated with industrial activity and discharges from large and medium municipal separate storm sewer systems. The notice of proposed rulemaking will present EPA's proposed control program and will solicit public comments on many important issues. The success of the final rule depends in part on thorough public reviews of the proposal and constructive public comments.

## Discussion of Mr. Gallup's Paper

Question:

Isn't the legal definition of MEP "maximum extent practicable" sort of terrifying, in that it has been interpreted previously in other contexts as "... to the maximum extent not prohibited by low?"

Answer:

We don't see MEP as being defined or restrained that way.

Comment:

It is a little scary, in that it can be interpreted "... in the eye of the beholder."

Comment:

It is analogous to an arbitrary standard. It will be a more difficult management standard for some than for others. We need more guidance--believe it or not--from EPA.

Answer:

We are really only looking at 57 cities across the county. And there **will** be EPA guidance to the states.

Question:

With respect to monitoring, will EPA be specifying types of monitoring techniques, i.e. particular measurement equipment, to be used?

Answer:

We have tried to replace the monitoring burden with a focus on management practices. We recognize that monitoring is expensive; and when we have run around monitoring everything before in every outfall, we've gotten a lot of "nondects." So, we think you should focus on illicit connections. Go to the outfalls that are, say, greater than 36" in diameter, and see if there is any dry-weather flow (DWF) there. If there is, measure some things. And remember, cities of 100,000 to 200,000 are the ones for which the permit application begins to be relevant.

Question:

A comment on the 36" pipe criterion: In our city, we have a watershed that is 50.4 acres, just as an example, from which there are

1142 storm drain outfalls and 300 combined sewer overflow sites to sample. That situation just seems overwhelming on its face.

Answer:

That's not really a right interpretation. Again, we think you should find the ones that are truly causing the problems and sample there.

Question:

When we last counted our outfalls in Houston, more than 20 years ago, we had 10,000; we've added a million people and an uncounted number of outfalls since then. Also, these 57 cities, like Houston, are large places that may include whole counties inside them, plus hundreds of small "cities," other jurisdictions, and special districts.

Answer:

We have counted 57 "cities" that are independent.

Question:

Is the enforcement of the Water Quality Act an absolute "strict-liability" situation?

Answer:

(Minor introductory, prefacing remarks; but, finally and essentially:) yes.

A View From The Middle - Water Quality Act of 1987:
Problems and Possibilities for the Design Professional

Comments by Michael T. Llewelyn, Chief
Wisconsin Nonpoint Source Program

Milwaukee, WI

Two major revisions to the Clean Water Act will create new initiatives in the area of stormwater management. Stormwater Management permits and the nonpoint source program will expand the scope of state and local government roles in urban runoff control. For many states, the issue of stormwater management will be addressed for the first time as a water quality issue. Perhaps this is the most fundamental change in engineering philosophy that will occur as a result of the Water Quality Act of 1987.

In Wisconsin, a comprehensive nonpoint source control program has been in existence since 1978. This program addresses both rural and urban nonpoint pollution. In the past 5 years major effort has been expended in development of an urban nonpoint source program. The issue of stormwater management as a water quality vs. water quantity problem has created a need to rethink basic engineering philosophy of viewing stormwater as a flood control concern. Many communities and their engineering staff have not accepted the premise that urban runoff is a major water quality problem for both surface and groundwater quality.

The Wisconsin Nonpoint Source Program is primarily a cost-share program. It is based on identifying the most critical nonpoint sources in "priority watersheds", developing best management practices for those problems and providing 50 - 70% of the cost of those practices. In urban areas two distinct approaches are being taken. One relies on preventive measures in developing areas which emphasizes stormwater management and construction site erosion ordinances. The other relies on structural retrofitting and "housekeeping" practices such as street sweeping.

Our experience in Wisconsin is showing that there is a different expectation and perception regarding stormwater management between the engineering community and state water quality agency staff on the prioritization of water quality vs. water quantity. Stormwater must be managed for both impacts. The designer of stormwater management practices will have to address both water quality impacts in terms of pollutant loads and hydrologic modifications to urban waterbodies as well as the more traditional flood prevention issues. By necessity, more emphasis on land use planning in developing areas for minimizing stormwater impacts must occur.

## PROBLEMS FACING THE DESIGN ENGINEER

- lack of uniform standards which emphasize water quality component of stormwater management

- reluctance of local communities to retrofit in established urban areas without state and federal funding - Congress did not intend to have nonpoint source program become next Construction Grant program

- traditional approach to flood control will not satisfy state water quality agency concerns

- increased scrutiny over the quantifiable impacts of structural or nonstructural stormwater management practices related to water quality will require comprehensive planning approaches prior to design

## OPPORTUNITIES FACING THE DESIGN ENGINEER

- the relationship between point source and nonpoint source pollution control is becoming more integrated which presents a professional challenge to the water resource professional

- greater needs to develop innovative, multi-purpose stormwater management practices will exist

- land use planning in developing areas will require design engineer participation to incorporate stormwater management practices into plan

### Summary

Stormwater management has traditionally been viewed as a water quantity concern. The Clean Water Act requirement for future permitting of stormwater discharges will dramatically alter the engineering communities approach to stormwater management design since an increased emphasis will be placed on minimizing the physical, chemical and biological impacts of stormwater discharges to surface and ground water. State water quality agencies will be required to develop new approaches to permitting systems to accommodate stormwater management into municipal and industrial discharger's permits. Stormwater will not be viewed as a nonpoint source problem in the future since, from a regulatory standpoint, NPDES permits have been developed for point source discharge systems. This new approach will demand new and innovative design of both structural and non-structural best management practices for stormwater management in urban areas.

A View from the "Bottom": Challenges and Prospects

L. Scott Tucker*
Member, ASCE

Abstract

This presentation addresses the Water Quality Act of 1987 (WQA) as it pertains to National Pollutant Discharge Elimination System (NPDES) permits for municipal separate storm sewers. Regulations are being prepared by EPA that will implement the intent of Congress. Key requirements of the WQA, likely requirements of regulations how the regulations might impact local governments (the "bottom"), and cost implications are discussed.

Introduction

Passage of the Water Quality Act of 1987 (WQA) by Congress signalled a new era in urban stormwater management. Congress in Section 405 of the WQA clearly stated that National Pollutant Discharge Elimination System (NPDES) permits will be required for municipal separate storm sewers. Congress setting at the top of the pyramid has set in motion a program that will eventually wind up at the bottom with local government in terms of implementation. The "bottom" will have no choice, but to comply - what are the prospects?

Background information, requirements of the WQA, and current rulemaking activities of EPA are addressed in the presentation entitled "Federal Requirements for Storm Water Management Programs" by Gallup and Weiss. Some elements of the WQA and rulemaking activities will be restated in this presentation in order to comment on how they may impact local governments, and to place the issues in perspective as far as local government is concerned.

NPDES permits are not new. They have been required since passage of the Clean Water Act in 1972 which made it unlawful to discharge any pollutant to surface waters without a permit. Typically NPDES permits have been granted for point discharges of pollutants such as from municipal or industrial waste treatment plants or other readily definable

* Executive Director, Urban Drainage and Flood Control District, 2480 West 26th Avenue, Suite 156B, Denver, Colorado, 80211, USA

point sources.  The application requirements for standard NPDES permits are substantial and are oriented toward end-of-the-pipe, technology based controls.  The numbers of NPDES permits issued or currently in effect are limited. Colorado, for example, currently has about 870 active permits.

Technically, it has been a requirement to obtain NPDES permits for storm sewer discharges since 1972 when the original Clean Water Act was passed.  Because of the large number of storm sewers in the United States, and the intermittent and unpredictable nature of stormwater, EPA was never able to implement an effective NPDES permit program for storm sewers.  Congress settled the issue, however, in the WQA.  They made it clear that NPDES permits would have to be obtained for municipal storm sewer discharges, but they established a somewhat different set of rules for municipal separate storm sewers vis a vis the other more typical point sources.  It is now clear that Congress intends that a national effort be initiated to reduce pollutants from municipal stormwater discharges.

The process of implementation is now underway.  The EPA is writing regulations to that will be used to administer law passed by Congress.  Once EPA adopts regulations, the next step will be to administer the regulations.  Many states, 39 as of July, 1987, have NPDES programs approved by EPA.  These states will have to adopt regulations similar to what EPA will adopt in order to include the storm sewer permitting activity into their programs.  In the non-NPDES states EPA would administer the new program along with current NPDES activities.  Applicants will be required to submit applications to the state or EPA, as the case may be, for NPDES permits for storm sewer discharges.  The issuance of permits will be based on the applicants submitting all the required information and their commitment to fulfill requirements set forth in the permit.

The financial burden of the municipal storm sewer permitting program will fall primarily on local governments, i.e., cities, counties, flood control districts, etc.  The permitting program is not voluntary and there are no grants or federal funding support to assist with meeting permit requirements.  Also, in some states, Colorado for instance, the NPDES programs are primarily cash funded, and the cost of processing applications will very likely be passed on to the applicants, local government.

## Water Quality Act of 1987

The "law" is set forth in Section 405 of the WQA.  The requirements of Section 405 are discussed by Gallup and Weiss in their presentation.  Section 405 of the WQA defines who needs to get a permit, when they need to have the permit, basic conditions of a permit, and a requirement for special studies by EPA leading to development of regulations for other stormwater sources not initially covered.

Who needs to get a permit? Section 405 requires that permits be obtained for: discharge from municipal separate storm sewer systems serving populations of 250,000 or more (large system); discharges from municipal storm sewer systems serving populations between 100,000 and 250,000 (medium system); and discharges that the state or EPA in non-NPDES states determines to contribute to a violation of a water quality standard or is a significant contributor of pollutants.

Congress established deadlines for preparing the regulations, applying for permits, issuing permits, and complying with permit conditions. The deadlines Congress set are as follows:

|  | Large Systems | Medium Systems |
|---|---|---|
| EPA shall promulgate regulations by: | 2/89 | 2/91 |
| Applications submitted by | 2/90 | 2/92 |
| Permits issued or denied by | 2/91 | 2/93 |
| Compliance with permit requirements within 3 years of issuance | 2/94 | 2/96 |

Some of these deadlines will apparently not be met. The EPA has indicated that final regulations would be promulgated in late 1989, which is after the February, 1989, deadline for large systems.

Some local governments may be in the position of not having permits because of the lack of adopted regulations, thereby being in violation of the WQA. Any time delay involved in the states adopting appropriate regulations will further extend the likely time period of the permitting process. It is the writer's opinion that Congress' deadlines were unrealistic. The stormwater management problem is complex and it will take many years to force reform from the top down. Whether or not this results in civil actions against local governments for being in violation will remain to be seen.

The WQA also requires that by October, 1992, EPA is to have regulations covering those discharges for which permits were not originally required. This will likely include municipal storm sewer systems serving populations less than 100,000. In other words, for those areas not having to obtain permits initially, their time is coming.

The WQA states that permits for discharges from medium and large municipal separate storm sewers:

"1) may be issued on a system- or jurisdiction-wide basis;

2) shall include a requirement to effectively prohibit non-stormwater discharges into the storm sewers; and

3) shall require controls to reduce the discharge of pollutants to the maximum extent practicable, including management practices, control techniques and system, design and engineering methods, and such other provisions as the Administrator or the State determines appropriate for the control of such pollutants."

It is significant that permits may be issued on a system- or jurisdiction-wide basis. Before the WQA, each individual storm sewer outlet needed a permit; now an entire system can be permitted. This should reduce significantly the monitoring, data, and information requirements for permit application, but permit application requirements will still be substantial.

The second requirement means that any illegal or non-stormwater connections to storm sewers will have to be found and eliminated. There is no choice or negotiation, but this requirement makes sense.

The third requirement is a big unknown. A new term has been introduced - "maximum extent practicable" (MEP). What will it mean to reduce discharges of pollutants to the maximum extent practicable? Take note of this term as we may be hearing a lot about it in the future.

## EPA Regulations

The WQA establishes some broad guidelines, but the meat of the actual requirements will be included in the regulations. The EPA is preparing proposed regulations at the time of this conference, July, 1988, and final regulations will most likely not be published until at least late 1989. What will be required in the regulations is critical to local governments as all costs associated with preparing applications and complying with permit requirements will be borne by local governments.

At this writing it is not known what requirements the proposed regulations will contain. An early draft dated November, 1987, that circulated within EPA, however, may provide a good "feel" of where we are headed and what permit applications might be like. Being considered was a two-part application process, each part taking one year to complete. Part 1 would include development of existing information plus field inspection and screening of outfalls. Included would be:

- definition of legal authority and needed changes to meet permit requirements

- description of funding mechanisms and administrative capabilities

- locate on a map known outfalls larger than 36-inches in diameter or its equivalent (drainage area of 50 acres or more), outfalls 12 inches and larger or its equivalent (drainage area of 2 acres or more) if there is industrial activity in the drainage basin

- For each outfall estimate the boundaries of the drainage basin, identify the land use classifications, identify where there may be stormwater runoff from industrial activity, and identify any NPDES permitted discharges to storm sewers.

- Assemble existing rainfall/runoff data.

- Conduct a field inspection of each outfall for dry weather discharge and describe, sample and test any dry weather discharges found.

- Describe any existing management programs and controls.

Part 2 of the application would consist primarily of the development of a management plan and the initial characterization of stormwater discharges. Included would likely be:

- Demonstration that legal authority exists, and provision of fiscal analysis to show that local funding exists to meet regulatory requirements.

- Collect and analyze data needed to characterize discharges from storm sewers.

    1) conduct limited testing for outfalls suspected of illicit discharges for both wet and dry weather flows,

2) Conduct wet weather testing program of one storm event from five to ten outfalls and of several storms at one outfall,

3) Estimate the pollutant concentration in stormwater discharges, and

4) Estimate the pollutant load from stormwater discharges.

Prepare management plans that show how applicant will implement various programs to control pollutants being discharged from storm sewers to the maximum extent practicable. Plans will cover the following sources of stormwater discharges:

1) Program to control illicit discharges,

2) Program to control stormwater discharges associated with industrial activity,

3) Program to control urban runoff,

4) Program to control runoff from construction sites, and

5) Program to assess effectiveness of controls.

A management plan would be a key component of a permit. The management plan will describe what the applicant is going to be required to do in the future. It is the compliance plan, and will likely be the subject of negotiations between the applicant and state or EPA.

The standard to be applied to municipal separate storm sewers is to establish controls to reduce the discharge of pollutants to the maximum extent practicable (MEP). This is the standard referred to earlier and is the language used in the WQA. The emphasis of EPA appears to be in forcing the development and implementation of stormwater management programs at the local level for the four types of pollutant sources listed above: urban runoff, stormwater discharges associated with industrial activity, runoff from construction sites, and non-stormwater discharges into storm sewers. Applicants will be required to propose MEP control measures for each of these components of pollutant discharge. Where other sources, such as land disposal, contribute significant amounts of pollutants to a municipal storm sewer system, control measures will be required on a site specific basis.

Potential controls that could be part of a management plan include maintenance of structural facilities;

enforcement of measures to control pollutant discharges from new development after construction is completed; modified operation and maintenance practices for public highways; requiring flood management projects to assess the impacts on the water quality of receiving water bodies; development of controls for commercial pesticide and herbicide applicators and distributors; development of structural and non-structural measures for controlling stormwater runoff at construction sites (erosion control); initiation of programs to detect and eliminate illicit discharges and improper discharges into storms sewers; development of educational activities such as for proper disposal of used oil and toxic materials; and establishment of programs to monitor and control discharges from industrial activity that discharge into storm sewers.

Included in the requirements of the management plan will most likely be an ongoing effort to monitor stormwater discharges from representative storm sewers and representative storm events. An annual report will have to be submitted that would include the status of implementing the stormwater management plan, proposed changes to the stormwater management plan, and revisions to the assessment of the effectiveness of controls and fiscal analysis.

The impact of these regulations will be significant on those involved at the local level in the control and management of urban stormwater runoff. There will soon be a federal mandate in place to develop and implement comprehensive stormwater quality management plans at the local level. The program is a regulatory program and failure of "municipalities" to apply for and obtain permits will subject them to fines and penalties.

One of the most basic issues faced by EPA is the definition of a municipal separate storm sewer system. This definition will determine who will have to get permits and when. One option will be to define a municipal storm sewer system as a county containing the threshold populations. This would mean a large system would include any county with a population in excess of 250,000, and all incorporated entities and the county would be required to obtain permits for storm sewers within their respective jurisdictions. This definition would tend to encourage the development of regional stormwater management plans in the larger metro areas. For example in the Denver area four counties exceed 250,000 population. These four counties include some 25 incorporated jurisdictions, and only one, the City and County of Denver, exceeds 250,000, but all cities, towns, and unincorporated areas would have to apply for permits.

Another option would be to use a city or unincorporated county population as the criteria, but give the option of using the other definition. This would reduce the number of communities initially impacted by the regulations. For example, in the Denver metro area only Denver exceeds 250,000. Only three entities exceed 100,000; Aurora (223,200), Lakewood (121,400) and unincorporated Arapahoe County (104,800). The numbers in parenthesis are the 1986 estimated populations. This would reduce the number of local governments requiring a permit for the first go around from 28 to 4, but it will tend to encourage Balkanization of the permitting process and create other problems and inconsistencies. For example watersheds generally do not coincide with political boundaries. Thus there would likely be runoff originating outside the boundaries of the large or medium sized communities, but getting into their storm sewers where it is regulated. A local government has no regulatory power outside its jurisdiction, and thus may not be able to comply with permit requirements. Also, this definition orients the program toward already developed cities and counties with large populations where it will be harder to achieve reductions in pollutant discharges, <u>vis a vis</u> developing communities with smaller but growing populations where controls such as retention can be incorporated into new developments.

Since permits may be issued on a system- or jurisdiction-wide basis it may be possible to get one permit for an entire metro area with <u>each</u> local government being a co-permittee. Another possibility would be for a permit to be given to a county with the cities in the county and the county being co-permittees. A problem with this approach would occur where cities overlap more than one county.

Deciding who is to obtain a permit will vary considerably from area to area. In some cases it may be necessary to obtain additional authority and/or raise additional revenues to implement the approved management plans. Changes in institutional authorities and arrangements and raising local taxes are difficult to achieve and can not be readily accomplished. If institutional change and significant new revenue are indeed required it may take many years to achieve actual implementation.

Cost of Program Implementation

No one has any idea what this program is going to cost. The primary burden of implementation, and associated costs, however, will fall on local government. There are several phases of the permit program with various cost impacts to different parties. These include writing regulations, preparing permit applications, reviewing permit applications and issuing permits, complying with permit requirements, and reviewing and evaluating permit compliance.

The EPA is responsible for initially writing the regulations, which is a federal cost. Once EPA promulgates regulations states will have to adopt regulations similar to that of EPA. Staff will have to be added by the states, or EPA, to administer the new requirements. In Colorado these costs are passed on to applicants through fees which means local governments will end up paying for administering the program as well.

A certain real dollar cost to local governments will be for the development of the permit application. An estimate was prepared for the Urban Drainage and Flood Control District (UDFCD) of the cost of developing all the information required in the 11/87 EPA draft permit application process for the Denver metro area. The cost estimate is based on all, a total of 36, local governments in the Denver metro area obtaining a permit. The cost would be spread over a two year period for the two-part permit. Costs were estimated for two different scenarios. Scenario 1 assumed that each city, town, and county (for unincorporated areas) would be individually responsible for all application requirements. Scenario 2 assumed that one regional entity would be the primary applicant with 36 co-applicants (cities and counties). The cost estimate includes staff time, consultant fees, and data collection and monitoring. For Scenario 1 the total cost estimate is $4.8 million and for Scenario 2 it is $3.2 million. The cities and counties would be actively involved in both scenarios. The primary reason for Scenario 2 being less costly is due to lower data monitoring costs because the number of storm sewer monitoring locations could be significantly reduced.

The biggest unknown as far as costs are concerned is for compliance with permit requirements. This cost will depend on the adopted management program and continuing monitoring and reporting requirements determined during the permit application process. The costs will probably be spread between the public and private sector. Added requirements for new development will add to developer costs, and continued operation and maintenance of any structural controls will most likely be borne by cities and counties. Additional local government staff will be required to administer all the non-structural measures and continued reporting requirements. A long term monitoring program will require continued lab analysis and maintenance of the data collection system, which includes both staff demands and out-of-pocket expenditures. Since a goal of EPA appears to be to create local stormwater quality management programs, this infers the establishment of a federal, state, and local bureaucracy to maintain the programs. Such programs will probably be long term in nature extending over many years. In terms of cost keep in mind that term maximum

<u>extent practicable</u> (MEP), which will be the operating standard.

## Summary

Congress in passing the Water Quality Act of 1987 (WQA) left little doubt that it is their intent that a national effort be undertaken to reduce the discharge of pollutants from municipal storm sewers. The program will fall under the NPDES program administered by the states or EPA in non-NPDES states. The program is to be carried out at the local level, is regulatory in nature, and no federal financial assistance is likely.

The EPA is responsible for writing the regulations that will implement the "intent" of Congress. It is important for local governments to be involved in the regulation writing/review process. This opportunity will be available only once after EPA publishes proposed rules in the Federal register. Once regulations are published, it is a matter of complying with the requirements of the regulations.

The long term goal of EPA appears to be to force the development of comprehensive stormwater management plans at the local level that will in time reduce the discharge of pollutants into and from stormwaters. This will take many years to accomplish throughout the Nation and passage of the WQA and adoption of regulations is just the beginning of a long journey.

Practically all costs associated with the effort will be borne by local governments. Costs will be incurred in connections with permit application, permit compliance, and performance monitoring.

Local governments will be required to prepare management plans as a part of the permit application process. The management plans will become the compliance schedule upon which the permits will be based. The management plans will have to include controls to reduce the discharge of pollutants to the maximum extent practicable (MEP). The term MEP will be the operative word for judging whether or not management plans will be accepted by the states or EPA.

Long term monitoring may be a part of a compliance schedule. It is only through long term monitoring that the impacts of the program can be measured.

Unless the law is changed by Congress local governments will be required to develop comprehensive stormwater management plans. It is unlikely that Congress will modify the standards or exempt municipal storm sewers from the

NPDES permit process. Given this basic premise local governments are now in the position of reacting as best they can to the forthcoming EPA and state regulations. The challenge will be to develop cost effective programs that will do some good yet be affordable. The prospects are for the development of better stormwater management plans at the local level. The problem is that it is no longer a matter of choice.

Basis for Design of Wet Detention Basin BMP's

John P. Hartigan, M. ASCE*

Introduction

Two different approaches have typically been used to formulate design criteria for wet detention basin BMP's. One approach relies upon solids settling theory and assumes that all pollutant removal within the BMP is due to sedimentation (1,2). The other approach views the wet detention basin as a lake achieving a controlled level of eutrophication, in an attempt to account for biological and physical/chemical processes that have been documented as the principal nutrient removal mechanisms (3,4,5,6). Both approaches suggest that pollutant removal efficiency should be positively related to hydraulic residence time, although the controlled level of eutrophication approach results in greater storage capacities and longer residence times to achieve the same estimated nutrient removal efficiencies.

This paper evaluates the different design methods for wet detention basin BMP's. General design criteria for facilities designed primarily for nutrient removal are outlined. The costs associated with alternate design criteria are presented for a typical regional BMP master plan.

When Should Wet Detention Basin BMP's Be Used?

Two different detention options are currently used for runoff pollution control: wet detention and extended dry detention. In wet detention basins, pollutant removal occurs primarily within a permanent pool during the period of time between storm events. The "extended dry" method provides increased detention times for captured first-flush runoff in order to enhance solids settling and the removal of suspended pollutants.

In comparison with extended dry detention basins, wet detention basin BMP's offer the advantage of pollution removal mechanisms for dissolved phosphorus and dissolved nitrogen. Whereas dry detention systems can only rely upon solids settling processes for phosphorus and nitrogen removal, wet detention can achieve removal of dissolved nutrients through other physical/chemical and biological
-------------------
*Vice President, Camp Dresser & McKee (CDM), 7535 Little River Turnpike, Suite 200, Annandale, VA  22003.

processes in the permanent pool (e.g., uptake of nutrients by free-floating algae and wetland vegetation around the edge of the pool). As a result, monitored average pollutant removal efficiencies (1,7) for wet detention basin BMP's are on the order of 2 to 3 times greater than extended dry detention BMP's in the case of total P (50%-60% vs. 20%-30%) and 1.3 to 2 times greater in the case of total N (30%-40% vs. 20%-30%*). The increased removal rates for total P and total N in wet detention basins can be attributed in large part to average removal rates on the order of 50%-70% for dissolved nutrients, the nutrient fraction that is most readily available for biological activity and of greatest interest from a water quality management standpoint.

For other pollutants, the average removal rates for wet detention basins and extended dry detention basins are very similar (1,7): 80%-90% for total suspended solids; 70%-80% for lead; 40%-50% for zinc; and 20%-40% for BOD or COD.* Since the major difference between the performance of wet and dry detention basins is the greater removal of nutrients in the former, wet detention basins are most appropriate for areas where the receiving water quality problem is caused by nutrient loadings. Since nutrients typically require extended hydraulic residence times to cause a serious receiving water quality problem, watersheds where free-flowing streams are the critical receiving waters are least likely to require wet detention basin BMP's for water quality management. Examples of situations where nutrient control needs will tend to foster the use of wet detention basin BMP's are as follows:

o  Watersheds of reservoirs and lakes: Wet detention basins should be the preferred BMP for watershed management programs designed to protect lakes and reservoirs, particularly water supply reservoirs. The use of wet detention basins to control urban runoff pollution can help achieve eutrophication management goals in downstream reservoirs.

o  Watersheds of tidal embayments and estuaries: Reductions of nutrient loadings into estuarine systems is becoming a growing concern not only in coastal areas but in upland areas that drain into tidal waters (e.g., Chesapeake Bay drainage area (8)). In these areas, wet detention basin BMP's can help maximize the control of nutrient discharges.

As shown in Table 1, which summarizes design criteria for a recent regional BMP master plan for Fairfax County, Virginia (9), the permanent pool of a wet detention basin can require anywhere from 2 to 7 times more storage than an extended dry detention basin depending upon the land use.

---
*Assumes average hydraulic residence time of 2 weeks or greater for permanent pool of wet detention basin, and 12-24 hr detention time for extended dry detention basin with a storage capacity of 1.0 inch of runoff per impervious acre.

In addition to greater storage requirements, wet detention basins also require larger structures and more land, meaning that they tend to require greater capital costs than extended dry detention. Cost increases associated with wet detention are on the order of 50%-150% for the nonpoint pollution control storage alone, although this differential gets smaller if the detention basins require overlying storage for peak-shaving purposes (e.g., flood control, erosion control).

Table 1. Comparison of Detention Storage Requirements in Northern Virginia: Permanent Pool of Wet Detention Basin vs. Extended Dry Detention

| Land Use | Percent Impervious | Wet Detention (inches) | Extended Dry Detention (inches) |
|---|---|---|---|
| Low Density Single Family | 20% | 0.7 | 0.1 |
| Medium Density Single Family | 35% | 0.8 | 0.2 |
| Multifamily Residential | 50% | 1.0 | 0.4 |
| Industrial/Office | 70% | 1.2 | 0.5 |
| Commercial | 80%-90% | 1.3 | 0.8 |
| Forest/Undeveloped | 0% | 0.5 | 0.0 |

NOTES:

1. Wet detention storage volume is based upon an average hydraulic residence time of 2 weeks.

2. Extended dry detention storage volume is based upon capture of first flush runoff (0.8-0.9 inches of runoff per impervious acre).

Given the greater capital costs associated with a wet detention basin, this type of BMP will be most cost-effective in areas where nutrient control is a primary concern and least cost-effective where it is not. For areas, where control of other pollutants (e.g., toxicants) is the principal concern or where the major objective is to show some progress with BMP implementation, extended dry detention basin BMP's are likely to be more cost-effective than wet detention systems. For example, this general guideline would apply to areas where BMP implementation is intended to satisfy permitting requirements (e.g., upcoming NPDES stormwater discharge permitting program) that are not tied to a specific receiving water quality problem.

Although they may be less cost-effective for situations where nutrient control is not a serious concern, wet detention basin BMP's do offer some other advantages which should be considered in BMP selection. Wet detention basins are usually more attractive looking than dry basins, particularly if there is extensive wetland vegetation around the perimeter of the permanent pool. Wet detention basins are actually considered as property value amenities in many areas. Also, wet detention basins offer the advantage that sediment and debris accumulate within the permanent pool. Since these accumulations are out-of-sight and well below the basin outlet, wet detention basins tend to require less frequent cleanouts to maintain an attractive appearance and prevent clogging. Therefore, if regular BMP maintenance cannot be guaranteed (e.g., cases where the responsibility for maintenance resides with the property owner rather than the local government), there may be some advantage to requiring more costly wet detention basins rather than dry basins.

Other issues to consider when choosing between wet and dry detention basin BMP's include the following:

a. Whether the drainage area is large enough to sustain a permanent pool;

b. Whether the receiving waters immediately downstream are particularly sensitive to increased water temperatures in the effluent from the wet detention basin; and

c. Whether existing wetlands at the detention basin site restrict the use of a permanent pool.

If the drainage area is too small, storm runoff and dry weather inflows into the wet detention basin may be too small to maintain a permanent pool during "dry" seasons. While excessive drawdown of the permanent pool does not pose a nonpoint pollution control problem, it will cause aesthetic problems. Suggested guidelines for minimum drainage areas of wet detention basins are presented elsewhere in this paper.

Because the summer temperature of waters stored in the permanent pool tends to be greater than downstream receiving waters, there may be some concern about using wet detention basins along streams where beneficial uses are very temperature-sensitive (e.g., trout streams). Extended dry detention basins may be a preferable BMP where water temperature impacts are a concern.

The potential impacts of stormwater management structures on wetlands are coming under increasing scutiny from regulatory agencies. While it can be argued that wet detention basins can be designed to produce new wetland systems and that the additional water quality protection justifies potential wetlands impacts, extended dry detention systems are likely to be the preferable BMP for sites with a significant amount of existing wetlands.

## Alternate Design Methods

The most important feature of a wet detention basin is the permanent pool. Urban runoff detained in the permanent pool following a storm event is subjected to physical/chemical and biological processes which achieve removal of selected pollutants. During the next storm event, urban runoff inflows displace "treated" waters in the permanent pool followed by treatment after the storm ends. This means that the size and shape of the permanent pool is an important design criterion. For example, the larger the permanent pool storage volume in comparison with design runoff conditions (e.g., first 0.5 inch of runoff), the lower the outflow of urban runoff inflows and the higher the retention and treatment between rainstorms.

### Solids Settling Design Method.
The solids settling design method (1,2) relies upon rainfall/runoff statistics, settling velocities for assumed particle size distributions, and the assumed percentage of pollutant mass attached to sediment (particulate fraction) in order to calculate suspended pollutant removal for specified overflow rates. Separate efficiency calculations for dynamic conditions during storm events and for quiescent conditions following storm events are weighted by the duration of each condition to determine a long-term average pollutant removal rate. The method assumes an approximate plug flow system in the detention basin, with all pollutant removal resulting from Type I sedimentation. While this assumption may be reasonable for dynamic conditions during storm events, completely mixed conditions which account for longitudinal dispersion may be more likely under quiescent conditions. Pollutant removal under quiescent conditions is based upon a capture/pumpout model originally developed for evaluations of combined sewer overflow (CSO) interceptors. For permanent pool storage volumes typically considered for wet detention basins, the solids settling method usually assigns more than 90% of the total pollutant removal to quiescent conditions and less than 10% to the dynamic conditions which are probably best represented by the plug flow assumptions of the design model.

Design curves for the solids settling method for wet detention basin design are shown in Figure 1. As may be seen, these curves relate average TSS removal to the size of the permanent pool. For these particular design curves, the permanent pool size is expressed in terms of the ratio of its surface area to the BMP drainage area. Since these curves are based upon a mean depth of 3.5 ft for the permanent pool, the x-axis can easily be converted to a permanent pool storage volume (i.e., product of surface area and mean depth). The design curves shown in Figure 1 assume that the land use pattern in the BMP drainage area is restricted to low density single family residential (runoff coefficient (RV) = 0.2). Separate design curves must be developed for other land use patterns.

Permanent pool designs can be based upon the performance standard for average TSS removal. For example, if an 80%

# WET DETENTION BASIN DESIGN

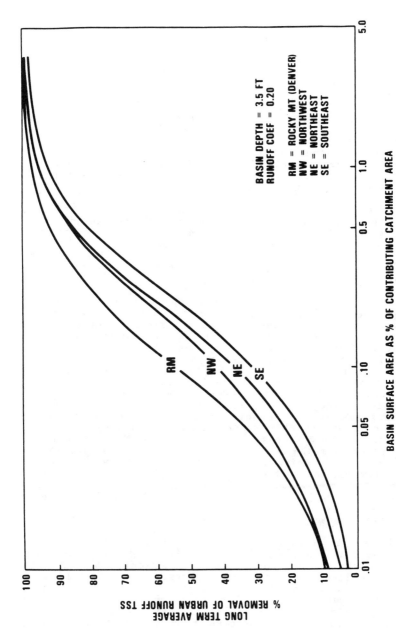

Figure 1. Geographically Based Design Curves for Solids Settling Model (1)

removal of TSS is desired for areas in the southeastern U.S. (SE), the permanent pool surface area should be at least 0.5% of the BMP drainage area. For a 40-acre drainage area comprised of low density single family land use, this translates into a permanent pool with a minimum surface area of 0.2 ac and a storage volume of 0.7 ac-ft (0.2 inch runoff) assuming a mean depth of 3.5 ft.

Average removal rates for other pollutants may be derived by multiplying the TSS removal rate by the average particulate fraction of the pollutant of interest. For example, previous applications of this method have relied upon an average particulate fraction of 67% for total P, based on the nationwide database of the EPA Nationwide Urban Runoff Program (NURP) (2).

To test the solids settling design model, it was applied to nine of the wet detention basin BMP's monitored during the NURP program. Based upon goodness-of-fit plots, it was concluded that the method does a reasonably good job of predicting the average pollutant removal efficiencies at nine NURP sites based upon wet detention basin design characteristics (2).

The method appears to be most appropriate for handling suspended solids and constituents such as heavy metals (e.g., lead) which tend to appear primarily in suspended form, since sedimentation should be the dominant pollutant removal process. However, it would appear to be less appropriate for the evaluation of nutrient removal efficiencies since monitoring data at several NURP wet detention basin sites indicate that the majority of total P and/or total N mass removal was in the form of dissolved P and dissolved N. This is illustrated in Table 2 which summarizes average removal of total P and dissolved P monitored at the NURP wet detention basin sites and other testing sites. As may be seen, dissolved P removal represents a major component of total P removal at seven of the ten sites which reported dissolved P removal rates. For example, the portion of the total P efficiency attributable to dissolved P removal ranged from 62% to 86% for four of the NURP sites. This suggests that solids settling theory alone does not account for the most important nutrient removal mechanisms in wet detention basin BMP's.

<u>Lake Eutrophication Model Design Method.</u> This approach assumes that a wet detention basin BMP is a small eutrophic lake which can be represented by empirical models used to evaluate lake eutrophication impacts. The intent of this approach is to use lake eutrophication models to account for the significant removal of dissolved nutrients observed in the field and attributable to biological processes such as uptake by algae and rooted aquatic vegetation. Using this design method, a wet detention basin can be sized to achieve a controlled rate of eutrophication and an associated removal rate for nutrients.

The design method is restricted to nutrients. However, since wet detention basin BMP's that achieve significant

Table 2. Summary of Monitored Wet Detention Basin Efficiencies: Total P and Dissolved P

| Location | Site | Average Hydraulic Residence Time (weeks) | Average Total P Removal (%) | Average Dissolved P Removal (%) | Dissolved P Fraction of Total P Removal (%) |
|---|---|---|---|---|---|
| A. NURP | | | | | |
| Lansing, MI | Grace N. | <1 | 0% | 0% | 0% |
| Lansing, MI | Grace S. | <1 | 12% | 23% | 25% |
| Ann Arbor, MI | Pitt | <1 | 18% | 0% | 0% |
| Ann Arbor, MI | Swift Run | <1 | 3% | 29% | 65% |
| Ann Arbor, MI | Traver | <1 | 34% | 56% | 62% |
| Long Island, NY | Ungua | 5 | 45% | N/A | N/A |
| Washington, DC | Burke | 5 | 48% | 53% | 86% |
| Washington, DC | Westleigh | 7 | 54% | 71% | 73% |
| Glen Ellyn, IL | Lake Ellyn | 6 | 34% | N/A | N/A |
| Lansing, MI | Waverly Hills | 14 | 79% | 70% | 19% |
| B. USGS (1986) | | | | | |
| Orlando FL | Highway Pond | 1-2 | 29% | 54% | 63% |
| C. Minnesota | | | | | |
| Twin Cities, MN | Fish | 5 | 44% | 32% | 47% |
| Roseville, MN | Josephine | 6 | 62% | 69% | 75% |

SOURCE: Reference 4

nutrient removal also achieve removal rates for other pollutants that are similar to other BMP's, it is probably not necessary for the design method to address other constituents besides nutrients. Likewise, wet detention basin BMP's may not be cost-effective unless nutrient control is the principal water quality management objective.

The recommended lake eutrophication design model is the phosphorus retention coefficient model developed by Walker (4,10). Like most input/output lake eutrophication models, the Walker model is an empirical approach which treats the permanent pool as a completely mixed system and assumes that it is not necessary to consider the temporal variability associated with individual storm events. Unlike the solids settling model which accounts for temporal variability of individual storms, the Walker model is based upon <u>annual</u> flows and loadings. Because it does not consider storm to storm variability, the Walker model is much simpler to apply than the solids settling model. The model is applied in two parts:

$$K2 = (0.056)(QS)(F)^{-1}/(QS + 13.3) \qquad \text{(Equation No. 1)}$$

where

$K2$ = second order decay rate ($m^3$/mg-yr)
$QS$ = mean overflow rate (m/yr) = $Z/T$
$F$ = inflow ortho P/total P ratio
$Z$ = mean depth (m)
$T$ = average hydraulic residence time (yr)

and

$$R = 1 + (1 - (1 + 4N)^{0.5})/(2N) \qquad \text{(Equation No. 2)}$$

where

$R$ = total P retention coefficient = BMP efficiency
$N$ = $(K2)(P)(T)$
$P$ = inflow total P (ug/L)

As may be seen, the model relies upon a second order reaction rate which means that the total P removal per unit volume is proportional to the concentration squared. The second order decay rate ($K2$) is calculated from the mean overflow rate and the ortho P fraction of total P. The average total P removal rate ($R$) is then calculated from the decay rate, the inflow total P concentration and the average hydraulic residence time. The model was developed from a database for 60 Corps of Engineers' reservoirs and verified for 20 other reservoirs.

To test how well the model represents wet detention basin BMP's, Walker (4) applied it to 10 NURP sites and 14 other wet detention systems and small lakes. The goodness-of-fit assessments yielded an $R^2$ of about 0.8, indicating that the model does a very good job of replicating monitored average total P removal from detention basin design characteristics.

<u>Comparison of Design Methods</u>. A key parameter in the eutrophication modeling approach is the average hydraulic residence time (T). The average hydraulic residence time (in units of "years") may be computed from the ratio of permanent pool storage volume (VB) to the product of mean storm runoff (VR) times the average number of storms per year. For example, for eastern U.S. locations which average about 100 storm events per year, the average hydraulic residence time (in years) is equal to VB/VR divided by 100.

To compare the two design methods, average removal rates for total P are plotted against average hydraulic residence time in Figure 2. These design curves are based on northern Virginia hydrologic conditions and a mean depth of 1.0 m for the permanent pool of the wet detention basin. For the solids settling design model, the particulate fraction of total P is set at 67%, the nationwide mean of the NURP database used in previous testing of the model (2). For the lake eutrophication model, the mean total P in the inflow to the wet detention basin is set at 380 ug/L, the mean total P for residential land use in the national NURP database, and the ortho P/total P ratio, F, is set at 0.33. The curves in Figure 2 are generally applicable for a range of runoff coefficients (RV) for the BMP drainage area, so long as the same runoff coefficient is used for both design methods. As may be seen, the solids settling design curves result in considerably shorter average hydraulic residence times to achieve a specified total P removal rate. For example, to achieve an average total P removal rate of 50%, the solids settling model design method results in a permanent pool with a T of about one week, while the lake eutrophication model design method results in a T of about three weeks. In other words, the lake eutrophication model design method results in a permanent pool which is about three times larger than the solids settling method. The greater storage requirements for the lake eutrophication modeling method can be attributed in large part to the smaller reaction rates associated with the biological processes (e.g., algal uptake and sinking) that remove dissolved phosphorus, compared to the higher reaction rates assumed for solids settling processes under approximate plug flow conditions.

The relationships in Figure 2 may be used to select an optimum T for wet detention basin designs. For the lake eutrophication model design curves, the optimum T is on the order of 2 to 3 weeks compared to an optimum range of 1 to 2 weeks for the solids settling design curves. Both design curves become assymptotic for T's greater than these values, indicating diminishing marginal benefits for incremental increases in the size of the permanent pool. Based upon these types of relationships, a design T of 2 weeks has been recommended for wet detention basin design standards for major portions of Florida (6), Virginia (9), and Maryland (5).

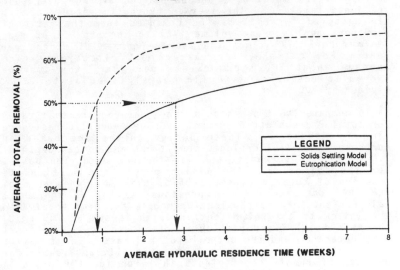

**FIGURE 2
COMPARISON OF WET DETENTION BASIN DESIGN MODELS: NORTHERN VIRGINIA
(MEAN DEPTH = 1.0 M)**

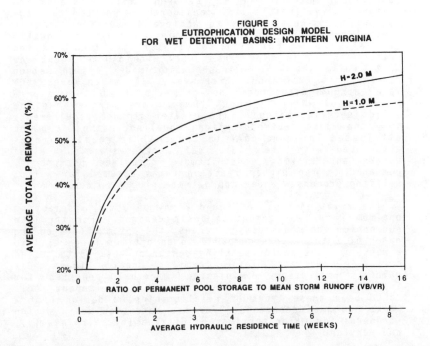

**FIGURE 3
EUTROPHICATION DESIGN MODEL FOR WET DETENTION BASINS: NORTHERN VIRGINIA**

Figure 3 shows lake eutrophication model design curves for permanent pool mean depths (storage volume/surface area) of 1.0 m and 2.0 m. These curves are based upon the same assumptions about runoff inflow as Figure 2. As may be seen, this design method indicates that, for relatively large T's (e.g., greater than 4 weeks), average total P removal rates can be increased by about 10% by doubling the mean depth of the permanent pool. By comparison, the solids settling design model indicates that increasing the mean depth has no impact on total P removal for values of T greater than 2 to 3 weeks.

Table 3 summarizes the characteristics of two of the most intensively monitored wet detention basin BMP's in the NURP study, Burke and Westleigh in the Washington, DC region (7). Both wet detention basin sites were monitored over the same 12-month period in 1981. Data from these two BMP testing sites were used to evaluate how well the two design methods represented total P removal rates, both long-term averages and efficiencies for individual storm events. As may be seen, both wet detention basin sites drained single family residential development and both permanent pools were characterized by relatively large T's, 5 to 7 weeks, and relatively large ratios of surface area (As) to drainage area (Aw), 2.5% to 3.3%. Average total P removal rates for the monitoring period are on the order of 50% at the two sites, while average dissolved P removal rates are as high as 71%. For the observed mean total P

Table 3. Characteristics of Wet Detention Basins: Metropolitan Washington, DC NURP Study

|   |   | Burke | Westleigh |
|---|---|---|---|
| A. | **Watershed** | | |
|   | Land Use | S.F. Res. | S.F. Res. |
|   | Drainage Area (Aw) | 27.1 ac | 47.9 ac |
|   | Density (DU/Ac) | 3.0 | 1.2 |
|   | % Impervious | 34% | 24% |
| B. | **Permanent Pool** | | |
|   | Storage (VB) | 1.4 in | 1.5 in |
|   | Surface Area (As) | 0.9 ac | 1.2 ac |
|   | Mean Depth | 3.5 ft | 5.0 ft |
|   | VB/VR | 9.4 | 12.8 |
|   | Average Residence Time | 4.9 wks | 6.7 wks |
|   | Overflow Rate (QR/As) | 0.06 ft/hr | 0.06 ft/hr |
|   | As/Aw Ratio | 3.3% | 2.5% |
| C. | **Average P Removal** | | |
|   | Total P | 48% | 54% |
|   | Dissolved P | 53% | 71% |

inflow (290 ug/L), mean ortho-P fraction (0.50), and mean particulate fraction (21%) at the Burke BMP site, the lake eutrophication model projected an average total P removal rate (43%) reasonably close to the measured mean (48%), while the solids settling model projected an average removal rate (20%) which was almost 60% lower than the measured value. The poor performance of the solids settling model can undoubtedly be attributed to its inability to handle the large dissolved P inflows to the Burke site. For the observed mean total P inflow (397 ug/l), mean ortho-P fraction (0.46), and mean particulate fraction at the Westleigh site, the lake eutrophication model also produced average total P removal projections (52%) which agreed more closely with the observed mean (54%) than did the solids settling model projection (42%).

Evaluations of total P removal efficiencies for the range of storm events at Burke and Westleigh indicate a significant decline in pollutant removal efficiencies for relatively low VB/VR ratios. For example, at the Burke wet detention basin site, total P removal rates were on the order of 50% to 80% for storms producing less than 0.4 inch runoff (VB/VR > 4), but only about 10% to 25% for three storms with runoff volumes of 0.6 to 1.2 inches (VB/VR < 2). Since the VB/VR ratio is an indicator of average hydraulic residence time (T), these storm-to-storm comparisons also highlight the importance of incorporating a relatively long residence time into the permanent pool design. Likewise, at the Westleigh wet detention basin site, total P removal rates were on the order of 50% to 90% for most storms producing less than 0.1 inch runoff (VB/VR > 15.0), but were only about 20% to 40% for three storms producing 0.15 to 0.6 inch runoff (VB/VR of 2.5 to 10).

<u>Cost Comparisons for Different Sizing Criteria.</u> To evaluate the cost impacts of different sizing criteria for the permanent pool of wet detention basin BMP's, cost estimates were developed for three different design approaches. The different sizing criteria were applied to 34 wet detention basins recommended in a regional BMP master plan for the 49.4 sq mi Cub Run watershed in Fairfax County, Virginia. The regional BMP plan strategically locates offsite control facilities which drain about 150 acres on the average and serve a total area of about 7.9 sq mi with a total imperviousness of 22%. The three different sizing criteria are as follows:

o <u>VB/VR = 4.0</u>: This is equivalent to an average hydraulic residence time of 2 weeks. Based upon applications of the lake eutrophication design model (see Figure 2), a T of 2 weeks is the recommended design criterion for the regional BMP plan for the Cub Run watershed (9).

o <u>VB/VR = 2.5</u>: This is equivalent to an average hydraulic residence time of about 9 days. Based upon applications of the solids settling design model, this is one of the design criteria recommended by the State of Maryland.

o **Permanent Pool Storage Equivalent to 0.5-inch Runoff:** This is the design criterion used for Montgomery County, Maryland. The T achieved by this criteria will depend upon the level of urban development in the BMP drainage area. For single family residential development, T will be on the order of 1.5 to 2.5 weeks; while for industrial and commercial development, it will typically be less than 1.0 week.

The characteristics of wet detention basins sized in accordance with this criteria for the Cub Run regional BMP master plan are summarized in Table 4. This table lists total storage requirements, surface area and cost for each design approach. The storage requirements include peak-shaving storage for 10-year and/or 2-year design storms as well as freeboard provisions. Table 4 also presents cost projections assuming that the 34 facilities were designed as extended dry detention basins (see Table 1) rather than wet detention basins. As may be seen, total storage, land area and cost projections are very similar for criteria based upon a VB/VR ratio of 2.5 and the 0.5-inch runoff criteria. Storage requirements for the two-week residence time criteria are 12%-14% greater than the other two design approaches for wet detention basins, while land requirements are about 10% greater. Compared to extended dry detention, the wet detention basin design based on a two-week residence time averaged about 31% more storage and about 20% more land area. Total cost for the design based on a T of 2 weeks is $15.6 million for the Cub Run master plan. This cost projection is only about 10% greater than the cost projections for the less conservative wet detention designs and 23% greater than the costs to design the 34 facilities for extended dry detention. Compared to the other wet detention basin criteria, the design for a T of 2 weeks will achieve greater total P removal for a relatively small increase in cost. Likewise, compared to extended dry detention, the wet detention basins designed based on a T of 2 weeks should achieve average nutrient removal rates which are about two times greater with a relatively small increase in cost.

Unit costs for the regional wet detention basin system are $3,080 per acre of drainage area for a T of 2 weeks and about $2,800 per acre for the less conservative designs. By comparison, the unit costs for the extended dry detention facility are $2,500 per acre.

## General Design Criteria

Table 5 summarizes general design criteria for the design of wet detention basin BMP's, including: storage volume, mean depth, and surface area of the permanent pool; minimum drainage area; shoreline side slopes; length/width ratio; and soil permeability. Separate criteria are presented for onsite BMP's (e.g., drainage area less than

Table 4. Comparison of Costs of Alternate Regional BMP Facility Plans for Cub Run Watershed (Fairfax County, VA)

| Parameter | Alternate Wet Detention Designs | | | Extended Dry Detention |
|---|---|---|---|---|
| | $V_B/V_R = 4.0$ | $V_B/V_R = 2.5$ | 0.5 IN. R.O. | |
| 1. No. of Facilities | 34 | 34 | 34 | 34 |
| 2. Avg. Drainage Area | 149 ac | 149 ac | 149 ac | 149 ac |
| 3. Avg. Total Storage (with peak shaving) | 29.1 ac-ft | 25.5 ac-ft | 26.0 ac-ft | 22.2 ac-ft |
| 4. Avg. Land Area (including buffer) | 6.1 ac | 5.6 ac | 5.6 ac | 5.1 ac |
| 5. Total Cost (including land cost) | $15.6 mill. | $14.1 mill. | $14.3 mill. | $12.7 mill. |
| 6. Cost per acre of drainage area | $3,080 | $2,780 | $2,820 | $2,510 |

Notes:

1. All facilities are designed for peak-shaving (10-yr and/or 2-yr)
2. Total drainage area is 7.9 sq mi with average imperviousness of 22%
3. Land costs are assumed to average $25,000/ac
4. $V_B/V_R$ = 4.0 is equivalent to "T" of 2 weeks
5. $V_B/V_R$ = 2.5 is equivalent to "T" of 9 days
6. Extended dry detention BMP storage is 0.8-0.9 in per impervious acre

Table 5
## GENERAL DESIGN CRITERIA: WET DETENTION BASIN BMP'S

| Design Parameter | Recommended Criteria | |
|---|---|---|
| | Onsite BMP | Regional BMP |
| 1. Storage Volume (Permanent Pool) | • $T \geq 2$ weeks<br>• $V_B/V_R \geq 4$ | • Same as Onsite |
| 2. Depth (Permanent Pool) | • $Z = 1$ to $3$ m<br>• MAX. $= 4$ to $6$ m | • Same as Onsite |
| 3. Surface Area | • $\geq 0.25$ ac | • $\geq 3$ to $5$ ac |
| 4. Drainage Area | • 20 to 25 ac min. | • 100 to 300 ac (depending on % imp.) |
| 5. Shoreline Side Slopes | • 5H:1V to 10H:1V | • Same as Onsite |
| 6. Length/Width Ratio | • L/W $\geq$ 2:1 (goal)<br>• Max. I/O travel times | • Same as Onsite |
| 7. Soil Permeability | • B, C, and D soils<br>• Compaction | • Same as Onsite |

50 acres) and regional BMP's (e.g., drainage area of 100 to 300 acres).

Storage Volume. The lake eutrophication model design method is recommended for use in sizing wet detention basins for nutrient control. Because this method accounts for the biological uptake of dissolved nutrients which appears to be an important factor in wet detention basin performance, it produces a more conservative design. Given the fact that wet detention basins cannot yet be engineered like water/ wastewater treatment plants to guarantee a specified effluent limit, it makes sense to use the more conservative design procedure that has been shown to provide a fairly accurate representation of long-term BMP performance.

Based upon the design curves for the lake eutrophication model method shown in Figures 2 and 3, the permanent pool storage volume should be based upon an average hydraulic residence time (T) greater than or equal to 2 weeks (i.e., 0.0385 yr). For most sections of the eastern U.S., this T criteria corresponds to a ratio of permanent pool storage to mean storm runoff which is greater than or equal to 4 (VB/VR $\geq$ 4). An alternate approach for regional BMP's is to include both surface runoff (VR) and baseflow in the hydraulic residence time calculations. This typically results in VB/VR ratios greater than 4 for eastern U.S. areas (e.g., see Table 1 which is based on VB/VR ratios on the order of 5).

While enhanced sedimentation is likely to result from the use of a T which is much greater than 2 weeks, such a BMP facility would have a greater risk of thermal stratification and anaerobic bottom waters. As a result, there would be increased risk of short-circuiting and significant export of nutrients from bottom sediments subject to anaerobic conditions. Consequently, it is advisable to maintain the average residence time at the lowest level which can ensure adequate nutrient uptake.

Depth of Permanent Pool. Mean depth of the permanent pool is calculated by dividing the storage by the surface area. The mean depth should be low enough to minimize the risk of thermal stratification, but high enough to ensure that algal blooms are not excessive and to minimize resuspension of settled pollutant during major storm events. The prevention of significant thermal stratification will help minimize short-circuiting and maintain the aerobic bottom waters that should maximize sediment uptake and minimize the release of nutrients from bottom sediments into the water column. A mean depth of about 1 to 3 m should be capable of maintaining an acceptable environment within the permanent pool for the average hydraulic residence times recommended herein, although separate analyses should be performed for each locale. The mean depths of the more effective wet detention basins monitored by the NURP study typically fall within this range as do the recommendations of recent Florida monitoring studies of retention basins (11).

The maximum depth of the permanent pool should be set at a level which minimizes the risk of thermal stratification. Based upon typical thermal profiles for different impoundment sizes and geographical regions (12), a maximum depth of no greater than 4 to 6 m should be acceptable for most regions assuming a T of 2 weeks.

Extended Detention Zone Above the Permanent Pool. In some areas, wet detention basin regulations require the provision of extended detention of a specified runoff volume in a storage zone above the permanent pool. For example, a requirement of some jurisdictions in the State of Maryland is that a runoff volume equivalent to 0.5 inch should be dewatered over a 40-hr period. This requirement is intended to minimize short-circuiting and to enhance solids settling. In addition, it should also achieve some peak-shaving benefits for minor design storms (e.g., 2-year or less), particularly in watersheds with a relatively low level of imperviousness. While it is likely that this design approach will further improve solids settling, the incremental increase in sedimentation will probably be relatively small since the recommended permanent pool designs (e.g., T of 2 weeks) should already be achieving average TSS removal rates in excess of 90%. It is conceivable that adding the same storage volume (e.g., 0.5 inch) to the permanent pool may result in a greater incremental increase in nutrient removal, with the additional removal taking the form of increased uptake of dissolved nutrients which are more readily available for biological activity than suspended nutrients. In addition, a larger permanent pool should also enhance solids settling under quiescent conditions.

Minimum Surface Area of Permanent Pool. For onsite BMP's, factors to be considered in establishing a minimum surface area are topography, aesthetics, desired mean depth, and solids settling guidelines. A minimum surface area of 0.25 ac is probably not unreasonable for an onsite BMP given the typical drainage areas required to sustain a permanent pool during summer months.

For regional BMP's which typically drain a few hundred acres, minimum surface areas have been set at levels which facilitate maintenance activities. For example, a recent regional BMP master plan for Fairfax County, Virginia (9) relied on a 3-acre minimum standard for the permanent pool surface area based on maintenance considerations. Assuming drainage aceas up to 300 acres, a 3-acre minimum surface area requirement will result in an As/Aw ratio of 1% and average TSS removal rates on the order of 90%-95% for residential watersheds based on the solids settling design model. For nonresidential watersheds with relatively high levels of imperviousness, As/Aw ratios in excess of about 3% are desirable to achieve high levels of sedimentation based on the solids settling design model.

Minimum Drainage Area. For onsite BMP's, the minimum drainage area should permit sufficient flow to prevent severe drawdown during the dry season and the associated aesthetic problems. Drainage area is also a factor in the sedimentation efficiency achieved by the detention basin, in that design curves based on As/Aw ratios have been developed with the solids settling model (see Figure 1). A minimum drainage area of 20 to 25 acres should be adequate for many areas to ensure adequate dry weather inflows. Further, in conjunction with a 0.25-acre minimum for surface area, a 20- to 25-acre minimum drainage area provides an As/Aw ratio of 1.0% or greater and an average TSS removal rate of 85% to 95% for most residential land uses. For nonresidential land uses with relatively high levels of imperviousness, smaller drainage areas (e.g., As/Aw ratio on the order of 3% to 5%) are preferable to achieve average TSS removal rates of 90% or greater based on solids settling model design curves.

For regional BMPs, the maximum drainage area should be set at a level which minimizes the exposure of upstream channels to erosive storm flows and also minimizes public safety hazards associated with the dam height. A rule of thumb that has been applied to regional detention basin master plans in Virginia and Maryland is to restrict the maximum drainage area to 100 to 300 acres depending upon the amount of imperviousness in the BMP watershed. This drainage area range is considered adequate to prevent excessive streambank erosion upstream of the regional BMP. Highly impervious drainage areas will typically be restricted to the lower end of the range (100 acres) and vice versa. Another factor is limiting the drainage area so as not to exceed the thresholds (usually expressed in terms of storage and/or dam height) for state dam safety permits. Detention basin designs which do not require a dam safety permit are desirable from the standpoint of reduced safety hazards and reduced lead time requirements. A recent master plan (9) for a regional detention basin system in Fairfax County, Virginia found that a 100- to 300-acre limit of drainage area adequately addresses concerns about exposure of upstream channels and minimizing dam safety hazards.

<u>Side Slopes Along the Shoreline.</u> The slope of the littoral zone around the perimeter of the permanent pool should be gradual enough to minimize safety hazards, to promote the growth of wetland vegetation along the shoreline, and to facilitate maintenance (e.g., grass mowing). Side slopes in the range 5H:1V to 10H:1V are recommended. The side slopes should also be topsoiled, nurtured or planted from 2 ft below to 1 ft above the permanent pool control elevation to promote vegetative growth. Wetland vegetation will not only improve the aesthetic qualities of the detention facility, but they will also help minimize the proliferation of free-floating algae. The nutrient uptake achieve by wetland vegetation will help keep the algae concentrations in check by limiting the amount of nutrients available for phytoplankton. Additional guidelines for using wetland vegetation within shallow sections of the permanent pool have been published by the State of Maryland (13).

Length:Width Ratio. Since it is generally not an important factor in achieving flooding/erosion control performance standards, length:width ratios are rarely considered in the design of peak-shaving detention basins. However, relatively high length:width ratios can help minimize short-circuiting, enhance sedimentation, and also help prevent vertical stratification within the permanent pool. A minimum length:width ratio of 2:1 is probably a reasonable planning goal for the permanent pools of wet detention basins. In addition, the location of the outlet structure within the basins should maximize travel time from the inlet to the outlet. Baffles or islands can also be installed within the permanent pool to increase the flow path length and minimize short-circuiting.

Soil Permeability. Highly permeable soils may not be acceptable for wet detention basins due to the greater potential for excessive drawdown of the permanent pool during dry periods and associated aesthetic problems. However, successful operation of a wet detention basin BMP without severe drawdown have been demonstrated at NURP testing sites underlain by relatively permeable soils. Examples include the Burke and Westleigh wet detention basins which were both underlain by relatively permeable soils classified as hydrologic soil group B. In cases where relatively permeable soils are encountered, drawdown rates can be minimized by compacting the permanent pool soils during construction.

## Summary and Conclusions

Given the greater costs and increased land requirements compared to dry detention BMP's, wet detention basin BMP's are most cost-effective where the control of nutrient loadings is the primary concern and least cost-effective where nutrient control is not a critical issue. Wet detention basins cannot yet be engineered like water/wastewater treatment facilities to guarantee a specified effluent limits. Typical design methods involve somewhat empirical approaches which can approximate long-term pollutant removal rates by lumping together several physical/chemical and biological unit processes. Two different design methods were evaluated. One of the methods bases design criteria solely on solids settling, while the other method treats the wet detention basin as a eutrophic lake which achieves removal of both dissolved and suspended nutrients. Because 60% to 80% of total P removal monitored at wet detention basin BMP's is in the form of dissolved P, the solids settling model may not be appropriate for designs based on nutrient removal requirements. Instead, the solids settling model may be most appropriate for designs which address TSS and constituents which tend to appear primarily in suspended form (e.g., certain heavy metals). Because the lake eutrophication model design method accounts for the biological uptake of dissolved nutrients, it produces a more conservative design which is more appropriate for nutrient control than the solids settling design method. The permanent pool storage resulting from a eutrophication model design is on the order of three times larger than a design based on the solids settling model. Average hydraulic residence time is a very important design parameter for the permanent pool, and a minimum T of 2 weeks is recommended as an optimum design standard. Other general criteria for enhancing nutrient removal include mean and maximum depth of the permanent pool and shoreline side slopes that foster wetland vegetation.

References

1. U.S. Environmental Protection Agency, "Results of the Nationwide Urban Runoff Program: Volume I: Final Report," Water Planning Division, Washington, DC, December 1983.

2. Driscoll, E.D., "Performance of Detention Basins for Control of Urban Runoff Quality," prepared for 1983 International Symposium on Urban Hydrology, Hydraulics and Sediment Control, University of Kentucky, Lexington, KY, 1983.

3. Hartigan, J.P., "Regional BMP Master Plans," Urban Runoff Quality - Impact and Quality Enhancement Technology, Urbonas, B. and Roesner, L.A., eds., American Society of Civil Engineers, New York, NY, 1986, pp. 351-365.

4. Walker, W.W., "Phosphorus Removal by Urban Runoff Detention Basins," Lake and Reservoir Management: Volume III, North American Lake Management Society, Washington, DC, 1987, pp. 314-326.

5. Camp Dresser & McKee Inc., "Use of Stormwater Infiltration Practices for Water Quality Management: Minimum Criteria and Planning Guidelines," prepared for Maryland Water Resources Administration, Annapolis, MD, July 1985.

6. Camp Dresser & McKee Inc., "An Assessment of Stormwater Management Programs," prepared for Florida Department of Environmental Regulation, Tallahassee, FL, December 1985.

7. Northern Virginia Planning District Commission, "Washington Metropolitan Area Urban Runoff Demonstration Project," Annandale, VA, April 1983.

8. "Chesapeake Bay: A Framework for Action," U.S. Environmental Protection Agency, Chesapeake Bay Program, Annapolis, MD, September 1983.

9. Camp Dresser & McKee, "Regional Stormwater Management Plan: Final Report," prepared for Fairfax County Department of Public Works, Fairfax, VA, June 1988.

10. Walker, W.W., "Empirical Methods for Predicting Eutrophication in Impoundments--Report 3: Model Refinements," Technical Report E-81-9, U.S. Army Corps of

Engineers, Waterways Experiment Station, Vicksburg, MS, March 1985.

11. Yousef, Y.A., et al., "Fate of Pollutants in Retention/Detention Ponds," <u>Stormwater Management: An Update</u>, Publication #85-1, University of Central Florida, Environmental Systems Engineering Institute, Orlando, FL, July 1985, pp. 259-275.

12. Mills, W.B., et al., "Water Quality Assessment: A Screening Procedure for Toxic and Conventional Pollutants," <u>EPA-600/6-82-004</u>, U.S. EPA, Environmental Research Laboratory, Athens, GA, 1982.

13. "Guidelines for Constructing Wetland Stormwater Basins," Sediment and Stormwater Division, Water Resources Administration, Maryland Department of Natural Resources, Annapolis, MD, March 1987.

## Discussion of Mr. Hartigan's Paper

Question:

Is 50 to 70% phosphorous removal--that you say is achievable in these wet detention ponds--enough?

Answer:

It may not be.

Question:

What are some maintenance criteria?

Answer:

Use regionally large-enough facilities wherever possible, so the responsible agency will have sufficiently organized and competent maintenance staff and resources. Also, expect to perform major clean-outs of accumulated solids or noxious plants or whatever, every two or three years.

Long Term Performance of
Water Quality Ponds

Eugene D. Driscoll *

## Introduction

A report developed under EPA's Nationwide Urban Runoff Program (NURP) (1) describes an analysis methodology and presents graphs and example computations to guide planning level evaluations and design decisions on two techniques for urban runoff quality control. The control techniques addressed, infiltration devices and wet pond detention devices (basins that maintain a permanent pool of water), were shown by the NURP studies to be effective techniques for reducing pollutant discharges in urban stormwater.

This presentation provides a condensed summary of the analysis methodology, and an overview of the performance that can be expected from stormwater detention basins. The equations involved can be programmed for execution on personal (micro-) computers, for evaluations on a case-by-case basis An alternate approach, on which this presentation is based, solves the equations for the range of values the controlling parameters can assume, and presents pertinent results in a series of graphs which are then used to illustrate the general relationship between design size and performance. The presentation also includes a discussion of several issues relating to water quality ponds, including data on storm runoff settling velocities, the effect on the analysis of small storms that do not produce runoff, and use of the analysis method to size sediment forebays.

## Background

A detention device installed at a specific location will necessarily have a fixed size or capacity. Storm runoff, on the other hand, is highly variable. A particular basin will exhibit variable performance characteristics, depending on the size of the storm being processed. In general, it will perform more poorly for the larger storms than for the smaller ones. For detention devices such as wet ponds, which maintain a permanent pool of water, there is a further consideration that influences the ability to characterize performance based on monitoring data. For many storms in all basins, and for virtually all storms in large basins, the effluent displaced during a particular event represents, in fact, a volume contributed by the runoff of some antecedent event.

---

\* Senior Consultant, Woodward Clyde Consultants
101 Manito Ave Oakland, NJ 07436 (201) 337-2217

The methodology presented in this report is based on a probabilistic technique that accounts for the inherent variability of the situation it addresses. The analysis has a planning orientation that provides a basis for establishing "first order" design specifications (size, detention time), in terms of a long-term average removal of urban runoff pollutants. It is sufficiently simple, fast, and economical to apply, so that a large number of alternative scenarios are practical to examine. Comparison with actual performance data suggest that the procedure provides sufficiently accurate performance projections for the intended purposes.

The wet pond performance analysis presented below is based on and adapted from probabilistic analysis procedures conceived and formulated by DiToro, and developed by DiToro and Small (2,3,4). These procedures provide a direct analytical solution for the long term average removal of stormwater pollutants for several different modes of operation of a control technique. The variable nature of storm runoff is treated by specifying the rainfall and the runoff it produces in probabilistic terms, established by an appropriate analysis of a long-term precipitation record for an area. The equations that compute long-term average performance include a normalization that ratios the size or treatment capacity to the mean runoff event volume or flow, so that the solution is a general one, independent of local differences in rainfall and runoff. The equations are solved for a range of values for coefficient of variation to produce the general performance curves shown by Figures 1 through 4.

These plots depict long term average performance based on the mode of operation of a device processing variable stormwater inflows. Figure 1 relates to a device (such as porous pavement) in which removal (capture) of runoff is strictly a function of flow rate. In this, as in the other cases, the removal efficiency is seen to increase with size or treatment capacity. Since for any location or condition, the mean runoff flow (QR) will be a constant, higher ratios of QT/QR represent higher treatment capacities, which will result from greater percolating area when the percolation rate remains the same. On the other hand, efficiency degrades with increasing variability (here the coefficient of variation of runoff flow rates). A device with a ratio of 1.0 would capture 100% of the applied flows if there was no variability and all runoff reached the unit at the mean flow, QR.

Figure 2 indicates the performance relationships for a device, such as a sedimentation basin, that doesn't capture any runoff, but whose removal efficiency is a function of flow rate. Figures 3 and 4 apply to devices for which removal is controlled by the storage volume provided. An anology would be a recharge pit or basin. The volume ratio in the basic performance plot, Figure 3, is based on "effective" storage volume, which would be the same as the physical volume if the basin were always empty at the start of a storm. This will not usually be the case, and the actual amount of storage capacity available will depend on the emptying rate (e.g., by infiltration or pump out). Figure 4 is used in association with Figure 3 to estimate the relationship between physical capacity (VB), effective capacity (VE), and emptying rate.

# WATER-QUALITY PONDS

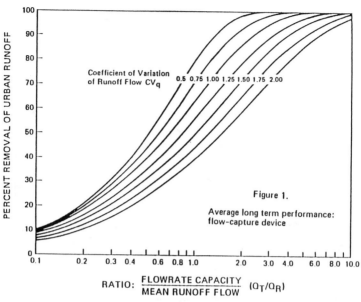

Figure 1. Average long term performance: flow-capture device

RATIO: $\dfrac{\text{FLOWRATE CAPACITY}}{\text{MEAN RUNOFF FLOW}}$ $(Q_T/Q_R)$

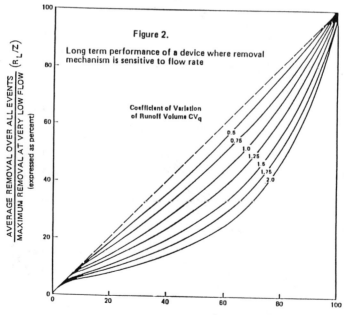

Figure 2. Long term performance of a device where removal mechanism is sensitive to flow rate

$\dfrac{\text{REMOVAL AT MEAN RUNOFF FLOW}}{\text{MAXIMUM REMOVAL AT VERY LOW FLOW}}$ $(R_M/Z)$
(expressed as percent)

# URBAN RUNOFF QUALITY CONTROLS

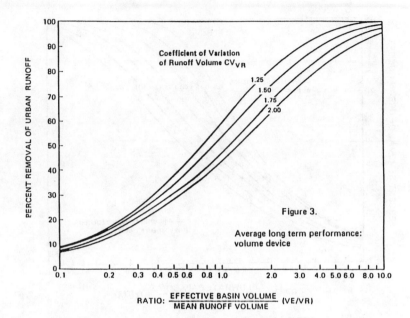

Figure 3.

Average long term performance: volume device

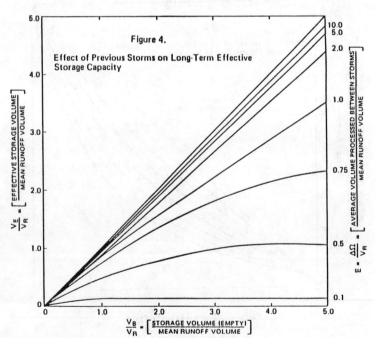

Figure 4.

Effect of Previous Storms on Long-Term Effective Storage Capacity

Efficiency is computed as a long-term average reduction in runoff volume or mass loading and is considered to be an appropriate measure of performance. For the procedures used in this report, variable rainfall/runoff rates and volumes are specified as a mean and coefficient of variation (CV = standard deviation / mean). A meaningful measure of device size or capacity is then the ratio of its volume or flow capacity to the volume or flow rate for the mean storm runoff event. This permits a convenient generalization of the analyses performed and allows results to be readily applied to various combinations of local conditions.

Wet Pond Analysis Method

Detention basins that receive storm runoff, but that have negligible losses through infiltration, can be considered to rely principally on sedimentation processes for pollutant removal. Under some conditions, reductions attributable to other processes can result in removal of specific pollutants. For example, substantial reductions in coliform bacteria have been observed and attributed to natural die-off, and algal uptake of nutrients is considered to be responsible for observed reductions in soluble forms of nitrogen and phosphorus.

A basic aspect of such a system is that part of the time (while runoff inflows occur), stormwater is moving through the basin, and sedimentation takes place under dynamic conditions. During the considerably longer dry periods between storm events, sedimentation takes place under quiescent conditions. The other important factor influencing pollutant removal by sedimentation is the settling velocity of the particulates present in the urban runoff.

*Removal Under Dynamic Conditions* - Characterization of the performance of sedimentation devices has been extensively analyzed over the years because of the important role such devices play in both water treatment and wastewater treatment systems. A method of analysis which is particularly suitable is presented by Fair and Geyer (5). Removal due to sedimentation in a dynamic (flow through) system is expressed by the following equation:

$$R = 1 - \left[ 1 + \frac{1}{n} \cdot \frac{v_s}{Q/A} \right]^{-n} \qquad (1)$$

where:

- $R$ = fraction of initial solids removed ($R * 100 = \%$ Removal)
- $v_s$ = settling velocity of particles
- $Q/A$ = rate of applied flow divided by surface area of basin (an "overflow velocity," often designated the overflow rate)
- $n$ = turbulence or short-circuiting parameter

One value of this model is that it provides a quantitative means of factoring into the analysis an expression for impaired performance due to short-circuiting (since many stormwater retention basins will not have ideal geometry for sedimentation). Fair and Geyer suggest an empirical relationship between performance and the value of "n," which is: n = 1 (very poor); n = 3 (good); n > 5 (very good). In addition, when a value of n = ∞ is assigned (ideal performance), the equation reduces to the familiar form wherein removal efficiency is keyed to detention time.

$$R = 1 - \exp\left[-\frac{v_s}{Q/A}\right] = 1 - \exp\left[-kt\right] \quad (2)$$

where:

$k$ = $v_s / h$ (sedimentation rate coefficient)
$h$ = average depth of basin
$t$ = $V / Q$ residence time
$V$ = volume of basin

The two expressions are equivalent. To use them, one must be able to identify an appropriate value for either settling velocity, or for the rate coefficient (k), which will ultimately depend on the settling velocity of the particulates present.

Solving equation 1 for a range of overflow rates and particle settling velocities and plotting the results as shown by Figure 5, indicates the wide range in removal that can be expected either (a) at a constant overflow rate for particles of different size, or (b) at different rates of flow for a specific size fraction. Both of these variable factors are present in urban runoff applications. The effect of a range of particle settling velocities is addressed by performing separate computations for a number of settling velocities and then using the weighted average of the mass fractions to compute net removal.

Removal/flow relationships for a range of settling velocities representative of urban runoff, computed by equation 1, are presented in Figure 6 as a semi-log plot on which an exponential approximation, equation 2, would plot as a straight line. For a site-specific analysis (for each settling velocity separately), the straight line approximation would match the exact solution at the point corresponding to the mean overflow rate (QR/A), and the slope would be adjusted to give the best match over the range of rates expected to span the bulk of the important storms. The intercept of this fitted line (Q/A = 0) provides the estimate for the factor Z shown earlier in Figure 2. Over the range of overflow rates of interest, the exponential approximation will usually be within about 10%.

Long-term average removal of a pollutant under dynamic conditions can, therefore, be estimated from the statistics (mean and coefficient of

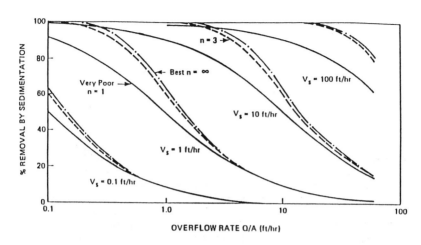

Figure 5 Effect of settling velocity and overflow rate on removal efficiency

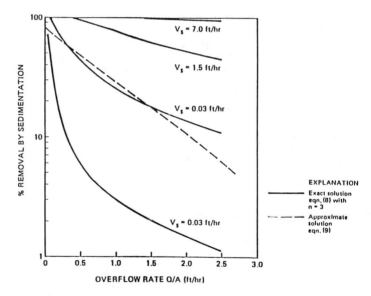

Figure 6 Flow-removal relationships for exponential approximation

variation) of runoff flows, the basin surface area, and representative particle settling velocities for urban runoff.

Removal Under Quiescent Conditions - For much of the country, the average storm duration is about 6 hours, and the average interval between storms is on the order of 3 to 4 days. Thus, significant portions of storm runoff volumes may be detained for extended periods under quiescent conditions, until displaced by subsequent storm events. The volume of a basin relative to the volumes of runoff events routed through it is the principal factor influencing removal effectiveness under quiescent conditions.

The probabilistic computation summarized by design performance curves in Figures 3 and 4, estimates the removal of physical volume from the basin under quiescent conditions during the dry periods between storms. However, for sedimentation devices that maintain a permanent pool of water, a modification is required because there is no loss of stored volume between runoff events. Instead, it is the particulates in the detained volume that settle out under quiescent conditions. The modification required is to express this condition in terms of the parameters of the design performance curves.

The term $\Omega$ may be thought of as a "processing rate." For a recharge device, it is the rate at which volume is removed from the basin by percolation through the bottom and sides. For a sedimentation device, it may be thought of as a particle removal rate. Using this interpretation, the term $\Omega \Delta$ used in Figure 4 can be considered to represent that portion of the basin volume from which solids with a selected settling velocity have been completely removed. Instead of the TSS concentration of the entire volume diminishing with time under quiescent settling, the concentration is assumed to remain constant, while the remaining volume with which this concentration is associated diminishes with time. The solids removal rate is then:

$$\Omega = v_s * A \qquad (3)$$

where:

$v_s$ = particle settling velocity (ft/hr)
$A$ = basin surface area (square feet)

Long Term performance

Since both dynamic and quiescent conditions prevail in a detention basin at different times, the overall efficiency of a basin is the result of the combined effect of the two processes at work. The simple model used to integrate these effects is illustrated by Figure 7. Five identical storms with an interval between event midpoints ($\Delta$) of 3.5 days are routed through a basin, assuming plug flow. Each storm has a duration of 12 hours (0.5

# WATER-QUALITY PONDS

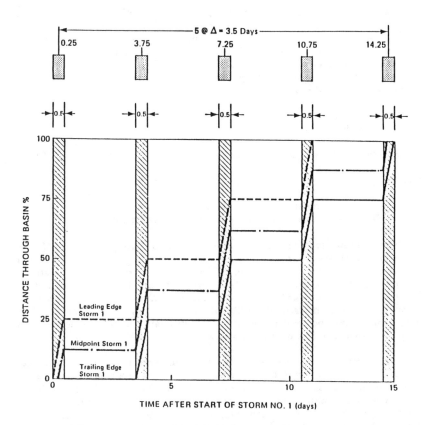

For Storm Midpoint Volume

Total Residence Time = 14.0 Days
Dynamic Time: (0.25) + (3 x 0.5) + 0.25 = 2.0 Days     2/14 = 0.14
Quiescent Time: 14.0-2.0 = 12 Days                     12/14 = 0.86

$D/\Delta = 0.5/3.5 = 0.14$
$(1-D/\Delta) = 0.86$

**Figure 7**    Illustration of quiescent vs. dynamic residence time in a storm detention basin

day), and a volume that is 25% of the basin volume (VB/VR = 4). The plotted lines track the residence / displacement pattern in the basin for the leading edge, midpoint, and trailing edge of Storm #1. The shading highlights the fraction of the total residence time when dynamic conditions prevail. For this simplified case, and for actual conditions where both storm volumes (VR) and intervals ($\Delta$) fluctuate, the fraction of time under dynamic conditions is estimated by:

Fraction of residence time
under dynamic conditions = $D / \Delta$ (4)

Fraction under quiescent conditions = $1 - (D / \Delta)$ (5)

where:

$D$ = mean storm duration
$\Delta$ = mean interval between storm midpoints

This simple schematic illustrates several relevant features of the operation of this type of device. When the basin is as large as that indicated (which is not uncommon for current practice), the outflow volume during an event represents a different parcel of water than that for the storm that causes it to be displaced. Assessing performance by comparing paired influent and effluent loads for individual storms is less appropriate than the comparison of overall influent and effluent loads for a long-term sequence of storm events.

All runoff volumes that enter the basin undergo the dynamic removal process one or more times before discharge. For the large basin illustrated, this is broken up into four different periods of displacement. For a basin with a volume small enough that the runoff passes all the way through, there would be only one such period of dynamic removal, and only a fraction of the runoff would remain in the basin to undergo quiescent removal. For most detention basins, the analysis procedure suggests that the dominant influences on performance efficiency are the basin volume and the settling velocity of the pollutants in the runoff.

Settling Velocity of Particulates in Urban Runoff

The settling velocity of particulates in urban runoff is estimated from data obtained from settling column tests performed by a number of the NURP projects. Settling column tests were conducted by a number of NURP projects on samples of urban runoff. Results from these tests, and from a similar set of tests reported by Whipple and Hunter (6), were analyzed to derive information on particle settling velocities in urban stormwater runoff. The analysis procedure used for developing settling velocity distributions from test data is summarized by Figure 8. The overall results of the analysis of 46 separate settling column tests are tabulated below.

# WATER-QUALITY PONDS

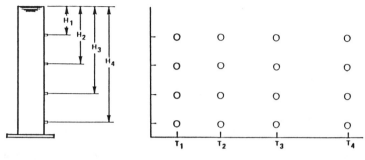

SETTLING COLUMN        ELAPSED TIME TO SAMPLE WITHDRAWAL

O = Data Point - Record % removed based on observed vs. initial concentration

Settling velocity ($V_s$) for that removal fraction is determined from the corresponding sample depth (h) and time (t)
$$V_s = H/T$$

Observed % removed reflects the fraction with velocities equal or greater than computed $V_s$

A probability plot of results from all samples describes the distribution of particle settling velocity in the sample

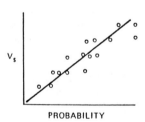

PROBABILITY

Figure 8    Estimating settling velocity distributions from settling column tests

| Size Fraction | % of Particle Mass in Urban Runoff | Average Settling Velocity (ft/hr) |
|---|---|---|
| 1 | 0 - 20% | 0.03 |
| 2 | 20 - 40% | 0.30 |
| 3 | 40 - 60% | 1.5 |
| 4 | 60 - 80% | 7. |
| 5 | 80 - 100% | 65. |

One limitation of the NURP analysis results tabulated above, is that the analyses were restricted to TSS data, and it was necessary to infer the settling velocity distribution of other pollutants from their soluble fraction together with assumptions regarding how many of the above TSS size fractions the pollutant is expected to be associated with. For example, in the NURP analysis it was assumed that most pollutants would be associated only with the four smaller size fractions.

Settling velocities for specific pollutants of interest can be developed by the use of the procedure described above wherin the samples withdrawn are analyzed for one or more pollutants other than (or in addition to) suspended solids. There have been a few studies that have taken this approach recently, but for which published results are not yet available.

Applications of Analysis Method

One of the advantages to the availability and use of a relatively simple-to-apply probabilistic analysis, given always the caveat that a "simple model' in fact provides a simplified representation of reality and approximately correct results. The principal one is that it becomes practical to make large numbers of analyses to generalize results, examine the sensitivity of predicted results to uncertainty in any of the input parameters, or to explore the effect of modified concepts for water quality pond design. An illustration of each of these possibilities is presented below.

1. Generalizing Results -

An example of this type of output is shown by Figure 9. The detention basin's average depth, and the runoff coefficient of the contributing drainage area are both held constant. For these conditions, the generalized performance plot indicates removal efficiency as a function of basin size and regional rainfall characteristics.

This figure illustrates the order of differences in basin performance characteristics which can result from regional differences in rainfall patterns, all other things being equal. The effect of design size is also shown. Basin size is expressed as a (percentage) ratio between the surface area of the basin and the contributory urban drainage area. For example, an area ratio of 0.10% on the horizontal axis reflects a basin with a surface area of 0.64 acres serving a 1-square-mile (640-acre) urban drainage area. The performance relationships could alternatively be expressed in terms of

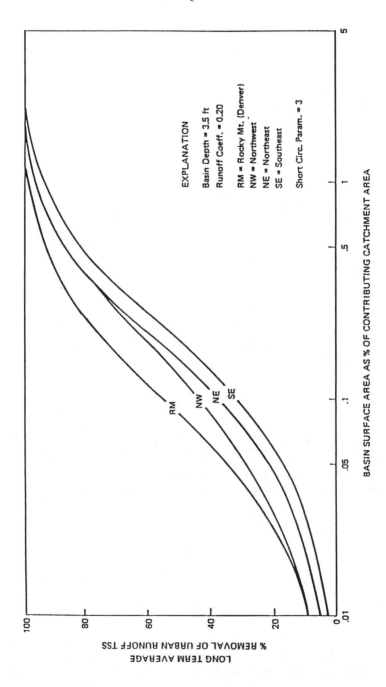

Figure 9  Regional differences in detention basin performance

basin volumes, although depth would also have to be shown in such a case because performance depends on both area and volume provided.

Alternative detention basin performance charts (similar in general appearance) could be developed using the same approach, that would provide a more useful format as a working guide for local planning decisions. For example, the local rainfall properties would be held as a constant, and the size-performance curves would be developed to show the relative effect of basin depth and/or the impervious fraction of the contributing area.

2. Sensitivity to Input Estimates --

A useful example of a sensitivity analysis is illustrated by the following examination of the concern that applying a runoff coefficient to the complete rainfall record provides a biased estimate of the statistics for runoff flows and volumes. Most of the very small storms (e.g., those that produce total depths less than about 0.1 inch) are completely absorbed and generate no runoff at all. Schueler has made this observation for the Washington D.C. area, as has Urbonas for the Denver area (7).

At my firm, Woodward-Clyde Consultants, we are nearing completion of a PC (microcomputer) version of the SYNOP rainfall analysis program. Several useful features are being added to enhance flexibility and usefulness, compared with earlier versions. One of the added features is the ability to filter out all storms smaller than a user-specified minimum volume, and compute the statistical parameters of the larger storm events expected to produce runoff.

The general effect of the removal of a group of small values from a data set is that the value of the mean will increase, and the coefficient of variation will decrease. The higher the cutoff level selected, the greater will be the percentage change in the statistic. Based on some preliminary results from the SYNOP program, some analyses using artificial data sets, and suggestions offered by Urbonas (7), an initial assumption was made that the mean values of the rainfall statistics will increase by 30 percent, and the coefficients of variation will decrease by 15 percent. These conditions were employed in a sensitivity analysis to assess the degree to which elimination of the small storms that produce no runoff might affect predictions of basin performance.

Inspection of the general performance relationships shown by Figures 1 and 3, will indicate that the changes tend to compensate. The higher mean values reduce efficiency, but the lower coefficients of variation operate to increase efficiency. The wet pond analysis methodology was applied routing runoff from a range of completely impervious catchments through a 3 foot deep basin. Table 1 compares the performance predictions for the two sets of rainfall/runoff statistics. It is seen that for the assumed changes, there is no significant effect on long term performance efficiency. Additional testing, using better estimates of the actual changes in the rainfall/runoff statistics are necessary before concluding that the result is generally true, but it is probably safe to say that the effect will not be substantive.

## TABLE 1. EFFECT OF ELIMINATION OF SMALL STORMS ON PERFORMANCE PREDICTIONS

### RAINFALL - RUNOFF EVENT STATISTICS

|  |  | ALL STORMS | | W/O SMALL STORMS | |
|---|---|---|---|---|---|
|  |  | MEAN | COV | MEAN | COV |
| VOLUME | inch | 0.40 | 1.50 | 0.52 | 1.28 |
| INTENSITY | in/hr | 0.07 | 1.30 | 0.09 | 1.11 |
| DURATION | hours | 6.0 | 1.10 | 7.8 | 0.94 |
| INTERVAL | hours | 90 | 1.00 | 117 | 1.00 |

### WET POND PERFORMANCE ESTIMATES

| *BASIN SIZE RATIO | % REMOV DYNAMIC | % REMOV QUIESCENT | % REMOV COMBINED | % REMOV DYNAMIC | % REMOV QUIESCENT | % REMOV COMBINED |
|---|---|---|---|---|---|---|
| 0.10% | 10 | 6  | 15 | 10 | 5  | 15 |
| 0.20% | 12 | 12 | 22 | 12 | 10 | 21 |
| 0.33% | 13 | 18 | 29 | 14 | 16 | 27 |
| 0.50% | 14 | 25 | 36 | 15 | 22 | 34 |
| 1.00% | 16 | 45 | 54 | 18 | 42 | 52 |
| 2.00% | 19 | 69 | 75 | 20 | 67 | 73 |

NOTES -

* Value shown is the ratio of the basin surface area to the area of the catchment that contributes runoff (expressed as a percentage.

Computations assume an average basin depth of 3 feet, and a completely impervious catchment.

3. Investigation of Alternate design Concepts -
The settling velocity data for urban runoff indicate that a large fraction of TSS removed by sedimentation, will settle out very rapidly. As a general approximation, on the order of 50 or 60% of the particulates have settling velocities that exceed 1 or 2 feet per hour. This is consistent with Yousef's observation that "a majority of the particulate metals settle out quickly and are deposited near the point of input" (8). His observation is in turn confirmed by independent data from one of the NURP basins (Lake Ellyn, IL), that also examined sediment accumulation patterns in detail.

Consider the foregoing information in context with Wiegand's observation that "the one-time costs for sediment removal can be staggering"... even though carried out only once every 10 to 20 years (9). The evidence suggests that there should be significant practical and economic value to the development of a basin design that has a two stage operation with the following features.

(a). A relatively small inlet section with size and design features that will permit capture of the easily separated particulate fraction, and convenient and economic removal of accumulated sediment. More frequent cleaning will be required, but there will be a practical benefit if this activity becomes a semi-routine element of a maintenance program. The likelihood of long term maintainence of effectiveness should be enhanced if sediment removal can be converted from a 20 year event with staggering cost and difficulty.

(b). A final section that provides the bulk of the storage volume determined by applicable local criteria. The effective life of these larger, more costly to dredge sections should be extended substantially.

The analysis methodology is applied here in a preliminary examination of the practical feasibility of this concept. A 1 acre, completely impervious catchment is used together with a set of rainfall statistics typical of the eastern US. The analysis examines the projected performance of a forebay that has a surface area of 43.5 square feet and an effective depth of 1 foot. Additional depth for solids accumulation would, of course, have to be provided. Table 2 lists computed removal efficiencies for each of five size fractions of solids in urban runoff. The results indicate that even a section this small could frmove 25 to 30 percent of all influent solids, and about 80 percent of the largest fraction. A tacit assumption is that accumulated solids would be removed frequently enough to maintain the forebay's effectiveness.

The actual removal efficiencies listed should considered approximations. It appears valid to conclude that even relatively quite small forebay sections can be designed to trap appreciable fractions of influent solids.

## TABLE 2. PRELIMINARY ANALYSIS OF PERFORMANCE CAPABILITY OF A FOREBAY

### RAINFALL - RUNOFF EVENT STATISTICS

|  |  | MEAN | COV |
|---|---|---|---|
| VOLUME | inch | 0.40 | 1.50 |
| INTENSITY | in/hr | 0.07 | 1.30 |
| DURATION | hours | 6.0 | 1.10 |
| INTERVAL | hours | 90 | 1.00 |

### FOREBAY PERFORMANCE ESTIMATES

| SIZE FRACTION | SETT VELOC FT / HR $V_s$ | EFFECTIVE VOL RATIO $V_E/V_R$ | % REMOV DYNAMIC | % REMOV QUIESCENT | % REMOV COMBINED |
|---|---|---|---|---|---|
| 1 | 0.03 | 0.01 | 1.4 | 0.4 | 1.0 |
| 2 | 0.30 | 0.02 | 5.0 | 1.2 | 6.1 |
| 3 | 1.50 | 0.03 | 12.5 | 2.2 | 14.5 |
| 4 | 7.0 | 0.03 | 32.5 | 2.4 | 34.2 |
| 5 | 65 | 0.03 | 80.6 | 2.4 | 81.0 |

NOTES –

Computation is based on the following conditions.

1 acre urban catchment, completely impervious ($R_v = 1.0$)

Forebay surface area = 43.5 square foot, and depth = 1 foot

**REFERENCES**

1. USEPA. 1986, <u>Methodology for Analysis of Detention basins for Control of Urban Runoff Quality</u>], Office of Water, NonPoint Source Division, Washington, D.C., September.

2. Hydroscience, Inc. 1979. <u>A Statistical Method for the Assessment of Urban Stormwater,</u> for USEPA NonPoint Sources Branch, EPA 440/3-79-023, May.

3. DiToro, D.M. and M.J. Small. 1979. <u>Stormwater Interception and Storage</u>, Journal of the Environmental Engineering Division, ASCE, Vol. 105, No. EE1, Proc. Paper 14368, February.

4. Small, M.J. and D.M. DiToro. 1979. <u>Stormwater Treatment Systems</u>, Journal of the Environmental Engineering Division, ASCE, Vol. 105, No. EE3, Proc. Paper 14617, June.

5. Fair, G.M. and J.C. Geyer. 1954. <u>Water Supply and Waste Water Disposal</u>, John Wiley and Sons.

6. Whipple, W., Jr. and J.V. Hunter. 1981. <u>Settleability of Urban Runoff Pollution,</u> Water Pollution Control Federation Journal, <u>53</u> (12), pp. 1726-1731, December.

7. Personal Communications with Thomas R. Schueler (Metropolitan Washington Council of Governments), and Ben Urbonas (Urban Drainage and Flood Control District, Denver CO). 1988

8. Yousef, Y.A., et al. 1986. <u>Design and Effectiveness of Urban Retention Ponds</u>, Engineering Foundation Conference, Urban Runoff Quality, It's Impacts and Quality Enhancement Technology.

9. Wiegand, C., et al. 1986. <u>Comparative Costs and Cost Effectiveness of Urban Best Management Practices</u>, Engineering Foundation Conference, Urban Runoff Quality, It's Impacts and Quality Enhancement Technology.

## Discussion of Mr. Driscoll's Paper

Question:

With respect to the wet pond forebay you mentioned, wherein you computed a 27% solids removal, that computation depends, does it not, on having a fully mixed, representative sample of all the solids being carried by the runoff?

Answer:

Right. I got my solids' distribution data from the NURP studies, and I have to presume those investigators did the sampling correctly.

Question:

I am prepared to assume that certain "laws" or theories, like Stokes Law, apply and can be used to derive these mathematically, theoretically based design criteria; and I assume we can make physical, mechanical measurements of particle sizes and hence surface areas to estimate the availability of sites where heavy metals and other things can adhere. But when we get to providing the forebay thus designed ahead of a wet detention pond, are we doing that to protect the quality of the water in the pond, or the downstream receiving water quality?

Answer:

It is for downstream protection against heavy metals and solids, primarily. But I recognize that quality in the ponds becomes important as well to the people who live next to them.

## Mixing and Residence Times of Stormwater Runoff in a Detention System

By Edward H. Martin,[1] Member, ASCE

**Abstract:** Five tracer runs were performed on a detention pond and wetlands system to determine mixing and residence times in the system. The data indicate that at low discharges and with large amounts of storage, the pond is moderately mixed with residence times not much less than the theoretical maximum possible under complete mixing. At higher discharges and with less storage in the pond, short-circuiting occurs, reducing the amount of mixing in the pond and appreciably reducing the residence times. The time between pond outlet peak concentrations and wetlands outlet peak concentrations indicate that in the wetlands, mixing increases with decreasing discharge and increasing storage.

### Introduction

The detention of urban stormwater runoff for water-quality enhancement is becoming more widely practiced. Among the most important factors in the overall performance of detention facilities are the mixing and residence times of runoff as it flows through the system.

The purpose of this paper is to examine the mixing characteristics and residence times of stormwater runoff in an urban detention system in central Florida. The results of 5 tracer runs under near steady-state discharge conditions were used to investigate mixing characteristics and residence times of runoff in a detention pond and wetlands system. The tracer runs were made at various inlet discharges and levels of detention storage. The constituent-removal performance of the system is described in Martin and Smoot (1986) and Martin (1988).

Two mixing idealizations that have been widely used in evaluating the characteristics of basins are the models of (1) completely mixed flow, and (2) plug flow. These idealizations represent two extremes in mixing. In practice, most detention basins will exhibit deviations from either mixing ideal. The deviations tend to reduce mean residence times and can result from short-circuiting, recycling, or the presence of stagnant zones. In this paper, flows exhibiting these mixing characteristics are referred to as moderately mixed flows.

In an idealized completely mixed flow, the influent is assumed to be completely and instantaneously mixed with the contents of the basin. Concentrations are uniform throughout the basin. The steady-state mean residence time for a completely mixed flow basin is VOL/Q (Weber 1972) where VOL is the volume of water stored in the basin and Q is the discharge.

---

[1]Hydro., U.S. Geological Survey, 224 W. Center Street, Altamonte Springs, FL 32714.

For idealized plug flow, the flow is assumed to proceed through the basin in an orderly manner, as discrete units of water traveling with equal velocity. Concentrations are uniform in the cross section, and there is no longitudinal mixing because of concentration gradients in the direction of flow. The steady-state residence time for a plug flow basin is VOL/Q (Weber 1972), the same as for a completely mixed flow basin. In contrast to the steady-state residence time of a completely mixed flow basin, which is a mean for the flow particles, the steady-state residence time for a plug flow basin applies to each individual particle.

Moderately mixed flow basins are characterized by incomplete mixing, short-circuiting, recycling, and dead zones. Moderately mixed flow basins do not represent an intermediate point between a completely mixed flow and plug flow basin mixing, but a departure from either ideal that will produce residence times lower than either a completely mixed flow or plug flow basin. It is usually necessary to experimentally determine the mixing and residence times in a moderately mixed flow basin.

The technical literature to date indicates that plug flow is the recommended type of mixing in stormwater detention basins. This recommendation seems to be based not on a detailed analysis of completely mixed or plug flow mixing in a stormwater detention basin, but on the analogy of a stormwater detention basin to a wastewater plant settling tank, which is designed to be a plug flow basin.

The choice of mixing type for a detention basin should depend upon the ratio of a design-storm runoff volume to pond storage volume. If the ratio is less than one, plug flow is the obvious choice because plug flow would ensure that all the storm runoff would be retained until the next storm and be subject to the chemical and biological changes that occur between storms. When the ratio of storm runoff to pond storage is greater than one, the choice of mixing type is not as obvious. The choice will depend upon several factors, including the reduction rate of constituents during a storm and the reduction rate of constituents occurring in the pond's water between storms.

A completely mixed detention basin will reduce peak constituent concentrations to a composite mean concentration of water in storage and water that flows into the basin during the storm. Runoff from a large storm entering a plug flow basin could cause the water containing the peak inlet constituent concentrations to be subject to only dynamic reduction during the storm and to be completely discharged out of the basin relatively quickly. In contrast, for the same large storm, a completely mixed basin could reduce peak inlet constituent concentrations below some threshold receiving water standard or toxicity level through mixing with the basin's between-storm-storage-volume water and water that flows into the basin during the remainder of the storm. This point, alone, may dictate the choice of a completely mixed basin in areas where water-quality discharge standards are specified in terms of constituent concentrations. The most appropriate mixing design for a detention system is worthy of further study.

This paper describes the mixing and residence times for a stormwater detention pond and wetlands system that experience moderately mixed flows and for which the constituent-removal performance has already been described. These data, along with performance data of other detention systems, could begin to provide an indication of which mixing types are most appropriate under different circumstances for stormwater detention basins.

## Description of the Detention Pond and Wetlands

The stormwater detention facility is part of an urban drainage system located in Orlando, Florida; Fig. 1 is a plan view of the system. The drainage area above the detention facility is approximately 41.6 acres. The land uses are: urban roadway, 33%; forest, 28%; high-density residential, 27%; and low-density residential, 13%. About one-third of the rainfall on the drainage area runs off (Martin and Smoot 1986). Runoff moves by overland flow to a system of curbs and gutters that direct the flow to a series of drop inlets connected to a reinforced concrete drainage pipe that discharges into the pond.

The pond was built into, and is surrounded by, a relatively impermeable clay layer. The sides of the pond have a 2:1 slope and are protected by sand-cement riprap. The pond shoreline is surrounded by a growth of cattails and small trees. The pond is roughly rectangular in shape with an area during dry weather of 8,600 sq ft (800 m$^2$). The pond's width to length ratio is approximately 1:2. Depths range from about 8 ft (2.4 m) during dry weather to as much as 11 ft (3.4 m) during storms. Dead storage, the volume of water retained between storms, is 53,040 ft$^3$ (1,500 m$^3$) or 0.35 in. (0.89 cm) of runoff volume from the drainage area and available live storage, the additional volume of water that is stored during a storm, is about 18,500 ft$^3$ (520 m$^3$) or 0.12 in. (0.30 cm) of runoff, so that the maximum available storage, dead plus live, is 71,540 ft$^3$ (2,030 m$^3$) or 0.47 in. (1.19 cm) of runoff volume. The terms dead and live storage refer to volumes retained between and during storms respectively, not to inactive (no mixing) or active (mixing) zones.

Water discharges from the pond into the wetlands over an earthen spillway into a shallow channel that runs the length of the wetlands, (Fig. 1); however, at high pond stages, water overflows the length of the east side of the pond. The bed material of the wetlands is composed of sand and loamy silt. Large cypress trees grow in the wetlands with heavy undergrowth of hyacinths, duckweed, cattails, and small trees. Water leaves the wetlands over a compound weir built around a drop inlet, and enters a culvert leading to a drainage canal.

The wetlands are about 32,000 sq ft (2,970 m$^2$) in area with dry-weather depths ranging from 0 to 3 ft (0.91 m), and storm depths reaching as much as 5 ft (1.5 m). Dead storage can be as much as 22,700 ft$^3$ (640 m$^3$) or 0.15 in. (0.38 cm) of runoff. The available live storage is about 98,500 ft$^3$ (2,790 m$^3$) or 0.65 in. (1.65 cm) of runoff, with total storage reaching about 121,000 ft$^3$ (3,430 m$^3$) or 0.80 in. (2.03 cm) of runoff during storms.

## Data Collection

The experimental method used to determine the mixing and residence times of a basin involved injecting a conservative tracer at the pond inlet, and collecting samples at the pond outlet (wetlands inlet)

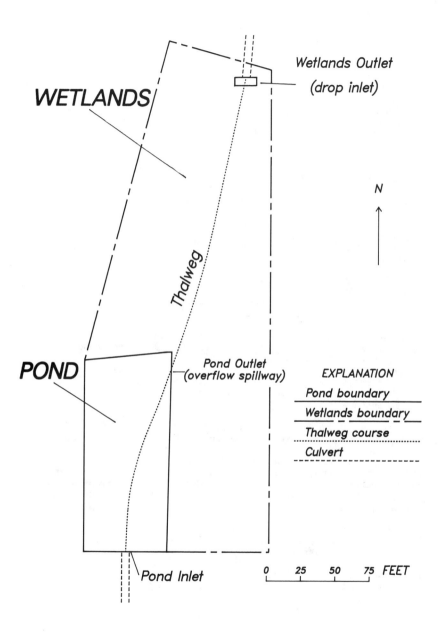

FIG. 1. Plan View of System

and the wetlands outlet. For a system subject to unsteady flows, data for different inlet discharges and volumes of storage are needed to adequately describe the mixing and residence time characteristics.

For purposes of this study, relatively steady discharges through the system were achieved by modification of the existing drainage system. With the use of a pumping station, water from a detention pond in an adjacent drainage system could be pumped into the study drainage system. High, relatively steady flows could be established in the system when desired. The major constraint for this procedure was the amount of water available for pumping; usually there was only enough water for one to two hours of pumping.

Discharge at the pond inlet was measured with an electromagnetic current meter mounted in the center of the inlet culvert. Flow at the pond outlet was affected by backwater from the wetlands; therefore, it was necessary to use a numerical solution of the standard storage equation and a record of pond stage to determine pond outlet flow. Discharge at the wetlands outlet was determined from the record of wetlands stage and a weir, calibrated using current-meter discharge measurements.

All of the tracer injections used the following procedure. The pumps were turned on and the flow at the pond inlet was established for 5 to 15 minutes. Rhodamine-wt was injected into the pond inlet culvert through a manhole, approximately 30 ft (9.1 m) upstream from the pond inlet (Fig. 1). The dye was put into a water-tight glass jar that was placed in a plastic pipe having a perforated end. The pipe containing the jar and dye was then inserted into the manhole so that the jar was in the center of the 5-ft (1.5 m) inlet culvert. The jar was then broken with a metal rod inserted into the plastic pipe. The travel time of the dye from the injection point along the 30 ft (9.1 m) of culvert was calculated using the pond inlet velocity, which ranged between 0.5 and 1.0 ft/s (0.15 and 0.30 m/s).

Tracer data were collected during three different pumping periods. During the first two pumping periods two dye injections were made, and on the last pumping period one injection was made, yielding a total of 5 tracer runs. The discharge was generally constant throughout each pumping period but the stage in the pond and wetlands, and consequently the storage, increased through the pumping period. Thus data were collected for the same discharge at various volumes of storage.

Dye samples were taken at the pond outlet and the wetlands outlet at a frequency necessary to define the passage of dye through the system. It was assumed that concentrations at the pond and wetlands outlet sampling points were representative of all the water discharging from the pond and wetlands outlets, respectively. Dye concentrations were determined by standard fluorometric procedures as described in Wilson, Cobb, and Kilpatrick (1986).

## Results

### Basic data

An example of the collected data, including two tracer runs, is shown in Fig. 2. The instantaneous discharges at the pond inlet, pond outlet, and wetlands outlet, shown in the top panel of Fig. 2, vary gradually with time, but are considerably more steady than storm-generated discharges at this site (Martin and Smoot 1986). The discharge from the pond and wetlands changes even more gradually than the pond inlet discharge because of the storage effects of the pond and wetlands. The volume of water in live storage in the pond and wetlands is also shown in the top panel of Fig. 2. The storage gradually increases as long as the discharge is maintained throughout both runs.

The dye concentrations at the pond outlet and wetlands outlet are shown in the bottom panel of Fig. 2. The pond outlet concentrations for run 3 are characterized by a sharp peak (about 150 $\mu g/L$) approximately 5 minutes after the injection. The pond outlet concentration for the second peak (about 40 $\mu g/L$), run 4, is much less in magnitude and occurred about 12 minutes after injection. Because the discharge was relatively constant, it is reasoned that the reduction and increased time to peak are because of increased storage and the resultant establishment of mixing currents in the pond. At high pond stages, water overflows the length of the east side of the pond (Fig. 1). It is possible that at higher stages, the pond's discharge water is not homogeneous. However, samples taken intermittently from points in the pond during the run do not indicate this.

Dye concentrations at the wetlands outlet generally follow the trend of the pond outlet showing the dissipation of the peak dye concentrations because of the effects of mixing and storage in the wetlands. The increased mixing because of increased wetlands storage is appreciable. Run 3 wetlands concentrations exhibited a high, (about 60 $\mu g/L$), but sharp peak, whereas run 4, with much increased storage, exhibited a gradual rise and even more gradual decline in wetlands outlet concentrations.

The maximum concentration that would have been found in the pond, if complete mixing had occurred, was calculated and plotted in the bottom panel of Fig. 2. Concentrations greater than the maximum, assuming complete mixing, may indicate that at least part of the dye is short-circuited across the pond, resulting in higher concentrations at the outlet. These higher concentrations might also be caused by plug flow with diffusion occurring in the pond. However, if plug flow were occurring, the time to peak concentration would be about the same as the mean residence time. As will be shown later, this is not the case. Run 4, with a peak concentration closer to the completely mixed flow concentration, seems to have had more complete mixing than run 3.

The concentration curves alone do not give a complete indication of the extent of mixing in the pond or wetlands. Analysis of mixing and residence times requires the calculation of the mass flux and cumulative mass of dye at the pond outlet and wetlands outlet.

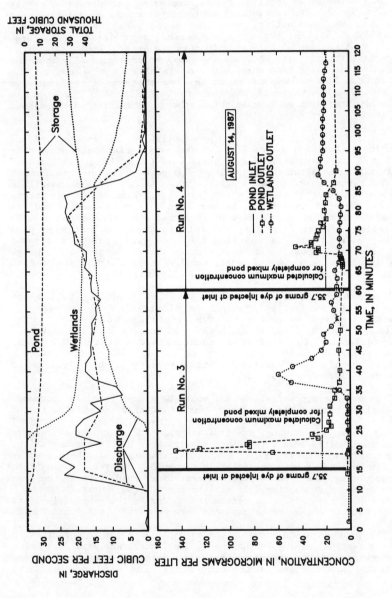

FIG. 2. Discharge and Dye Concentrations for Runs 3 and 4

Knowing the cumulative mass of dye transported past the pond outlet and wetlands outlet allows the determination of the amount of dye recovered and not retained in the pond and wetlands.

For analysis purposes, the dye runs were separated. The background concentration of dye at the sampling points at the time of each injection was subtracted from the measured concentrations and the mass flux calculated using the adjusted concentrations. In Fig. 3 the mass flux for run 3 is plotted. The area under the mass flux curve is the cumulative mass of dye recovered. This was calculated and also plotted in Fig. 3. The cumulative mass is shown in terms of both grams and percentage of total dye injected.

The mass flux curve for the pond shown in Fig. 3 indicates a steep rise and a corresponding steep recession until about 9 minutes, after which, the recession is more gradual. About 30% of the dye has been recovered up to this point. Field observations indicated some short-circuiting from pond inlet to pond outlet. This short-circuiting is probably responsible for the high mass-flux peak and the quick transporting of 30% of the injected dye across the pond. An additional 20% (total of 50%) is recovered in 47 minutes; the remaining 50% is released at a gradually decreasing rate.

The data seem to indicate that for this discharge and storage approximately 30% of the inlet flow particles are short-circuited directly across the pond within about 9 minutes, while the remaining 70% of the inlet flow particles are dispersed throughout the pond. Subsequent inlet discharge mixes with the pond water, transporting out a part of the dye and further diluting the remaining dye. This process would continue, if the inlet discharge were maintained, until all dye is transported out of the pond. This process is conceptually the same as the moderately mixed flow model.

After about 50 minutes, 50 percent of the dye had been recovered. The mass flux seems to be declining at a constant rate and the cumulative mass curve is increasing in a gradual and smooth manner (Fig. 3). To estimate the statistics of the residence time of all the injected dye, it was necessary to extrapolate the mass flux and cumulative mass curves to the point where most of the injected dye would have been recovered. The cumulative curves were extrapolated using the equation:

$$M = M_T(1-e^{-kt}), \quad \quad \quad \quad \quad \quad \quad \quad (1)$$

where
$t$ = time,
$M$ = cumulative mass transported out of the pond at time t,
$M_T$ = total mass of dye injected into the pond,
$k$ = rate constant, and
$e$ = base of natural logarithms.

This equation is similar to the equation for a conservative constituent discharged from a completely mixed flow basin (Weber 1972), except that the calculated rate constant, k, incorporates the effects of moderate mixing. The use of Eq. (1) requires the assumption that the inlet discharge is maintained and that the storage in the pond does not change. The extrapolated mass flux was calculated by taking differences in the extrapolated cumulative mass curve.

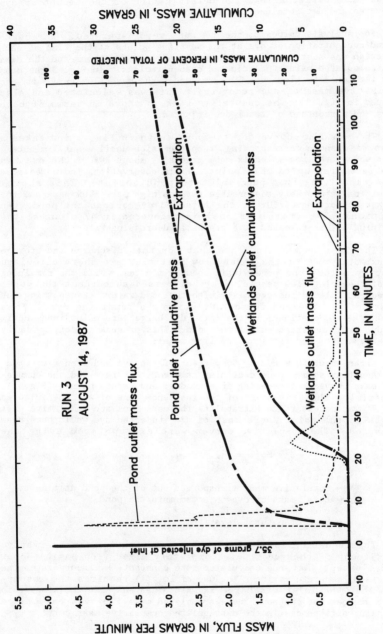

FIG. 3. Mass Flux and Cumulative Mass for Run 3

Eq. (1) has only two constant values ($M_T$ and k). The extrapolation was required to have an end point in common with the available data to prevent any discontinuities in the cumulative mass curve. Prevention of discontinuities allows calculation of statistics describing mixing and residence times. This requirement allows the use of only one other point determining the parameters in Eq. (1). The two points used were the last point for which data are available and an arbitrarily chosen point prior to the last data point. The point was chosen to yield a smooth cumulative and mass flux curve. A more elegant extrapolation technique is being considered for further analysis of this data, however, review of Fig. 3 indicates that the simple extrapolation technique used will provide adequate results, and allow comparisons of the results from different runs.

For purposes of calculations, the extrapolation was carried out until 90% percent of the injected dye was recovered or the mass flux rate became less than 1% of maximum mass flux. In reality, the high inlet discharge used in the dye runs probably would not be experienced for more than 30 minutes during a natural storm.

The estimated mean residence time and variance about the mean was calculated using the measured mass flux curve and the extrapolated mass flux curve. The equations used are:

$$\text{mean resident time} = \frac{\Sigma\ t(\text{mass flux})\Delta t}{\Sigma\ (\text{mass flux})\Delta t} \quad \ldots\ldots\ldots\ldots\ldots\ldots (2)$$

$$\text{variance about mean} = \frac{\Sigma\ (t - \text{mean residence time})^2(\text{mass flux})\Delta t}{\Sigma\ (\text{mass flux})\Delta t} \quad ..(3)$$

where

$\Delta t$ is a discrete time step.

## Detention Pond

The data for the 5 tracer runs are tabulated in Table 1. The average inlet discharges varied between about 19 and 16 cfs (0.54 and 0.45 m³/s) for runs 1-4 and about half that for run 5. The average discharges of runs 1-4 are a little more than one third of the peak discharge recorded at this site (48 cfs, 1.36 m³/s) (Martin and Smoot 1986). Storm pond inlet hydrographs are characterized by sharp peak discharges; the duration of discharge greater than about 10 cfs (0.28 m³/s) was less than 30 minutes for 11 of 13 previously monitored storms. The average live storage in the pond varied from 2,650 ft³ (75 m³) for run 5 to 10,600 ft³ (300 m³) for run 4. Utilization of the maximum available live storage, 18,500 ft³ (520 m³), varied then from 14% to 57%.

The theoretical maximum concentration of dye, if complete mixing instantaneously occurred, was calculated and is shown in Table 1 along with the actual measured peak concentrations found at the pond outlet. The actual concentration is higher for each run than the completely mixed concentration indicating that the pond is not completely mixed for these discharges and levels of storage. Similar results were found for the maximum mass flux assuming complete mixing and the actual measured peak mass flux (Table 1).

TABLE 1.--<u>Hydraulic parameters and residence times for the detention pond</u>

[Measured peak concentration at pond outlet are values that have been reduced by the background concentrations at the time of the injection. The abbreviations used in the following table are: ft$^3$/s, cubic feet per second; ft$^3$, cubic feet; µg/L, micrograms per liter; g/min, grams per minute; %, percent; min, munutes; and min$^2$, minutes squared]

|  | Tracer Run | | | | |
|---|---|---|---|---|---|
|  | 1 | 2 | 3 | 4 | 5 |
| **Discharge (average during run):** | | | | | |
| Pond inlet (ft$^3$/s) | 16.0 | 15.9 | 16.4 | 19.4 | 9.6 |
| Pond outlet (ft$^3$/s) | 14.8 | 16.6 | 16.3 | 19.1 | 8.8 |
| Mean of inlet and outlet (ft$^3$/s) | 15.4 | 16.2 | 16.4 | 19.2 | 9.2 |
| **Storage:** | | | | | |
| Live at time of injection (ft$^3$) | 580 | 7,770 | 570 | 9,800 | 2,310 |
| Average live (ft$^3$) | 4,930 | 8,590 | 5,610 | 10,600 | 2,650 |
| Average total (ft$^3$) | 58,000 | 61,600 | 58,600 | 63,600 | 55,700 |
| Dye injected (grams) | 23.8 | 23.8 | 35.7 | 35.7 | 23.8 |
| **Mixing:** | | | | | |
| Theoretical maximum concentration with complete mixing (µg/L) | 16 | 14 | 24 | 20 | 15 |
| Measured peak concentration at pond outlet (µg/L) | 155 | 68 | 142 | 38 | 69 |
| Theoretical maximum mass flux with complete mixing (g/min) | .50 | .44 | .61 | .59 | .22 |
| Measured mass flux at peak (g/min) | 3.8 | 2.0 | 4.4 | 1.3 | 1.0 |
| **Short-circuiting:** | | | | | |
| Time to peak, mode (min) | 5 | 5 | 5 | 11 | 17 |
| Time for inititial cloud to pass (min) | 6 | 7 | 9 | 12 | 25 |
| Percent of mass in initial cloud (%) | 40 | 16 | 30 | 18 | 35 |
| **Residence times:** | | | | | |
| Percentiles: | | | | | |
| T25 (min) | 5 | 20 | 7 | 23 | 18 |
| T50 Median (min) | 20 | 56 | 47 | 47 | 95 |
| T75 (min) | 69 | 117 | 168 | 88 | 282 |
| Theoretical mean with complete mixing (VOL/Q) (min) | 63 | 63 | 60 | 55 | 101 |
| Estimated mean (min) | 31 | 62 | 50 | 52 | 125 |
| Estimated variance about mean (min$^2$) | 1,140 | 2,690 | 3,020 | 1,230 | 17,000 |

The short-circuiting was quantified for each run as already described by arbitrarily picking the time after the peak at which there appears to be an abrupt change in slope in pond outlet concentrations and mass flux. The short-circuiting dye appeared visually as a cloud of dye crossing the pond. The amount of mass contained in the initial cloud was then determined from a plot of the cumulative mass curve.

The times for the short-circuiting initial cloud to pass varied from 6 to 12 minutes for runs 1 to 4 and more than 25 minutes for run 5. The time for the short-circuiting cloud to pass varies with discharge when storage is low as in runs 1, 3, and 5. As discharge increases, velocity of the short-circuiting increases and cloud-passage time decreases.

The amount of dye in the short-circuiting cloud (percent mass, table 1) seems to be inversely related to the amount of water in storage in the pond; however, the associated amount of mixing that the inlet particles are subject to is directly related to the amount of water in storage. Large amounts of water in the pond allow mixing currents to become established, reducing short-circuiting across the pond. When the live storage in the pond was about 5,600 ft$^3$ (160 m$^3$) or less (runs 1, 3, and 5), the initial cloud (short-circuiting) contained 30 to 40% of the initial injected mass. When the live storage was about 8,600 ft$^3$ (240 m$^3$) or more (runs 2 and 4), the mass in the initial short-circuiting cloud was about half that for the other runs. These data indicate that the greater the live storage, the greater the mixing that occurs to the inlet particles.

The median time for dye recovery, the time for recovery of 50% of the injected dye, ranges between 47 and 95 minutes for runs 2 to 5, and was 20 minutes for run 1. As previously described, run 1 had the largest amount of dye contained in the short-circuiting cloud. The estimated time to recover 75% of the injected dye ranged between 69 and 282 minutes.

The theoretical maximum mean residence time with complete mixing (VOL/Q) in the pond was calculated and is shown in Table 1. The completely mixed-flow mean residence time is remarkably close to the estimated mean residence times for runs 2, 3, and 4. The runs have been shown to have been made under fairly well-mixed conditions. The measured mean residence time for run 1 is about half the theoretical maximum residence time for a completely mixed flow basin, indicating the effects of short-circuiting. Run 5 had an estimated mean residence time considerably larger than the theoretical maximum residence time. This is most likely because of the relatively long tail of the extrapolated mass-flux curve and the resulting uncertainties. The variance about the mean for this run is much larger than those of the other runs. The median residence times, which is not as much affected by the extrapolation for the data, for run 5 is still less than the completely mixed flow mean residence time for this run.

Wetlands and System
---

In table 2 are shown the hydraulic parameters and residence time statistics for the wetlands and system. Because of the nature of dye transport through the pond, with part being quickly delivered and part

TABLE 2.--Hydraulic parameters and residence times for the system

[Measured peak concentration at pond outlet are values that have been reduced by the background concentrations at the time of the injection. VOL/Q (min) of pond plus wetlands storage was used to calculate the theortical mean residence time with complete mixing was the volume. The abbreviations used in the following table are: ft³/s, cubic feet per second; ft³, cubic feet; $\mu$g/L, micrograms per liter; g/min, grams per minute; %, percent; min, minutes; min², and minutes squared]

|  | Tracer Run | | | | |
|---|---|---|---|---|---|
|  | 1 | 2 | 3 | 4 | 5 |
| **Discharge:** (average during run) | | | | | |
| Pond inlet (ft³/s) | 16.0 | 15.9 | 16.4 | 19.4 | 9.6 |
| Wetlands outlet (ft³/s) | 11.4 | 13.2 | 10.4 | 13.8 | 3.9 |
| **Storage:** (average during run) | | | | | |
| Wetlands live storage (ft³) | 31,800 | 35,900 | 30,130 | 36,300 | 23,700 |
| System total storage (ft³) | 89,800 | 97,530 | 88,780 | 99,940 | 79,390 |
| Dye injected (grams) | 23.8 | 23.8 | 35.7 | 35.7 | 23.8 |
| **Measured concentration peak** | | | | | |
| at wetlands outlet ($\mu$g/L) | 23 | 25 | 58 | 19 | 7.3 |
| Measured mass flux at peak (g/min) | .52 | .62 | .95 | .44 | .06 |
| **Short circuiting:** | | | | | |
| Time to peak, mode (min) | 20 | 18 | 24 | 29 | 54 |
| Time pond outlet peak to wetlands outlet peak (min) | 15 | 13 | 19 | 18 | 37 |
| **Residence times:** | | | | | |
| Percentiles: | | | | | |
| T25 (min) | 36 | 35 | 35 | 53 | 219 |
| T50 Median (min) | 79 | 87 | 85 | 94 | 466 |
| T75 (min) | 153 | 174 | 190 | 165 | 888 |
| Theoretical mean with complete mixing (VOL/Q) (min) | 109 | 112 | 110 | 100 | 196 |
| Estimated mean (min) | 86 | 96 | 102 | 101 | 500 |
| Difference mean (pond outlet to wetlands outlet) (min) | 55 | 34 | 52 | 49 | 375 |
| Estimated variance about mean (min²) | 1,370 | 1,660 | 2,230 | 990 | 26,300 |

being gradually delivered to the wetlands, it is not possible to determine the wetlands response to a pulse input at the wetlands inlets (pond outlet) directly from the available data. However, it is possible to calculate time between peaks and means, and make inferences about the residence time of particles in the wetlands. Some of these differences have been calculated and are shown in table 2.

Average discharge at the wetlands outlet varied between about 10 and 14 cfs (0.28 and 0.40 m$^3$/s) for runs 1 through 4, and was about 4 cfs (0.11 m$^3$/s) for run 5. Average live storage in the wetlands was greater than 30,000 ft$^3$ (850 m$^3$) for runs 1 through 4 and less than 24,000 ft$^3$ (680 m$^3$) for run 5. Utilization of the maximum available live storage in the wetlands, 98,500 ft$^3$ (2,790 m$^3$), varied between 24% and 37%.

It was not possible to directly quantify short-circuiting in the wetlands, however, the times between pond and wetlands outlet peak concentration give an indication of short-circuiting and mixing in the wetlands. Longer times to peak concentration probably indicate greater mixing and less short-circuiting than shorter times to peak concentration. The time to peak ranged from 18 to 54 minutes, generally increasing with decreasing discharge and increasing storage. The time to peak between the wetlands outlet and pond outlet (between 13 and 19 minutes) is generally about the same for runs 1 through 4, and much greater for run 5 (37 minutes). The increased peak travel time of run 5 compared to runs 1 through 4 is probably due to the much reduced discharge of run 5 (less than 4 cfs, 0.11 m$^3$/s) compared to those of runs 1-4 (between 10 and 14 cfs, 0.28 and 0.40 m$^3$/s).

The median residence time, time for 50% of the mass to be recovered, in the system ranged from about 80 minutes to more than 400 minutes. Time to recover 75% of the injected dye varied between 153 minutes to almost 900 minutes.

As with the pond, the theoretical mean residence times with complete mixing are in close agreement with the estimated residence times of those runs with generally more complete mixing (runs 2 and 4).

Previous studies (Martin and Smoot 1986; Martin 1988) have shown that the system loads-removal efficiencies of total lead, zinc, and solids ranged between about 50 to 80%. Total nitrogen and phosphorus efficiencies were found to range between about 30 to 40%. As has been described, these efficiencies were obtained with a fairly well mixed system that exhibits some short-circuiting at high flows. Reducing short-circuiting and making the system approximate a completely mixed flow basin probably would increase this system's efficiency. Similarly configured and sized basins, designed to be completely or well mixed, should experience similar loads-removal efficiencies.

### Summary

Five tracer runs were performed on a detention pond and wetlands system to determine mixing and residence times in the system. The detention pond is 8,600 sq ft (800 m$^2$) in size with a width to length ratio of 1:2. The pond's dead storage is 53,040 ft$^3$ (1,500 m$^3$) and maximum available live storage is 18,500 ft$^3$ (520 m$^3$). The wetlands is 32,000 sq ft (2,970 m$^2$) in size. Wetlands dead storage is 22,700 ft$^3$ (640 m$^3$) and maximum available live storage is 98,500 ft$^3$ (2,790 m$^3$).

The data indicate that at low discharges and with large amounts of storage the pond is moderately mixed with residence times not much less than the maximum possible under ideal complete mixing conditions. At higher discharges and less storage, short-circuiting occurs reducing the amount of mixing in the pond and appreciably reducing the residence times. On one run, 40% of the inlet particles were short-circuited across the pond in less than 6 minutes.

It was not possible to directly quantify short-circuiting in the wetlands, however, the time between pond outlet peak concentrations and wetlands outlet peak give an indication of mixing in the wetlands. The times to peak concentrations at the wetlands indicate that mixing increases with decreasing discharge and increasing storage.

This system has been shown to reduce loads of total lead, zinc, and solids between 50 and 80% and total nitrogen and phosphorus loads between 30 and 40%. These efficiencies were obtained with a system that tends to be fairly well mixed at low discharges and moderately mixed with some short-circuiting at high flows. Reducing short-circuiting and making the system approximate a completely mixed flow basin probably could increase this system's efficiency. Similarly configured and sized basins, designed to be completely or well mixed, could probably experience similar loads removal efficiencies.

**Acknowledgements**

The study was conducted by the U.S. Geological Survey in cooperation with the Florida Department of Transportation, Mr. Gary Evink, project coordinator. The study would not have been possible without the cooperation and help of Orange County, Florida, especially Mr. William Masi and Mr. Tom Perrine.

**Appendix-References**

Martin, E. H. (1988). "Effectiveness of an urban runoff detention pond-wetlands system." J. Envir. Engr. ASCE, Paper no. 22649, July 1988.

Martin, E. H., and Smoot, J. L. (1986). <u>Constituent-load changes in urban stormwater runoff routed through a detention pond-wetlands system in central Florida</u>. U.S. Geological Survey Water-Resources Investigations Report 85-4310.

Weber, J. W. (1972). <u>Physicochemical Processes for water-quality control</u>, John Wiley and Sons, New York, 640 p.

Wilson, J. F., Cobb E. D., and Kilpatrick, A. F. (1986), <u>Fluorometric procedures for dye tracing</u>, U.S. Geological Survey Techniques of Water-Resources Investigations, Book 3, chap. A12.

## Discussion of Mr. Martin's Paper

Question:

Some of the results you showed for the 5th of your five short-circuiting dye tests appear to be out of line with the results of your other four runs. Can you explain that?

Answer:

The discharges involved in the 5th run were relatively low compared to the other runs. Hence, a larger theoretical residence time in the pond was predictable. The measured mass of dye that short-circuited through the pond, however, was about the same as the mass in the other four runs; and I am not able to explain why.

Question:

In addition to the dye you used, did you consider the use of labeled isotopes?

Answer:

No.

Design of Extended Detention Wet Pond Systems

Thomas R. Schueler * and Mike Helfrich **

ABSTRACT

This paper presents the concept of the extended detention wet pond system developed in the Washington metropolitan area (WMA) to fulfill a growing number of diverse stormwater management objectives. The system is oriented to maximize urban pollutant removal, reduce pond maintenance and to provide environmental benefits. The adaptable design consists of three complementary storage components: a permanent pool, extended detention storage and stormwater storage. The multiple storage components provide greater pollutant removal and downstream streambank erosion protection than can be achieved by an individual wet pond or dry extended detention pond alone. The paper presents guidance on the basic elements of hydraulic and hydrologic design for the system, as well as practical design tips for enhancing pollutant removal, reducing maintenance costs, providing environmental amenities, and improving pond safety and aesthetics.

INTRODUCTION

The extended detention pond system concept has gradually evolved over the last two decades in the Washington metropolitan area in response to a growing number of stormwater management objectives.

During the early 1970's, the major regional objective in stormwater management was flood control. In terms of design criteria, this meant that the post development peak discharge rate was to be reduced to predevelopment rates for runoff events associated with the two year return frequency storm. The primary practice used in the region to meet the criteria was the dry detention basin.

Water quality emerged as a second stormwater management objective in the early 1980's in response to growing concerns

---

\* Principal Environmental Engineer, Metropolitan Washington Council of Governments, 1875 Eye St., N.W., Washington, D.C. 20006

\*\* Senior Planner, Montgomery County Dept. of Environmental Protection, 101 Monroe Street, Rockville, Maryland 20850

over the water quality impacts of urban runoff. Local monitoring studies demonstrated that urban areas produced significant pollutant loads, and that the traditional dry stormwater pond did not effectively remove them (1,2). Other pond designs, however, were shown to have moderate to high urban pollutant removal capability (2). Consequently, regional stormwater management objectives were modified to emphasize pollutant removal. Design criteria were developed to provide extra treatment of the first-flush of runoff, using the extended detention dry pond and the wet pond, respectively.

Maintenance reduction became another important stormwater management objective in the WMA during the 1980's. Surveys indicated that most stormwater ponds were not functioning as intended (3), and that most routine and non-routine pond maintenance requirements were being ignored (4). To address the growing maintenance problem, local governments developed specific pond design regulations to reduce the cost and frequency of future maintenance activities.

Recently, local governments have emphasized the need to provide greater streambank erosion protection. A growing body of evidence indicated that the traditional criteria of two year storm control was insufficient, by itself, to provide an adequate level of protection for streambank and channel erosion (5,6,7). Greater control of intermediate sized storms was needed to control the frequency of the pre-development two year storm, in addition to the change in its absolute magnitude. In the WMA, the increase in two year runoff frequency is controlled by extended detention (7). This realization again led to a shift in stormwater management objectives, causing local governments to recommend and/or require extended detention for storms less than the two year event but greater than the first flush.

In recent years, a fifth stormwater management objective has emerged in response to concerns about the environmental impacts of ponds. Local experience has shown that ponds exert both positive and adverse impacts on the environment and adjacent communities. Consequently, local governments have begun to require a more environmentally sensitive approach to design. To enhance pond appearance, habitat and pollutant removal values, developers often must prepare an aquatic and terrestrial landscaping plan for the site. Regulatory agencies, concerned about fishery habitat and wetland protection, have begun to require special pond designs that mitigate potential impacts to sensitive downstream aquatic life and habitats. These impacts include anoxic pond water releases, thermal loading and freshwater wetland disturbance.

As the number of stormwater management objectives multiplied in the WMA, it became clear that no individual practice could completely satisfy them. Instead, local engineers and planners have combined individual practices into a single pond system. The extended detention wet pond system

# DETENTION WET POND SYSTEMS

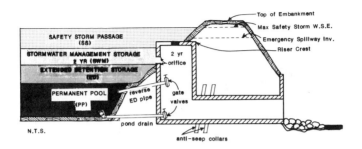

Figure 1. Cross-Section of the Extended Detention Pond System

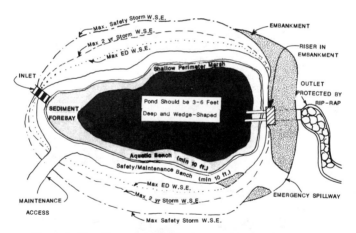

Figure 2. Plan View of the Extended Detention Pond System

Design No. 1: Standard Wet ED System

Design No. 2: Undersized Pool w/ED

Design No.3: Shallow Marsh w/Forebay

Design No.4: Oversized Pool w/no ED

Design No.5: Water Quality BMP w/no SWM

*Figure 3. Design Variants of the Extended Detention Pond System*

way of pollutant removal. Rather, the design relies exclusively on extended detention component for removal of particulate pollutants. Design No. 2 may be appropriate in areas where a large pool is not appropriate environmental reasons, where watershed area is a constraint, or at sites where nutrient management is not a priority.

The permanent pool is is replaced by a shallow marsh area in Design No. 3. The depth of the marsh ranges from zero to two feet, and also has two deeper pool areas to protect the ED pipe opening and to store deposited sediments, respectively. The design utilizes emergent wetland plants to enhance nutrient uptake from the under-sized permanent pool. The wetland area can either be intentionally planted or allowed to be colonized by volunteer plants.

Design No. 4 features a permanent pool with increased capacity, but no extended detention storage. This design relies exclusively on the permanent pool for pollutant removal, which can be substantial if the pool volume is large in relation to the runoff volume from the contributing watershed (11). The deep permanent pool allows for a sub-surface release of cooler water from the bottom of the pond, and can be managed to provide a recreational warm-water fishery. However, since the pond does not have extended detention, it may not adequately protect downstream channels from erosion.

Stormwater management storage is not included in Design No. 5. The pond functions solely as an urban BMP for pollutant removal, utilizing both the permanent pool and extended detention storage components. Design No. 5 may be applicable in situations where downstream areas are not subject to increased flooding (such as coastal areas and the lowermost portions of large watersheds).

## POLLUTANT REMOVAL CAPABILITY OF THE SYSTEM

While performance monitoring studies are not yet complete, the extended detention wet pond system is believed to provide a higher and more reliable level of pollutant removal than either the traditional dry extended detention pond and wet pond designs alone. Performance data for the system is currently being compiled through a two-year automated input/output monitoring program at three prototype facilities in Montgomery County, Maryland.

However, some of the expected water quality benefits of the extended detention wet pond system can be ascertained by analyzing the factors thought to limit pollutant removal evident in prior performance monitoring studies. Table 1 details the reported pollutant removal rates for dry, wet and extended detention dry ponds monitored in the Washington metropolitan area. The following discussion reviews the primary factors reported to limit pollutant removal in the field.

Table 2: Pollutant Removal Rates for Ponds in WMA

| POND TYPE | SITE NAME | DESIGN CRITERIA | TSS | TP | TN | COD | Zn | Pb | DATA SOURCE |
|---|---|---|---|---|---|---|---|---|---|
| DRY POND | Lake-Ridge | Two Year SWM Only | 14 | 20 | 10 | 0 | -10 | -- | (2) |
| DRY ED POND | Stedwick | 6-12 hrs of ED | 70 | 13 | 24 | 27 | 57 | 62 | (2) |
| | London Commons | 4-6 hrs of ED | 29 | 40 | 25 | 17 | 25 | 29 | (8) |
| | | 6-12 hrs of ED | 74 | 56 | 60 | 41 | 40 | 24 | |
| | Oak Hampton | 6-18 hrs of ED | 73 | 35 | -- | -- | -- | -- | (10) |
| | Settling Column Tests | 12 hrs | 68 | 42 | 25 | 35 | -- | 72 | (2) |
| | | 24 hrs | 75 | 45 | 32 | 39 | -- | 81 | |
| | | 48 hrs | 85 | 50 | 39 | 54 | -- | 85 | |
| WET PONDS | Burke/ Westleigh | VB/VR Ratio 8:1 | 54 | 66 | 28 | 30 | 51 | 65 | (2) |

## Pollutant Removal Limitations of Existing Pond Designs

As Table 1 shows, dry extended detention ponds are often moderately effective in removing particulate pollutants, such as TSS, trace metals and organic nutrients, even with relatively short detention times (6 to 12 hours). However, dry extended detention ponds have limited capability to remove soluble nutrients (2,7,8).

extended detention ponds have limited capability to remove soluble nutrients (2,7,8).

Removal rates in dry extended detention ponds appear to be limited by a series of factors: resuspension of previously deposited materials (2,8), the difficulty in achieving long detention times for relatively small but frequent storm events (8,10), the short detention time of runoff during the rising limb of the hydrograph (10), the lack of biological removal mechanisms for soluble nutrients (7), and the difficulty in getting extremely fine-grained particles to settle out.

Wet ponds, on the other hand, are more effective in removing both particulate and soluble pollutants (Table 1), provided that they have sufficient volume in relation to the contributing watershed (11). The higher removal is attributed to greater settling and biological uptake that occurs when incoming runoff is stored in the pool for a period of several days or more (2). Although most wet pond monitoring studies have indicated high long term removal rates, removal rates do vary significantly from storm to storm (2). Low removal has frequently been noted under two widely different cases: (a) when the incoming runoff volumes are greater than the permanent pool volume and (b) when the incoming runoff volumes are small in relation to the permanent pool. The low removal rates in the first case are expected and consistent with current models of wet pond performance (11). The low removal rates in the second case are believed to be due to the export of pond seston from the pond (i.e., fine-grained organic matter in suspension displaced out of the pond during storms--(7)

A review of shallow marsh performance monitoring studies indicate that wetlands can have a moderate to high capability to remove both particulate and soluble pollutants, if their surface area is large in relation to the contributing watershed area (12). However, low and sometimes negative removal rates have been reported during some storms, due to export of plant detritus and other organic matter, particularly after plants die back at the end of the growing season (12,13). Also, as would be expected, removal rates are often sharply lower during the non-growing season, as compared to the growing season.

## Improved Removal Capability of the System

The apparent redundancy of the multiple storage components is thought to enhance the removal effectiveness of each individual component. Thus, factors which limit removal in one component may be partially or completely eliminated by another component. Some examples are provided below:

- The standard design for the extended detention wet pond system treats a much greater range of runoff volumes than most conventional ponds designed under existing WMA criteria (see Table 2). The PP and ED components collectively treat one-inch of watershed runoff. The first half-inch of runoff is retained in the permanent pool; the second half-inch is subject to extended detention. The combined system provides some removal in cases where an individual wet pond or dry extended detention pond would provide little or none. For example, if storm runoff volumes are in excess of the PP, they still receive ED treatment. Similarly, smaller storm runoff events are first treated by the PP rather than the less effective method of extended detention.

- The effectiveness of the ED component is greatly enhanced by the permanent pool since the pool acts as an effective barrier to resuspension. Thus, pollutants that settle out of the pool to the pond sediments are protected from scouring, and cannot be easily be resuspended (as sometimes occurs in dry extended detention ponds).

- Perimeter wetland areas created by the extended detention wet pond system also improve the performance of the ED and PP components. While the total wetland area is not sufficient, by itself, to provide high levels of nutrient removal, the additional uptake should contribute to better overall performance. In particular, the perimeter wetland areas provide an ideal substrate for bacterial degradation of pollutants (13). Wetland plants also help to stabilize the sediments and improve the settling characteristics of both the ED and PP components. Export of plant nutrients from the wetland during the non-growing season should be reduced since pond water is released at the much lower ED rate, thereby allowing the detritus to sink to pool sediments.

## Other Design Features to Improve Pollutant Removal

The sizing criteria for the ED and PP components presented in the standard design both assume ideal settling and retention conditions. These conditions may not occur in the field unless careful attention is paid to the overall geometry of the system.

In particular, the designer can avoid "short-circuiting" by making the pool as long, narrow and relatively shallow, as possible (7). For example, a minimum pool length to width ratio of 3:1 (measured from inlet to outlet) is frequently recommended (a 5:1 ratio is considered ideal for the design of gravitational settling tanks). If the ratio cannot be achieved given the natural topography at the site, berms and baffles can be used to increase the length of the flow path through the pool.

Optimal pool depth is between 3 to 6 feet. Greater pool depths can reduce removal rates, as deeper waters often stratify during the summer months. Stratification of the pool can produce conditions that increase the potential for both sediment nutrient release and thermal short-circuiting (7).

The location of the ED pipe opening can also affect pond performance. Limnological studies of reservoirs have concluded that pipes located near the surface of the pool will trap and store nutrients better than those situated near the bottom (20). Thus, as a general rule, it is recommended that the ED pipe opening should be above the midpoint of the normal pool elevation and the pool bottom.

## BASIC ELEMENTS OF HYDROLOGIC DESIGN

The hydrologic and hydraulic design for the extended detention pond system is accomplished in four basic steps, in which the storage volumes, water surface elevations and pipe (or weirs) sizes are determined for each major component. Table 2 presents a summary of the methods and criteria for the standard design. Alternative design criteria can also be found in (7,14).

## FEASIBILITY TESTS

Although the extended detention pond system is adaptable, it cannot be universally applied at all development sites. The following feasibility factors should be investigated prior to actual design phase:

### Watershed Area

A major constraint is insufficient contributing watershed area to maintain a permanent pool during dry weather. The minimum required watershed area depends on pond soil conditions (i.e., infiltration rate) and regional climate patterns (evaporation rates and typical interval between runoff events. Permanent pools cannot normally be maintained if the contributing watershed area is less than 30 acres in size, unless a perennial source of baseflow exists. Reference (14) presents procedures for determining if a permanent pool is feasible in small watersheds. If not, the pond designer may wish to consider alternative design No. 2 (see Figure 3).

### Geotechnical Constraints

A series of soil borings should be taken at the proposed pond site prior to final design to characterize bedrock and groundwater levels, soil infiltration rates, and the adequacy

**Table 2: Hydrologic and Hydraulic Design Criteria for the Standard Extended Detention Wet Pond System**

**1. PERMANENT POOL STORAGE**

| | |
|---|---|
| Design Criteria: | Treat First Flush of Runoff |
| Storage Volume: | One-watershed inch * Rv * Watershed area |
| | Rv= 0.05 + 0.009 (% Imperviousness) |
| Water Surface Elevation (W.S.E): | Established by invert of ED Pipe |
| Pipe Sizing: (pool drain) | Drain pool volume within 24 hours |

**2. EXTENDED DETENTION STORAGE**

| | |
|---|---|
| Design Criteria: | Provide minimum 24 hours of detention for next one-half inch watershed runoff |
| Storage Volume: | One-half inch * watershed area |
| W.S.E.: | Upper limit set at beginning of 2 year stormwater storage |
| Pipe Sizing: | $Qr = \dfrac{(0.5 \text{ in})(\text{acre})(43560 \text{ cf/acre})(\text{ft}/12 \text{ in})}{2 \ (24 \text{ hours})}$ |

**3. TWO YEAR STORM EVENT PEAK DISCHARGE CONTROL STORAGE**

| | |
|---|---|
| Design Criteria: | Maintain Pre-Development Peak Discharge for the Two Year Design Event |
| Storage Volume: | Obtained from SCS TR-55 Short Cut Method or TR-20 (Ref 18 or 19) |
| W.S.E.: | Upper Limit: bottom of 100 Year Storage  Lower Limit: top of ED storage |

**4. CALCULATE SAFETY STORM/SIZE EMERGENCY SPILLWAY**

Design Criteria:
  Safety Storm (SS): Design event depends on hazard class
  Emergency Spillway (ES): Must pass Safety Storm

Storage Volume:  SS: Obtained from TR-20 (Reference 19)
                 ES: From SCS Spillway Charts (Ref. 20)

of excavated soils for use as core or embankment fill. Borings showing the bedrock level to be within 2 or 3 feet of the planned bottom of the pool are a signal that expensive rock blasting might be required during excavation. Similarly, if soil borings reveal that the seasonally high water table is near or above the planned bottom pool, then it is probable that a) expensive de-watering systems may be required during construction and b) storage capacity in the permanent pool may be reduced by groundwater. The best solution to either problem is to set the bottom of the pool at least 3 feet above the water table or bedrock level.

### Other Constraints

The location of all utilities and associated easements should be determined prior to final design. In general, most utility companies are reluctant to allow their lines to be submerged or frequently inundated. The option of relocating existing utilities is often impractical or prohibitively expensive. If utilities cannot be avoided, it may be possible to protect them using earthen berms or dikes.

Freshwater wetlands are often situated in low lying areas that are also ideal sites for constructing ponds. This can be a major constraint, as wetlands are protected by federal and/or state wetland laws. Freshwater wetland areas are diverse in type and appearance, and may not always be readily evident to untrained personnel. If an area is recognized as a wetland, a permit is required before it is disturbed or inundated. Due to the high ecological value and rarity associated with some wetlands, unconditional permit approval is not always assured. To avoid delays, pond designers should consult with wetland permitting agencies early in the planning process.

## MAINTENANCE ASPECTS OF DESIGN

Historically, pond maintenance has often been ignored, and it is realistic to assume that this trend will continue in the future. Therefore, the pond designer must explicitly consider means of reducing the need and frequency of maintenance. Often, relatively inexpensive and simple design enhancements can significantly reduce both the cost and scope of future maintenance activities. Several maintenance reduction techniques applied in the WMA are discussed below:

### Methods of Reducing Sediment Removal Costs

Based on experience in the WMA, the sediment clean-out cycle for a pond in a stabilized watershed is on the order of ten to twenty years (15). The cost of sediment removal at each cycle can be 20 to 40% of the initial pond construction cost, depending on the size and accessibility of the pond and the methods used to remove, transport and dispose of the sediment (15). Sediment removal operations often exceed the financial and technical capabilities of homeowner association's and local governments. Thus, one of the major tasks facing the pond designer is to reduce the frequency and ease of cleanout operations. A number of design features are helpful in this regard:

- Perhaps the easiest way to reduce the frequency of cleanouts is to increase the storage volume of the pool. Excavating sediment during construction is approximately five times cheaper than dredging it later (15). If possible, the extra capacity should be provided in the form of a sediment forebay, constructed near the inlet of the pool where much of the incoming sediment load will drop out. Direct access should be available to the forebay to make clean-out operations easier.

- Sediment removal costs can be reduced as much as 50% if an on-site sediment disposal area is reserved (15).

- The riser should be equipped with a drain pipe sized to completely drawdown the permanent pool within 24 hours. The drain pipe should be controlled by a lockable gate valve inside the riser (Figure 1). If the drain pipe has an upward facing inverted elbow, clogging from bottom sediments and debris during draining operations can be minimized (Figure 1). The pond drain is important as it allows the the pool to be de-watered rapidly, so that pond sediments can be removed mechanically (rather than by the more costly method of dredging).

- Sediment removal costs increase sharply if the pond cannot be reached with heavy equipment. Thus, each pond should have a standard maintenance right of way (from a roadway) that leads to an access bench around the entire perimeter of the pond (Figure 2). The 15 foot wide bench is located one foot above the PP or at the ED pool elevation (whichever is higher) and is graded to provide positive drainage. The access bench also serves as a safety bench to reduce hazards.

### Pond Specifications to Assure Longevity

Most of the earliest stormwater ponds constructed in the WMA have experienced maintenance problems due to poor design and inadequate construction materials. Learning from these failures, local governments have adapted and modified their regulations and plan review criteria to prevent the problems from occurring in the future. Thus, the extended detention pond system incorporates the following specifications:

- <u>PIPE AND RISER MATERIALS.</u>

    Most area jurisdictions are shifting away from the use of corrugated metal pipe (CMP) in pond construction and are now specifying the use reinforced concrete risers and pipes. Concrete is estimated to have at least twice the longevity of CMP in a pond environment (16).

- REVERSE SLOPED ED PIPE

    The reverse sloped pipe is one of the few ED control devices that has been able to operate in the field for long periods without clogging and associated maintenance costs. Most dry ED control devices rely on filter cloth, gravel, or rock filtering devices that can be rapidly clogged by deposited sediments. By contrast, the reverse sloped pipe opening is situated above deposited pond sediments, yet is located below the pool surface where it could be clogged by floatable debris. Hooded risers can also be used to protect the ED pipe, but are not as resistant to clogging from floatable debris.

- ADJUSTABLE GATE VALVES

    Both the pond drain and the reverse slope ED pipe are equipped with adjustable gate valves within the riser (Figure 1). This feature makes it very easy to "fine-tune" pond release rates after pond construction. Experience in the WMA has shown that target extended detention times and release rates are seldom achieved in the field. The use of gate valves allows for maximum future flexibility in adjusting target detention times, thereby avoiding expensive riser/outlet modifications.

## Reducing Mowing Costs

The sideslopes, embankment, emergency spillway and buffer area of a pond must be regularly mowed to discourage woody growth and control weeds. Consequently, mowing operations are typically the single largest cost component associated with routine maintenance activities (15). Mowing costs can be sharply reduced if the the pond buffer is managed as a meadow rather than a lawn (mowing only in late Spring and late Fall, compared to an estimated 14 mowing operations per year). Aggressive ground cover species such as crown vetch can be used to stabilize the embankment, as they out-compete woody species for the first 5 to 10 years.

For easier mowing, pond side-slopes should be kept no steeper than 3:1 (h:v, for ease of access and safety) and not flatter than 20:1 (h:v, to prevent soggy conditions that impede mowing).

## Providing Access for Future Maintenance Needs

Lack of adequate access makes future maintenance activities more difficult and costly. Every pond should have an easement that grants direct access to the pond from a public or private roadway. This right-of-way should have a minimum width of 20 feet and a maximum slope of 5:1 (h:v), and be directly linked to the access bench around the perimeter of the permanent pool.

It is also strongly recommended that the riser be placed within the embankment rather than out in the permanent pool. This makes it much easier to get to the riser to make inspections and repairs.

## ENVIRONMENTAL ASPECTS OF DESIGN

The extended detention wet pond system has both positive and negative impacts on the immediate environment and downstream aquatic life. Positive impacts include the creation of aquatic and terrestrial wildlife habitat through pond landscaping, and the reduction in downstream aquatic habitat degradation caused by channel and bank erosion. On the negative side, permanent pools can alter the sensitive ecology of headwater streams by releasing heated and oxygen-depleted waters downstream (17). These potential impacts should be carefully investigated during preliminary design to determine if the configuration of the pond system needs to be altered.

### Basin Landscaping

Both aquatic and terrestrial landscaping are critical elements of the extended detention wet pond system. Plant species selected have a profound influence on a pond's appearance, habitat value, pollutant removal performance and maintenance requirements. In turn, the pond creates a wide gradient in moisture conditions that influences which plant species will grow best (i.e., due to differences in the frequency and depth of inundation during storms). These zones are shown schematically in Figure 4, along with several recommended native plant species.

Most WMA jurisdictions require that some kind of landscape treatment be provided in the terrestrial areas associated with the pond buffer (Zones 4 to 6) to stabilize slopes, improve the appearance and create wildlife habitat or scenic effects. Trees, shrubs and ground covers selected for this area need to be able to withstand extremely compacted soils, tolerate full sun and exposure and have few maintenance requirements. In this respect, the use of regionally native plants is highly recommended. The importance of terrestrial landscaping and natural pond contouring cannot be overstated as these factors often promote greater acceptance by adjacent property owners.

The standard design creates a series of concentric rings of permanently or regularly inundated shallow wetland areas (Figure 2). Aquatic landscaping serves several important design functions in the shoreline fringe area (Zone 3) and shallow aquatic bench area (Zone 2).

# DETENTION WET POND SYSTEMS

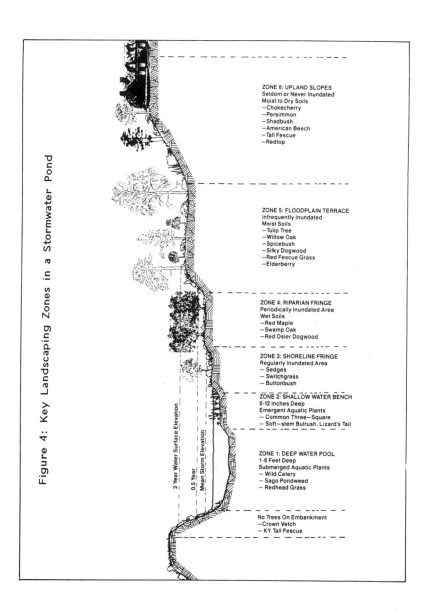

Figure 4: Key Landscaping Zones in a Stormwater Pond

The use of emergent wetlands plants in these zones helps to (a) enhance nutrient uptake, (b) reduce water velocities to promote better settling, (c) stabilize deposited pollutants and reduce resuspension, (d) allow a habitat for predaceous insects that serve as a check on mosquito populations, (e) conceal trash, debris and changes in water level, (f) protect the shoreline from erosion (g) provide food and cover for wildlife, waterfowl and shorebirds and (h) break up the engineered contours of the pond perimeter to make it more natural and pleasing in appearance. Useful guidance for establishment of wetland emergent plants in this zone is available in (7) and (13).

### Thermal Impacts of the Permanent Pool

Recent research has indicated that the release of warm-water from the permanent pool may adversely impact the thermal regime of downstream receiving waters (17). Simply put, the pool acts as a heat sink during the summer months in between storm events. When water is displaced from the pool and released downstream, it can be as much as 10 degrees F warmer than naturally occurring baseflow. Large impervious surfaces also act as a heat sink during the summer, and can warm surface runoff to 90 degrees F. (17). In either case, the increased water temperature can be critical for cold-water stream systems (e.g., trout streams) where fish and other aquatic life are constrained by maximum summertime water temperature. Most urban streams do not fall in this category, as prior urbanization and riparian disturbances have increased summer water temperatures and degraded aquatic habitat to the point that a cold-water fishery cannot be supported. However, in newly urbanizing watersheds, pond designers should pay close attention to potential thermal impacts of ponds.

Four basic approaches have been used in the WMA to alleviate temperature impacts in urbanizing watersheds that contain cold-water trout streams: (a) prohibiting the use of a permanent pool altogether (b) splitting out most of the incoming baseflow before it reaches the pool and bypassing the pool entirely, (c) using design alternative No. 2 that utilizes a drastically undersized permanent pool or (d) utilizing design alternative No. 4 (deep permanent pool) and positioning the ED pipe opening to release cooler water from deeper in the pool The actual impact of any of these measures on the downstream thermal regime has not yet been fully investigated.

### Low Oxygen Release From the Permanent Pool

Under some conditions, permanent pools can release waters very low in oxygen content to downstream receiving waters. This can occur in shallow pools with a high sediment oxygen demand or in deep stratified pools. The downstream dissolved oxygen sag tends to be less than a few hundred meters long due to natural reareation common in most streams. Pond designers can enhance the natural reareation process by creating a rip-rap cascade at the outfall of the pond, or placing surge-stone within the barrel to increase its roughness.

## SAFETY AND AESTHETIC DESIGN FEATURES

Concerns have been consistently been raised about the safety hazards and liability risks of ponds, although reported water accidents in ponds in the WMA are rare. While some local governments require fencing and warning signs around ponds, these measures may become an attractive nuisance, and are not encouraged in the extended detention pond system. Rather, efforts are concentrated on managing the contours of the pond to minimize possible hazard. Some of the basic safety features of the extended detention pond system are described below. In most cases, these features are required to achieve other design purposes, as well.

### Side-Slope Control

Side-slopes leading to the permanent pool should not be steeper than 3:1 (h:v), to reduce the risk of someone accidentally tumbling into the pool.

### Aquatic and Safety Benches

A 15 foot wide safety bench should be placed at the toe of the side-slopes leading to the permanent pool (this bench normally coincides with the maintenance access bench). A second bench is located around the perimeter of the pond at a depth of one-foot below the permanent pool. This aquatic bench is typically 10 to 15 feet wide and is intended to colonized or planted with wetland plants to aid in pollutant removal. In combination, the two benches make it extremely difficult to fall into the pool, and eliminate dangerous nearshore dropoffs. In addition, the relatively shallow depth requirements for the permanent pool (3 to 6 feet) or shallow marsh (0 to 1 foot) in the extended detention pond system, reduces the risk of accidental drowning.

### Placing Riser in the Embankment

For aesthetics, safety and maintenance reasons, it is important to place the riser within the embankment rather than in the middle of the permanent pool. Otherwise, the riser may become a diving and partying platform for teenagers. When designing a concrete riser, the basic principle is to keep access limited to lockable manhole covers (i.e., all large weir openings in the riser should be "fenced" with pipe. The pipe, set eight inches apart, also serves as an acceptable debris/trash rack).

## FUTURE REFINEMENT OF THE SYSTEM

The extended detention wet pond system presented in this paper remains in a constant state of refinement, and will be modified as more practical experience and technical data are obtained. Some priority areas for future research to improve the system's effectiveness are outlined below.

First, more data needs to be obtained on how the system works hydrodynamically, and how well it modifies the post-development hydrologic regime. Surprisingly, most of the field research to date has focused on the pollutant removal capability of ponds, and has not evaluated their hydrologic and hydraulic functions. Relatively inexpensive input/output flow monitoring and tracer dye studies could provide the needed data to develop design tools that can better imitate the natural predevelopment hydrologic regime over the entire range of runoff events.

Second, research needs to be performed to find the optimal combination of ED volume and ED duration for a pond that protects downstream channels from erosion and provides significant pollutant removal. Since the ED release rate can be manipulated by gate valves in the pond system, it is possible to perform a series of field experiments to vary the detention time and release rate to discover the optimal combination(s).

Third, more basic ecological research needs to be done to evaluate the downstream impacts of the pond system on aquatic life. The engineering community must acknowledge that stormwater ponds can adversely impact the aquatic ecosystem by regulating the flow regime, increasing thermal loads, and changing patterns of production in streams. Otherwise, it is quite likely that the engineers will become embroiled in continuing disputes with regulatory agencies concerned with water quality, fisheries and wetland protection. Only when the nature of the ecological impacts are fully known will it be possible to mitigate them through innovative design techniques.

## REFERENCES

1. Northern Virginia Planning District Commission. 1978. Land Use/Runoff Relationships in the Washington Metro Area. Prepared for Metro Wash Council of Govts. 73 p.

2. Metropolitan Washington Council of Governments. 1983. Urban Runoff in the Washington Metropolitan Area: Final Report Washington Area Urban Runoff Project. Prepared for U.S. EPA. 168 pp.

3. Tassone, J. 1984. Survey of Stormwater Management Maintainance Practices in Maryland. Report prepared for MD Water Resources Administration. 24 pp.

4. Maryland Water Resources Administration. 1986. Maintenance of Stormwater Management Structures: A Summary. Maryland Sediment and Stormwater Admin. 38 pp.

5. Berg, V., 1988. Sr. Engineer. Montgomery County Dept. of Environmental Protection. Rockville, MD.

6. Greenhorne and O'Mara, Inc. 1985. Initial Stormwater Management Investigations. Report prepared for Montgomery County Dept. of Environ Protection. 66 pp.

7. Schueler, T.R. 1987. Controlling Urban Runoff: A practical manual for planning and designing urban best management practices. Metro Washington Water Resources Planning Board. 224 pp.

8. Grizzard, T. 1988. Final Report: London Commons Extended Detention Pond Monitoring Project. Prepared for the Va. Department of Conservation and Historic Resources.

9. Grizzard, T. 1986. Efficiency of Extended Detention Ponds. pp.323-337 in Urban Runoff Quality: Impact and Quality Enhancement Technology. B. Urbonas and L. Roessner, (eds.), American Society of Civil Engineers.

10. Stack, W. 1988. Baltimore County Department of Public Works. Personal communication of preliminary results of Oakhampton Extended Detention Monitoring Project.

11. Driscoll, E.D. 1986. Methodology for Analysis of Detention Basins for Control of Urban Runoff Quality. U.S. EPA Nonpoint Source Branch. EPA 440/5-87-001. 72 pp.

12. Stack, W. 1988. Baltimore County Department of Public Works. Personal Communication of preliminary results of artificial ED marsh study.

13. Athanas, C. 1986. Wetland Basins for Stormwater Treatment: Analysis and Guidelines. Final report. Sediment and Stormwater Division. Maryland Dept of the Environment. Annapolis, MD. 222 pp.

14. Harrington, B. 1987. Feasibility and Design of Wet Ponds to Achieve Water Quality Control. Sediment and Stormwater Division. Maryland Dept of the Environment. Annapolis, MD. 32 pp.

15. Wiegand et al. 1986. Cost of Urban runoff Controls. pp. 366-380. in: Urban Runoff Quality-Impact and Quality Enhancement Technology. B. Urbonas and L. Roessner (eds.). American Society of Civil Engineers.

16. Maryland-National Capital Park and Planning Commission. 1984. Design guidelines for Park Stormwater Management Facilities. Dept. of Parks. Silver Spring, MD. 10 pp.

17. Galli, J. 1988. A Limnological study of an urban stormwater management pond and stream ecosystem. Masters thesis. George Mason University, Fairfax, Virginia.

18. U.S. Soil Conservation Service. 1986. Technical Release No. 55: Urban hydrolgy for Small Watersheds. National Technical Information Service. Springfield, Va. 162 pp.

19. U.S Soil Conservation Service. 1982. Technical Release No. 20. Hydrological Model Formulation. NTIS. 272 pp.

20. U.S. Soil Conservation Service. 1978. National Field Engineering Manual.

21. Petts, G.E. 1984. Impounded RIvers: Perspectives for Ecological Management. John Wiley & Sons. 326 pp.

## Discussion of Mr. Schueler's Paper

Question:

What spillway design flood do you use?

Answer:

For ponds fed by drainage areas less than 400 acres, the spillway must pass the 100-year event. The 2- and 10-year events are stored.

Question:

What are your sediment removal frequencies?

Answer:

Once every 5 to 10 years seems to be the practice (although once-an-infinity seems to be working for some people). It depends on the accumulations actually encountered.

Question:

Is mosquito control considered in your designs?

Answer:

No. That is not a big problem, except in swampy **dry** ponds.

Question:

Are you monitoring the extended detention ponds?

Answer:

Yes, there is a 2-year program underway in which we are monitoring 3 extended detention ponds against a wet pond as the control. Results so far indicate there may be a 5% removal improvement for extended detention storage.

Question:

How many of these ponds are now operating in the Washington area?

Answer:

Over 3,000. About 200 more are bing added each year. The maintenance that is done is being done largely by only one agency. Most local jurisdiction involvement is for construction inspection only.

Question:

Is the stratification you have observed all the result of temperature effects, rather than other fluid density differences?

Answer:

Yes.

Question:

Have you noted any design flaws with the reverse-slope orifice pipe?

Answer:

We think that they are best for treating smaller volumes of runoff; otherwise, you may want to go to multiple-port orifices.

Question:

With 3,000 of these facilities operating, have you spotted enormous or any downstream quality improvement?

Answer:

Excellent question. Unfortunately, the necessary studies haven't been done yet. We have seen the need to retrofit some of the existing facilities, first, before all the improvements we anticipated can be realized.

Question:

When doing maintenance for sediment removal, can trucks or loaders get **into** the ponds, or must this be done by draglines from the edges?

Answer:

It is best to dewater the pond, then wait 2 days for drying. Then, equipment **with wide tires** can get into the pond and scoop out the accumulated sediment or debris.

Question:

Given that some of your data indicate a decline in removal efficiencies over time, as we look down the road 5 years or more to successes at protecting receiving water quality, can we ever expect to spot improvements of a size that made it all worthwhile?

Answer:

Scott Tucker's people at "The Bottom" will have to answer that...

WATER-QUALITY PONDS --
ARE THEY THE ANSWER?

Patrick F. Mulhern, P.E.[1]
Timothy D. Steele, Ph.D.[2]

ABSTRACT

Water-quality "wet" ponds have been proposed to treat storm water for the removal of phosphorous in streams tributary to Cherry Creek Reservoir, a major recreational reservoir in the southeast Denver Metropolitan Area. No "wet" ponds for water quality treatment have been constructed in this area, but there are several "wet" ponds located on these streams that intercept stream sediments. Monitoring sites were established to collect inflow and outflow water-quality data at one of these ponds. The data were then briefly analyzed to attempt to determine the effectiveness of this pond in removing phosphorous from storm water runoff.

Introduction

The Denver Regional Council of Governments (DRCOG, 1983) has completed a Clean Lakes Study for the Cherry Creek Reservoir located in the southeast Denver Metropolitan Area. The Clean Lakes Study indicated developing eutrophic conditions in the Reservoir. The lake's coloration associated with the production of chlorophyll "a" was objectionable to the water-sports users, and high oxygen demands near the lake bottom, associated with bed sediments, threatened aquatic life. Cherry Creek Reservoir is one of the State Division of Parks and Outdoor Recreation's most popular facilities in terms of visitor-day use. Hence, the lake eutrophication presents a serious concern.

The Cherry Creek Basin Water Quality Management Master Plan (Master Plan) was also completed by DRCOG (1985). This study built upon the Clean Lakes study and presented a plan for fulfilling the recommended water-quality goals for the Reservoir. Phosphorous was targeted as the key nutrient to control, and a number of sources of phosphorous to the Reservoir were identified and evaluated. Nonpoint source runoff was the major phosphorous source by far, contributing approximately 90 percent of the total loading to the Reservoir.

Based upon the results of this study, an in-lake phosphorous concentration of .035 mg/l was established as the regulatory limit for Cherry Creek Reservoir. The Master Plan (DRCOG, 1985) then

1. General Manager, Inverness Water and Sanitation District, 317 Inverness Way South, Englewood, Colorado 80112.
2. Wate Resources Manager, In-Situ, Inc., 7401 W. Mansfield Avenue, Suite 114, Lakewood, Colorado 80235.

recommended treatment of runoff from the Basin to remove 50 percent of the phosphorous loading in storms producing up to one and one-half inches of runoff. The recommended method of treatment was "water-quality ponds" in conjunction with sand filtration or infiltration.

The proposed approach using water-quality ponds was recommended based upon research reported in the literature. The ponds could be "dry" or "wet", with "wet" implying a permanent pool of water. The reported literature suggested that a "wet" pond may be more effective in phosphorous removal. Never-the-less, the removal process was to be primarily from the deposition of sediments. In recommending this treatment methodology, the greater part of the phosphorous loading during a storm was judged to be attached to sediment particles.

To enhance the removal of sediments, and hence phosphorous, from storm runoff, the Master Plan (DRCOG, 1985) recommended the use of sand filtration of water discharging from the water-quality ponds. Sand beds underdrained by piping networks would tend to entrap additional sediment particles before releasing water back to the stream. The efficiency of removal of phosphorous using water-quality ponds with sand filtration or infiltration was estimated by DRCOG to be as high as 90 percent; thus, the target of 50 percent removal proposed under the Master Plan was judged to be readily attainable.

The Cherry Creek Basin Authority, an entity consisting of Counties, Cities, and Special Districts in the Basin, formed through intergovernmental agreement, has been given responsibility for regulating both the point and non-point source phosphorous allocation programs, and for establishing programs to protect the water quality of the Basin's streams as well as of the Reservoir. This entity has established an extensive monitoring program for the Reservoir and the major tributaries to the Reservoir (In-Situ, Inc., 1986). Part of that monitoring effort was targeted towards evaluating the effectiveness of water-quality ponds in achieving phosphorous removal.

## Procedure

No ponds are known to have been designed specifically for water-quality treatment in the Denver Metropolitan Area, and hence no prior data base was found locally. In the Inverness Business Park, which is located adjacent to Cottonwood Creek, a major tributary to the Cherry Creek Reservoir, a "wet" pond has been created as an aesthetic amenity for a golf course. This pond obviously tends to trap stream sediments, as evidenced by the need to dredge it on a periodic basis, specifically every three to four years. The volume of the pond is approximately 6.5 acre feet. Its surface area is 1.6 acres and average depth is about 4 feet. The pond is approximately circular in shape. This pond is controlled by a weir, and water historically has been spilling over the weir. Hence, surcharge storage in this pond during a storm event is very small.

Despite the fact that the pond was not designed specifically for water-quality treatment, it was judged that monitoring of the pond's inflow and out flow could provide data that might be beneficial to future water-quality pond design. Hence, three gaging station/water quality sampling sites were established as shown in Figure 1.

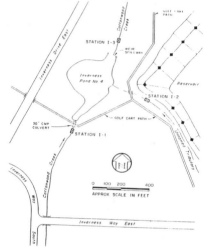

Figure 1. Monitoring Site Layout

Site I-1 measures the main inflow from Cottonwood Creek. Tributary drainage area is approximately 3.2 square miles. Development has occurred over about 8 percent of this basin all in commercial property. The remainder of the subbasin is agricultural or open range lands. Most of the commercial development occurs in close proximity to the pond.

Site I-2 measures inflow from a small tributary. Tributary drainage area is approximately 0.9 square miles. Approximately 8 percent of this subbasin is developed in commercial properties and airport runways. The remainder again is agricultural and range land.

Site I-3 then measures the outflow from the pond. The drainage area is approximately 4.2 square miles, consisting of areas tributary to Sites 1 and 2, and 0.1 square miles of intervening drainage.

Each sampling site consists of a Parshall Flume with stream-stage recording equipment activated by a pressure transducer. The equipment consists of Hermit data-logger recorders manufactured by In-Situ, Inc. Water-quality automatic samplers are manufactured by Manning; these samplers collect discrete samples at programmed intervals during a range of flow conditions for a given storm-runoff event. The samples then are composited prior to analysis.

The monitoring sites were established and operational beginning on April 29, 1987. Sampling was made for all storms exceeding a flow of approximately 0.5 cfs, which is the typical dry-weather flow rate. Twenty-four storm events were recorded during the period from April 29 through December 31, 1987. Water-quality analyses for the composite samples were completed identifying total and dissolved phosphorous, and total and dissolved orthophosphorous. Nitrogen species (ammonia, nitrate, nitrite, and Kjeldahl nitrogen) also were analyzed and periodic field measurements were made of pH, water temperature, specific conductance, and dissolved oxygen.

Regression functions were generated at each monitoring site, based upon storm-runoff and ambient low-flow sample analyses. In analyzing the data, all days of questionable data or no data were simply not included in calculations.

For this study, log-log plots of total phosphorous concentration versus stream discharge were developed at each monitoring site. Regression relationships then were calculated and were used to determine phosphorous loadings on a daily basis. The regression equations for each station are shown on Figures 2, 3, and 4. The correlation coefficient for site I-1 is 0.80 indicating 64 percent of the variance in the data is explained by this relationship. The correlation coefficient of site I-2 is only 0.48, but this was not considered a significant detriment to the analysis because of the small amount of flow from this site.

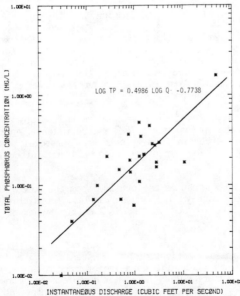

Figure 2. Total Phosphorous versus Discharge, Site I-1
Source: In-Situ, Inc., 1988

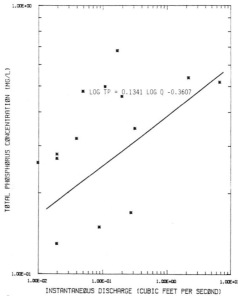

Figure 3. Total Phosphorous versus Discharge, Site I-2
Source: In-Situ, Inc., 1988

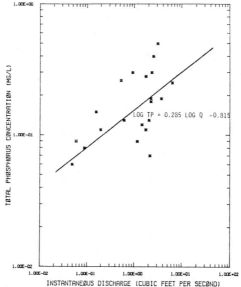

Figure 4. Total Phosphorous versus Discharge, Site I-3
Source: In-Situ, Inc., 1988

The correlation coefficient of site I-3 using all the data is 0.09 indicating that this relationship explains very little of the variance in the data. One data point during the May 14th storm event exhibited a very large flow rate with a relatively low phosphorous concentration. This point was judged to be an outlier, and was deleted from the data set. A regression relationship was then developed using all other data points and for this case, the correlation coefficient increased to 0.58. Because the data to be evaluated fell within limits that were well defined with a large number of data points, a decision was made to use the regression equation with the outlier deleted.

A summary of the loadings at each site during the month of June 1987 is shown in Table 1.

A summary of the month-by-month net phosphorous loadings through the pond for the period from May through December 1987, is shown in Table 2. Loadings for the five (5) major individual storms through the pond are shown in Table 3.

Table 1

Net Phosphorous Removal

June 1987

(discharge in cfs, loadings in lbs per day)

| JUNE | 1 | 2 | 3 | 4 | 5 | 6 | 7 | 8 | 9 | 10 | 11 | 12 | 13 | 14 | 15 |
|---|---|---|---|---|---|---|---|---|---|---|---|---|---|---|---|
| I-1 DISCHARGE | 0.23 | 0.17 | 0.09 | 0.19 | 0.37 | 0.51 | 0.52 | 0.45 | 2.40 * | | 0.09 | 0.10 | 0.09 | 0.10 | 0.28 |
| I-2 DISCHARGE | 0.00 | 0.00 | 0.00 | 0.00 | 0.00 | 0.00 | 0.00 | 0.00 | 0.06 * | | 0.00 | 0.00 | 0.00 | 0.00 | 0.00 |
| DISCHARGE IN | 0.23 | 0.17 | 0.09 | 0.19 | 0.37 | 0.51 | 0.52 | 0.45 | 2.46 * | | 0.09 | 0.10 | 0.09 | 0.10 | 0.28 |
| I-3 DISCHARGE | 0.23 | 0.17 | 0.09 | 0.19 | 0.37 | 0.51 | 0.52 | 0.45 | 2.46 * | | 0.09 | 0.10 | 0.09 | 0.10 | 0.28 |
| I-1 LOADING | 0.10 | 0.06 | 0.02 | 0.08 | 0.20 | 0.33 | 0.34 | 0.27 | 3.37 * | | 0.02 | 0.03 | 0.02 | 0.03 | 0.13 |
| I-2 LOADING | 0.00 | 0.00 | 0.00 | 0.00 | 0.00 | 0.00 | 0.00 | 0.00 | 0.10 * | | 0.00 | 0.00 | 0.00 | 0.00 | 0.00 |
| LOADING IN | 0.10 | 0.06 | 0.02 | 0.08 | 0.20 | 0.33 | 0.34 | 0.27 | 3.47 * | | 0.02 | 0.03 | 0.02 | 0.03 | 0.13 |
| LOADING OUT | 0.13 | 0.08 | 0.03 | 0.09 | 0.22 | 0.36 | 0.37 | 0.29 | 2.93 | 0.00 | 0.03 | 0.04 | 0.03 | 0.04 | 0.16 |
| NET LOADING | -0.03 | -0.02 | -0.01 | -0.01 | -0.02 | -0.03 | -0.03 | -0.02 | 0.54 | 0.00 | -0.01 | -0.01 | -0.01 | -0.01 | -0.03 |

| JUNE | 16 | 17 | 18 | 19 | 20 | 21 | 22 | 23 | 24 | 25 | 26 | 27 | 28 | 29 | 30 | 1 |
|---|---|---|---|---|---|---|---|---|---|---|---|---|---|---|---|---|
| I-1 DISCHARGE | 0.60 | 0.08 | 1.20 | 1.00 | 0.52 | 0.08 | 0.08 | 0.09 | 0.09 | 0.08 | 0.08 | 0.08 | 0.07 | 4.60 | 1.80 | 0.58 |
| I-2 DISCHARGE | 0.00 | 0.00 | 0.04 | 0.00 | 0.00 | 0.00 | 0.00 | 0.00 | 0.00 | 0.00 | 0.00 | 0.00 | 0.00 | 1.90 | 2.80 | 0.72 |
| DISCHARGE IN | 0.60 | 0.08 | 1.24 | 1.00 | 0.52 | 0.08 | 0.08 | 0.09 | 0.09 | 0.08 | 0.08 | 0.08 | 0.07 | 6.50 | 4.60 | 1.30 |
| I-3 DISCHARGE | 0.60 | 0.08 | 1.24 | 1.00 | 0.52 | 0.08 | 0.08 | 0.09 | 0.09 | 0.08 | 0.08 | 0.08 | 0.08 | 6.50 | 4.60 | 1.30 |
| I-1 LOADING | 0.42 | 0.02 | 1.19 | 0.91 | 0.34 | 0.02 | 0.02 | 0.02 | 0.02 | 0.02 | 0.02 | 0.02 | 0.02 | 8.93 | 2.19 | 0.40 |
| I-2 LOADING | 0.00 | 0.00 | 0.06 | 0.00 | 0.00 | 0.00 | 0.00 | 0.00 | 0.00 | 0.00 | 0.00 | 0.00 | 0.00 | 4.86 | 7.55 | 1.62 |
| LOADING IN | 0.42 | 0.02 | 1.25 | 0.91 | 0.34 | 0.02 | 0.02 | 0.02 | 0.02 | 0.02 | 0.02 | 0.02 | 0.02 | 13.79 | 9.74 | 2.02 |
| LOADING OUT | 0.44 | 0.03 | 1.22 | 0.88 | 0.36 | 0.03 | 0.03 | 0.03 | 0.03 | 0.03 | 0.03 | 0.03 | 0.03 | 9.89 | 6.14 | 1.27 |
| NET LOADING | -0.02 | -0.01 | 0.03 | 0.03 | -0.02 | -0.01 | -0.01 | -0.01 | -0.01 | -0.01 | -0.01 | -0.01 | -0.01 | 3.90 | 3.60 | 0.75 |

Table 2

Total Phosphorous Loading Summary

May through December 1987

|  | May (4) | June | July (2) | August (3) | Sept. | Oct. | Nov. | Dec. |
|---|---|---|---|---|---|---|---|---|
| Loading In (lbs) | 24.99 | 31.73 | N/A | 5.27 | .55 | 3.49 | 4.32 | 2.53 |
| Loading Out (lbs) | 20.73 | 23.25 | N/A | 4.39 | .90 | 3.93 | 5.03 | 3.14 |
| Net Removal(1) (lbs) | 4.26 | 8.48 | N/A | 0.88 | -0.35 | -0.44 | -0.71 | -0.61 |

| | |
|---|---|
| Total Phosphorous Loading In | 72.88 lbs |
| Total Phosphorous Loading Out | 61.37 lbs |
| Net Total Phosphorous Loading Removed By Pond | 11.51 lbs |
| Test Period Percent Removal | 15.8% |

Notes:

(1) Positive net loading indicates phosphorous removal by pond, negative net loading indicates phosphorous release by pond.

(2) Recording stations was not operational during the entire month of July and the first half of August.

(3) August loadings represent only approximately one-half of the month.

(4) Eight (8) days are missing due to equipment problems.

Table 3

Individual Storm Data

| Storm Event Record (1987) | Phosphorous Load In (lbs) | Phosphorous Load Out (lbs) | Net Phosphorous Removal (lbs) | Percent Removal (%) |
|---|---|---|---|---|
| May 3-7 | 8.30 | 7.11 | 1.19 | 14 |
| May 14-15 | 11.34 | 8.81 | 2.53 | 22 |
| May 24-27 | 3.69 | 3.37 | .32 | 9 |
| June 9 | 3.47 | 2.93 | .54 | 16 |
| June 29-30, July 1 | 25.55 | 17.30 | 8.25 | 32 |
| August 22-26 | 4.71 | 3.95 | .76 | 16 |

Table 2 shows that the net phosphorous removal for the 7-month period was approximately 11.5 pounds or about 16 percent of the loading entering the pond. This average daily loading result is supported by the data for the inflow and outflow stations. Twenty-four samples from I-1 gave an average concentration of .26 mg/l; of total phosphorous, and fourteen (14) samples collected at site I-2 showed an average concentration of 0.37 mg/l of total phosphorous. At the outflow site, twenty-three samples showed an average phosphorous concentration of 0.21 mg/l. Hence, based upon these average concentrations, values dropped significantly as the inflow (sites I-1 and I-2) passed through the pond to the outflow (site I-3).

Based upon actual data collected during the largest individual storm events of 1987, phosphorous removal ranged from about 9 percent to 32 percent (Table 3). A similar analyses of data for smaller storm events did not clearly indicate what was happening to the phosphorous load in outflow versus inflow.

During the large storm events, it was anticipated that the water-quality analyses would indicate relatively large percentages of phosphorous in suspended form. Although the data did not reflect this assumption in general for this case study, the analytical results for the May 14-15, 1987 storm event are notable. This storm event resulted in the highest instantaneous flow recorded during the 1987 monitoring year, 49 cfs, and an associated high phosphorous concentration of 1.67 mg/l. Dissolved phosphorous represented only 13 percent of the total concentration, and the corresponding measurement of phosphorous concentration downstream at site I-3 was .07 mg/l.

The amount of phosphorous removal during the months of May and June 1987 was estimated at a total of 12.7 pounds; whereas, during the months of September through December 1987, a net increase of 2.1 pounds of phosphorous was estimated from data at the outlet. In May and June, 5 significant storm events recorded relatively large phosphorous loads and removal by the pond. In August through November 1987, algal growth in the pond was substantial, and it appears that phosphorous was remobilized into the water column from pond bottom sediments during periods of high oxygen demand.

In October and November, with continued dense algal growth in the pond, additional water-quality samples were taken at sites I-1 and I-3, and in the pond itself. No flow was recorded at site I-2. Samples were collected on a weekly basis from October 23 to November 29. The results of these analyses are shown in Table 4. Note that the pattern for that period is for total phosphorous and dissolved phosphorous concentrations to increase slightly in water flowing out of the pond.

The regression curves from site I-1, the main inflow site and I-3, the outflow site, are shown on Figure 5. This suggests that for low flows, up to about 0.7 cfs, the concentration of phosphorous into the pond is lower than the concentration released from the pond. At discharges above 0.7 cfs, the trend reverses and phosphorous is removed by the pond. Hence, the data suggests that above 0.7 cfs, the inflow begins to carry sediment loads, a percentage of which is deposited in the pond. Site I-2 has minimal effects on the discussion since it was generally not flowing during periods of low-flow at site I-1.

Table 4

1987 Late Season Phosphorous Surveys

Inverness Pond No. 4

| Date | Site I-1 Total Phosphorous (mg/1) | Dissolved Phosphorous (mg/1) | Site I-3 Total Phosphorous (mg/1) | Dissolved Phosphorous (mg/1) |
|---|---|---|---|---|
| 10/23/87 | .03 | .03 | .11 | .15 |
| 10/30/87 | .07 | .01 | .10 | .03 |
| 11/06/87 | .01 | .01 | .11 | .02 |
| 11/13/87 | .06 | .01 | .09 | .02 |
| 11/20/87 | .09 | .05 | .09 | .04 |
| 11/29/87 | .07 | .05 | .08 | .05 |

Note: Site I-1 is upstream, and Site I-3 is downstream of pond.

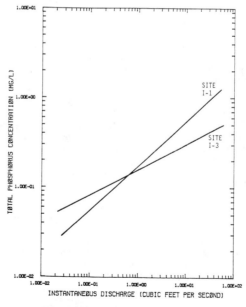

Figure 5. Comparison of Total Phosphorous versus Discharge Relationships at Sites I-1 and I-3

Although this comparison points out that phosphorous is released to the system from the pond presumably through releases from bottom sediments, it also shows that these releases are fairly insignificant in terms of total annual loadings. At a flow of 0.2 cfs, for instance, the net increase in concentration of phosphorous out of the pond is .02 mg/l, which corresponds to a net loading out of .02 pounds per day, or 0.6 pounds per month. This is fairly consistent with what happened during the month of September, 1987. At that rate, it would take over four (4) months of low flows to release the phosphorous entrapped during the storm event of May 14th and 15th. Hence, even though phosphorous is released from the ponds during low flows, the quantity of loading appears minor in comparison to removal during storm events.

Finally, a brief review of phosphorous concentrations versus time of year showed that phosphorous concentrations generally increased in late summer and fall versus concentrations in spring and early summer. This was true for both inflow and outflow from the pond.

## Summary

This paper presents some very preliminary findings regarding the functioning of an aesthetic golf course pond in removing phosphorous from storm water. The study is based upon 7 months of data and 6 significant storm events and 18 minor storm events. This represents a very small sample of data. However, because there is significant pressure to find a means of treating storm water in the Cherry Creek Basin, an attempt was made to complete an early evaluation of the limited data. Meanwhile, the monitoring program will continue this year and into future years. Hopefully, the additional data will present a more consistent removal characterization.

Based upon a broad overview of the data, there is a net reduction in phosphorous loading through the pond over the 7-month period of data collection during the 1987 monitoring year. This was determined by calculating the daily loading and accounting for the inflow and outflow loads over the period of data collection. Net phosphorous loading showed a removal of about 11.5 pounds of phosphorous, for a removal rate of approximately 16 percent over the 7-month period.

The 6 larger storm events sampled showed phosphorous removal rates of from 9 to 32 percent, and represented most of the phosphorous removal for the 7-month data period. This would support the conclusion that a primary mechanism for phosphorous removal is deposition of suspended sediments. The pond considered in this case study has not been designed for sediment removal for water quality treatment; hence, more efficient removal rates certainly chould be attainable.

Finally, it appears from the data that during base flow conditions, a net release of phosphorous generally occurs from the pond to the stream. This is assumed to relate to anaerobic conditions in the pond that allow the release of phosphorous from bed sediments. The release of phosphorous loading to the stream, however, appears minor in comparison with the amount of phosphorous deposited during the larger storm events.

References

Denver Regional Council of Governments, 1983, "Cherry Creek Reservoir Clean Lakes Study", Denver, Colorado, September.

Denver Regional Council of Governments, 1985, "Cherry Creek Basin Water Quality Management Master Plan", Denver, Colorado, September.

In-Situ, Inc,. 1986, Cherry Creek Reservoir and Cherry Creek Basin Monitoring Program, Southeastern Denver Metropolitan Area, Colorado: Prepared for the Cherry Creek Basin Authority, December 3.

In-Situ, Inc., 1987, Compilation and Graphical and Statistical Analysis of Historical Hdyrologic and Water-Quality Data for Cherry Creek Reservoir and the Cherry Creek Basin, Colorado: Prepared for the Cherry Creek Basin Authority, September 30.

In-Situ, Inc., 1987, Cherry Creek Basin Authority, 1987 Annual Report: Prepared for the Colorado Department of Health, Water Quality Control Commission, November 18.

In-Situ, Inc., 1988, "Cherry Creek Reservoir and Cherry Basin, Monitoring Report for the 1987 Calendar Year", Cherry Creek Basin Authority, Lakewood, Colorado, February.

# RAINFALL ANALYSIS FOR EFFICIENT DETENTION PONDS

T. Hvitved-Jacobsen*;
Y.A. Yousef**, MASCE;
and M.P. Wanielista**, MASCE

## ABSTRACT

Intensity-duration-frequency analysis of rainfall data is traditionally used for the selection of design storm required to size, storage capacity in detention/retention ponds without consideration of no rain periods before and after storm events. However storage utilization in the pond depends on incoming runoff event volumes and the inter-event dry periods. Inter-event dry periods between two successive storms should be sufficiently long to minimize cumulative effects of pollutants discharged to receiving streams.

This paper will discuss the concept of inter-event dry period and its application for designing urban retention/detention ponds.

## INTRODUCTION

Over the past 20 years, many investigators have documented the impact of urban runoff water on receiving streams. Also, various stormwater management systems such as detention/retention ponds, swales and others have been constructed and studied for their effectiveness on water quality improvement (Colston, 1974; Huber et al., 1977; Malmqvist, 1983; U.S. EPA , 1983; and Yousef et al., 1986). The methodologies available for the design and operation of these systems are generally empirical and the associated theoretical concepts are limited. The design is usually complicated by the large variability in runoff water quantity and quality. Therefore, there is a continuing interest in re-examination of the basic assumptions used for the design, including nature, temporal and special distribution of rainfall events.

Currently, it is essential to develop stormwater
-------------------
*Professor, Environmental Engineering Laboratory, University of Aalborg, Sohngaardsholmsvej 57, DK-9000, Aalborg, Denmark.
**Professors, Department of Civil Engineering and Environmental Sciences, University of Central Florida, Orlando, Florida 32816-0450, U.S.A.

management systems capable of providing both flood and pollution control in costly urbanized areas. Design of detention ponds should reduce the risk of flooding to a minimum and improve the water quality due to storage of water for a long enough period of time. Flooding is caused by extreme highly intense storm events, but water pollution problems are generally related to the more frequent storms and their temporal distribution effects (Arnell et al., 1983). For example, the intensity of 1 hour storm with a return period of once in 25 or 100 years is used for flood control. But another storm with a return period of one in six months or less may be used for water quality control. It is essential to understand the dynamic behaviour of the runoff control system and the most important characteristics of rainfall events to the system before developing meaningful design criteria.

Intensity-duration-frequency curves are developed from rainfall data supplied by the National Climatic Data Center (NCDC). Monthly tapes of rainfall data are available from NCDC, which contain hourly precipitation amounts and 24-hour totals for approximately 3000 stations in the U.S. Precipitation at most of the stations is recorded in tenths rather than hundredths of inches (Thorp, 1986). Consecutive hours of precipitation at a station are defined as a rainfall event. However, a storm is composed of one or more precipitation events separated by specified inter-event dry periods. Therefore, rainfall data are generally divided into separate, statistically independent storm events based on the time intervals of dry periods between two successive events. The duration of the dry period with minimal or no precipitation allowed within a storm can be varied. Volumes of rainfall storms separated by different inter-event dry periods can be developed from existing rainfall records and statistically analyzed. Rainfall volume - inter-event dry periods - frequency curves can be developed for desired locations and used to calculate storage volumes for pollution control in detention basins.

## INTER-EVENT DRY PERIODS

Analysis of rainfall data is generally based on statistically independent rainfall events with minimum time intervals of dry periods. These inter-event dry periods should be long enough to ensure independent flow events and to minimize cumulative effects of successive storm events. Intervals of 4-5 hours were used by DiToro (1980, Wenzel and Voorhees (1981), Arnell (1982), and Schilling (1983). Others such as Marsalek (1978) used a minimum of 3 hours and Johansen (1979) used one hour intervals to separate storm events. It is well understood that the total number of storm events in a specified rainfall record and the associated statistical analysis of this record will change if one changes the definition of a separate and independent storm event based on the length of minimum inter-event dry period separating two successive storms. Selection of different minimum inter-event dry periods may change the size of the design storm event, but this can be justified on the basis of potential downstream effects for pollutants of primary interest.

## DISTRIBUTION OF INTER-EVENT DRY PERIODS

Thorp (1986) studied hourly precipitation data from 89 stations in the northeastern United States to define seasonal storm (wet-period) and dry-periods statistics. He concluded that the largest contributors to regional winter precipitation were storms of 12 hour and 25 hour duration. In summer, the major regional precipitation came from storms of only 3 hour average durations. Storms of about 5 to 6 hours duration were the primary precipitation contributors for the spring and autumn season. The average dry-period duration was 56.2 hours for autumn, 51.5 hours for summer and 45.5 hours for winter and spring. These statistics were based on defining storm events to be separated by a minimum number of dry hours (NDHR) equivalent to two hours. A frequency distribution of dry periods for the northeastern states during 1977-1980 is presented in Figure 1. Approximately 50 percent of the

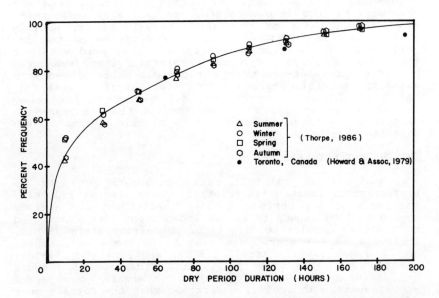

FIGURE 1: TYPICAL FREQUENCY DISTRIBUTION OF DRY PERIOD DURATION

storms are separated by dry periods ranging between 2 and 20 hours.

Similarly Howard & Associates (1979) studied the dynamics of rainfall data from 35 Canadian cities using storm events separated by a minimum inter-event dry period of one hour. They concluded that 76% of the storm events in Toronto are separated by dry periods ranging between one and 65 hours. These findings are in general agreement with data reported by Thorp (1976). They also concluded that practical engineering design should be concerned with more than one rainfall characteristics. It is obvious that a rainfall record can be statistically analyzed and different results are produced if one changes the definition of inter-event dry period.

## CASE STUDY

A 33 full years of rainfall record from the city of Odense on the island of Funen, Denmark between 1936 and 1979 is utilized for statistical analysis (Hvitved-Jacobsen and Yousef, 1988). The initial rainfall records contain 1571 storm events with rainfall depths greater than 3.0 mm and one hour minimum dry period in between storms. Thus initial rainfall record was rearranged and analyzed based on arbitrarily selected inter-event dry periods of 2, 12, 24, 36, 48, 72, and 96 hours. It is emphasized again that a storm event is composed of all rainfall events predeeded and followed by the desired minimum dry period. The total number of events declined from an initial 1571 to 716 events with a minimum of 96 hours without rainfall between two successive storms. However, the maximum depth increased from 59.4 mm initially to 87.2 mm respectively. Each set of data calculated for a specified inter-event dry period can be represented by a straight line on a logarithmic probability scale. The logarithmic mean ($\bar{X}_g$), standard deviation ($S_g$), coefficient of variance ($C_v$) and median ($X_{med}$) values were calculated.

The data indicate that the geometric mean values are lower than arithmetic mean values, and the median values are lower than geometric means for each set of data. Also, linear relationships are found to exist between an independent variable representing inter-event dry periods ($D_p$) and dependent variables of $X_{med}$, $\bar{X}_g$, $C_v$, and logarithmic values of total number of events ($\ln N_e$). The $D_p$ values strongly influences $\bar{X}_g$, $X_{med}$, $N_e$, $C_v$ and it is possible to predict these parameters for various minimum dry periods separating two successive storms.

Table 1 presents statistical parameters of 33 full

years of rainfall events in Odense, Denmark, for selected inter-event dry periods. All the parameters, $\bar{X}_g$, $X_{med}$, and geometric standard deviation ($S_g$) appear to increase gradually as the minimum dry period increases from 2 to 96 hours.

TABLE 1: STATISTICAL PARAMETERS OF RAINFALL EVENTS IN ODENSE, DENMARK

| PARAMETER | MINIMUM INTER-EVENT DRY PERIOD (h) | | | | |
|---|---|---|---|---|---|
| | 2 | 24 | 48 | 72 | 96 |
| $N_e$ | 1523 | 1178 | 973 | 809 | 716 |
| Range (mm) | 3-59.4 | 3-65 | 3-67.4 | 3-87.2 | 3-87.2 |
| $X_{med}$ (mm) | 5.60 | 6.80 | 8.00 | 9.60 | 10.80 |
| $\bar{X}_g$ (mm) | 6.52 | 7.86 | 9.04 | 10.36 | 11.35 |
| $S_g$ (mm) | 1.66 | 1.82 | 1.96 | 2.07 | 2.17 |

SIGNIFICANCE OF INTER-EVENT DRY PERIODS

Removal efficiencies in wet detention ponds are achieved by sedimentation, chemical interactions, and biological uptake. Stormwater detention for a period of twenty four hours or more may result in greater than 90% removal of suspended solids and associated pollutants carried by runoff water (Gizzard et al., 1986). However, soluble fractions, colloidal fractions and small size suspended solids concentrations in urban runoff are less than 45 microns in diameter. Also, approximately 20% of the particles in urban runoff have an average settling velocity of 0.03 ft/hr (Driscoll, 1983) which can be achieved by 2 um diameter particles with specific gravity 2.65 and type I sedimentation. Removal of these small particles and soluble fractions such as nitrate nitrogen, orthophosphorus and heavy metals is enhanced by increasing detention times in a wet detention pond by physical, chemical and biological processes within the permanent pool (Yousef et al., 1985 and Hartigan and Quasebarth, 1985).

The runoff storage of water in detention ponds is sized by different rules practiced in various parts of the country. They require a detention facility designed to detain on site either a minimum amount of runoff or the runoff from a design storm. The design storm varies from 1 inch (2.54 cm) to 2.5 inches (6.3 cm). In the state of Florida and others require that runoff volume must be bled down at a slow rate to allow a draw down within a minimum period of 72 hours. Other rules in the U.S. require the minimum size of the pond to be two and a half times the volume of runoff generated from the mean storm over the watershed areas, or the pond size to be adjusted to achieve an average of two weeks of detention within the pond (Hartigan, 1986) which may be equivalent to four times the volume of runoff generated by the mean storm.

Yousef et al., 1986, observed a decline in orthophosphorus concentration discharged into a retention pond receiving highway runoff to background level within 3 days following the rainfall event. Also in a recent study using model detention ponds (Yousef, 1988), it was concluded that a minimum detention time of at least 72 hours is needed to remove more than 95% of suspended solids and 30-70% of nutrients and heavy metals. Therefore, it may be desirable to design these ponds based on statistical analysis of rainfall records using storm events separated by a minimum dry period consistent with desired pollutant removal effectiveness in the pond. A minimum dry period between two successive events will allow a slow bleed off rate, minimize short circuiting, maximize detention periods and enhance removal efficiencies of pollutants. This dry period should be selected on the basis of time required to minimize cumulative effects. If design storm events are based on a minimum of 72 or 96 hours of dry period, sufficient time may be available for treatment of runoff events.

A clear and more specific design storm is lacking and it is important to emphasize basic criteria of inter-event dry periods and desired return periods. Therefore the rainfall data from Odense, Denmark are analyzed to calculate rainfall depth for inter-event dry periods between 2 and 96 hours and different return periods (Figure 2). It is proposed to make use of these graphs to design the volume of storage based on runoff produced from a storm event separated by 72 hour inter-event dry periods and return period of once in three or four month. Based on these assumptions the design storm will only be exceeded four or three times a year or less. Therefore, the design volume for wet detention ponds in Odense, Denmark can be estimated by the runoff volume produced form a storm event between 0.94 to 1.06 inch (24-27 mm) rainfall depth.

This is different from rainfall data collected for Orlando, Florida. The precipitations, inter-event dry period and frequency (PIF) for Orlando also are plotted in Figure 2. Preliminary analysis of rainfall data for ten years between 1978 and 1988, indicates that a storm event depth of 83 mm (3.25 inches) is separated by 72 hours and return period of once in three month. This analysis is based on eliminating all rainfall events of less than 0.04 inch (1.0 mm) depth from the record. The analysis is continuing.

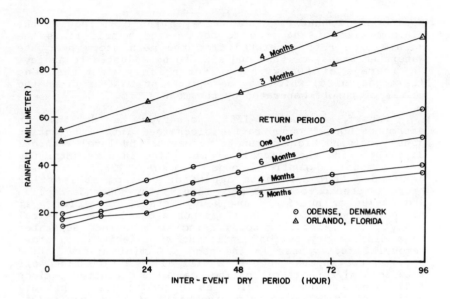

FIGURE 2: RAINFALL VOLUME FOR VARIOUS INTER-EVENT DRY PERIODS

CONCLUSION

The design storm for quality control in wet detention ponds and other stormwater management systems should be based on a desired return period and a minimum dry period duration between successive storms. This period should be consistent with length of time needed to minimize potential impacts on receiving water. Rainfall intensity-duration-frequency curves are traditionally used to develop design storms for flood control. However, rainfall volume - inter-event dry period - frequency curves can be developed and used to calculate design storms for quality control. A storm event with a recurrence interval of 3 or 4 months and a 72 inter-event dry period is recommended for design of wet detention ponds. Of course the recurrence period for a design storm should be based on the desired risk to be taken.

The design storm is the first step in calculating the storage volume in wet detention ponds. However, the pond efficiency is related to the surface overflow rating, average detention times and pond configuration (Yousef et al., 1988). The proposed approach will reduce the risk of cumulative effects from successive storms with short dry periods in between, and guarantee minimum detention times in the pond.

## APPENDIX

Arnell, V., "Rainfall Data for the Design of Sewer Pipe Systems," Report Series A:8, Department of Hydraulics, Chalmers University of Technology, Gothenburg, Sweden, 1982.

Arnell, V., Harremoes, P., Johansen, N.B. and Niemczynowicz, J., "Review of Rainfall Data Application for Design and Analysis," Proceedings of a Specialized Seminar on Rainfall - The Basis for Urban Runoff Design and Analysis," Copenhagen, Denmark, 1983.

Colston, N.V., Jr., "Characterization and Treatment of Urban Land Runoff," U.S. EPA-670/2-74/096, Washington, D.C., 1974.

DiToro, D.M., "Statistics of Receiving Water REsponse to Runoff," Proceedings of a National Conference on Urban Stormwater and Combined Sewer Overflow Impact on Receiving Water Bodies, Orlando, Florida, U.S. EPA-600/9-80-056, 1980.

Driscoll, Eugene D., "Performance of Detention Basins for Control of Urban Runoff Quality," Proceedings of International Symposium on Urban Hydrology, Hydraulics, and Sediment Control, University of Kentucky, Lexington, 1983.

Gizzard, T.J., Randall, C.W., Weand, B.L., and Ellis, K.L., "Effectiveness of Extended Retention Ponds," ASCE Proceedings of an Engineering Foundation Conference on Urban Runoff Quality - Impact and Quality Enhancement Technology," New England College, Henniker, New Hampshire, June 23-27, 1986.

Hartigan, J.P., "Regional BMP Master Plan," ASCE Proceedings of an Engineering Foundation Conference on Urban Runoff Quality - Impact and Quality Enhancement Technology," New England College, Henniker, New Hampshire, June 23-27, 1986.

Hartigan, J.P. and Quasebarth, "Urban Nonpoint Pollution Management for Water Supply Protection: Regional vs On-Site BMP Plans," Proceedings of International Symposium on Urban Hydrology, Hydraulics, and Sediment Control, University of Kentucky, Lexington, 1983.

Howard, Charles and Associates Ltd., "Analysis and Use of Urban Rainfall Data In Canada," Report No. EPS 3-WP-79-4 for the Water Pollution Control Directorate, Environmental Protection Service, Canada, July 1979.

Hvitved-Jacobsen, T., and Yousef, Y.A., "Analysis of Rainfall Series in the Design of Urban Drainage Control Systems," Water Research, Printed in Great Britain, MS 410,

1988.

Johansen, L., "Design Rainfall for Sewer Systems," Ph.D., Thesis, Department of Sanitary Engineering, Technical University of Denmark, Report 79-2, Lyngby, Denmark, 1979.

Malmqvist, P.A., " Urban Stormwater Pollutant Sources: An Analysis of Inflows and Outflows of N, P, Pb, Zn, and Cu in Urban Areas," Department of Sanitary Engineering, Chambers University of Technology, Gothenburg, Sweden, 1983.

Marsalek, J., "Research on the Design Storm Concept," ASCE Urban Water Resources Research Program, Technical Memorandum No. 33, 1978.

Schilling, W., "Univariate Versus Multivariate Rainfall Statistics: Problems and Potentials," Proceedings of a Specialized Seminar on Rainfall - The Basis for Urban Runoff Design and Analysis, Copenhagen, Denmark, 1983.

Thorp, J.M., "Mesoscale Storm and Dry Period Parameters from Hourly Precipitation Data: Program Documentation," Battelle Pacific Northwest Labs., Richland, WA, Journal of Atmospheric Environment, printed in Great Britain, Volume 20, Number 9, 1986.

U.S. Environmental Protection Agency, "Final Report the Nationwide Urban Runoff Program," Water Planning Division, WH-554, Washington, D.C., December 1983.

Wenzel, H.G. and Voorhees, M.L., "An Evaluation of the Urban Design Storm Concept," Water Resources Center, University of Illinois at Urbana-Champaign, Research Report UILU-WRC-81-0164, Ill., 1981.

Yousef, Y.A., Wanielista, M.P., and Harper, H.H., "Fate of Pollutants in Retention/Detention Ponds," Proceedings of a Conference on Stormwater Management: An Update," University of Central Florida, Environmental Systems Engineering Institute, Publication 85-1, July 1985.

Yousef, Y.A., Wanielista, M.P. and Harper, H.H., "Design and Effectiveness of Urban Retention Basins," ASCE Proceedings of an Engineering Foundation Conference on Urban Runoff Quality-Impact and Quality-Enhancement Technology, New England College, Henniker, New Hampshire, June 23-27, 1986.

Yousef, Y.A., Dietz, J.D., Wanielista, M.P. and Brabham, M.E., " Efficiency Optimization of Wet Detention Ponds for Urban Stormwater Management," Interim report submitted to FDER, Contract #WM 159, March 1988.

# Discussion of Dr. Jacobsen's Paper Presented by Dr. Yousef

Question:

In your talk, you referred to a one-cell basin and a three-cell basin. Do the results for the one-cell basin seem better?

Answer:

Only for solids, not for nutrients and metals. This is a case-specific result, however; our one-cell pond had exactly three times the volume of the three-cell pond.

Question:

Did you say that some U.S. cities use a detention volume criterion of 2.5 times the mean storm volume?

Answer:

Yes.

Question:

Did you do any work to determine if there was independence between storms?

Answer:

We used 5 hours as the separation period which we assumed meant independence.

Question:

How concerned are the Danes with urban runoff quality?

Answer:

They are beginning to see the area as a problem. They are beginning to install ponds.

Question:

Were all your rainfall data hourly data?

Answer:

No; 15 minutes was the measurement interval, and 0.6 mm was the smallest volume measured.

MONITORING AND DESIGN OF STORMWATER CONTROL BASINS

J. E. Veenhuis[1], J. H. Parrish[2], and M. E. Jennings[3],F.ASCE

Introduction

The City of Austin, Austin, Texas, has played a pioneering role in the control of urban nonpoint source pollution by enacting watershed and stormwater ordinances, overseeing detailed monitoring programs, and improving design criteria for stormwater control methods. The ordinances have resulted in widespread use of several methods of structural and nonstructural stormwater control and treatment. The most common of these methods include detention, retention, sedimentation or filtration basins, overland flow, limitations on impervious cover, street sweeping, spray irrigation of retained stormwater, and temporary controls on construction site runoff. Studies in other areas (Schueler, 1987) have indicated that these control methods result in some degree of stormwater quality improvement.

The effectiveness of the above methods used in Austin, and perhaps in other areas of the United States, to protect urban water resources has not yet been fully established. Therefore, detailed monitoring programs capable of quantitatively determining the effectiveness of control methods and of stormwater ordinances, are required.

The purpose of this report is to present an overview of the City of Austin's stormwater monitoring program (City of Austin, 1986), including previous monitoring programs with the U.S. Environmental Protection Agency and the U.S. Geological Survey, and to describe the relation of monitoring to design of stormwater control basins.

Previous Stormwater Monitoring

The City of Austin initiated the present Stormwater Monitoring Program (SMP) in 1983 to develop a data base needed to characterize the impacts of urban land use on the quality of stormwater runoff. In addition, the SMP is intended to establish the effectiveness of the various control measures currently used in the Austin area.

Previous to establishment of the SMP, the city participated in two other stormwater monitoring programs--the Nationwide Urban Runoff Program (NURP) with the U.S. Environmental Protection Agency (Environmental Protection Agency, 1983), and a cooperative study with the U.S. Geological Survey (USGS).

---
[1] Hydrologist, U.S. Geological Survey, Austin, Texas 78701
[2] Supervisor of Water Quality, City of Austin, Austin, Texas 78767
[3] Urban Hydrology Coordinator, U.S. Geological Survey, NSTL, MS 39529

The city's participation in the NURP was limited to three monitoring sites: a control watershed and two residential watersheds. Stormwater runoff from one of the residential areas passed through the Woodhollow retention/detention pond where inflow and outflow water quality was monitored. Because the NURP monitoring was conducted for only one year (1981) only a limited number of storm events were sampled. The Woodhollow site was subsequently included in the SMP.

### City of Austin and U.S. Geological Survey Cooperative Stormwater Monitoring Program

Since 1975, the City and USGS have been jointly involved in a productive cooperative stormwater monitoring program. A significant data base of rainfall, streamflow (both low flow and storm runoff), and water-quality information, on urban drainages and the Colorado River, has been compiled. In addition, ground-water monitoring and analysis of the Barton Springs segment of the Edwards Aquifer (Slade and others, 1985) has been performed. As a result of the cooperative monitoring program, the City has been able to develop mathematical models which describe stormwater runoff and pollutant loading for Austin urban streams. A report describing the study using data through 1982 is available (City of Austin, 1984) and is currently being updated using recent data. The studies are finding that water quality of Austin-area urban streams depends on the quantity of stormwater runoff which, in turn, depends on percent impervious cover of the urban watersheds. As percent impervious cover increases, pollutant loads increase for a given rainfall depth. The effect of increased impervious cover in terms of pollutant loads is more pronounced for higher rainfalls.

### U.S. Geological Survey Monitoring Program at Barton Creek Square Mall

The monitoring program with the U.S. Geological Survey included collection of precipitation, streamflow, and water-quality data collected during the period September 1982 to September 1984. A USGS report (Welborn and Veenhuis, 1987) describes the results of this monitoring program. Of particular interest are data and analyses connected with Barton Creek Square Mall (BCSM) where a detention and filtering pond is operated for stormwater control.

The buildings and parking lot of BCSM occupy about 100 acres. Most runoff from the mall flows into three detention and filtering ponds around the perimeter of the shopping mall. Figure 1 shows the relation of pond 1 and BCSM. Pond 1 drains about 46 acres (Section A) from the mall, plus an additional 33.5 acres (Sections B, C, D) adjacent to the mall. Sections B-D contain 13, 12, and 8.5 acres respectively and only runoff from Section C, which enters the pond as overland flow, is ungaged. Pond 1, with a storage capacity of about 3.5 acre-ft (equivalent to the first half inch of runoff from the drainage area), has a bed consisting of three layers of material that

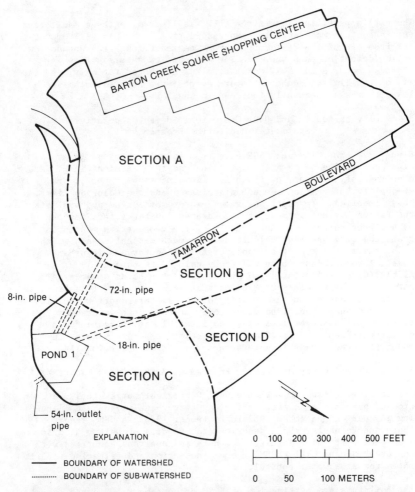

Figure 1.—Watershed of Pond 1 at Barton Creek Square Shopping Center.
See Figure 6 for location map of Austin, Texas

are used to filter water in the pond. The various layers of the pond are shown in Figure 2. Inflow from the 72-inch concrete pipe enters the pond through a control structure and is measured by a sharpcrested weir. Outflow is gaged by a v-notch weir. Influent flows less than the 3.5 acre-feet storage capacity leave the pond through the sand filter. Larger flows produce a discharge over a drop outlet that is not filtered. Both discharges pass through the outflow gage.

STORMWATER CONTROL BASINS 227

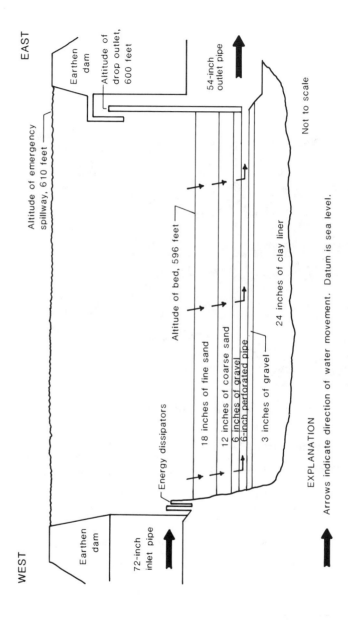

Figure 2.--Schematic cross section of Pond 1 at Barton Creek Square Shopping Center

A general view of pond 1 showing inflow and outflow structures is given in Figure 3.

Figure 3.--Pond 1 at Barton Creek Square Shopping Center and outflow control structure.

Instruments were installed at pond 1 to record rainfall and stage and to collect water-quality samples. A Manning UT "X" System Level Transmitter and Recorder[1] was used to measure stage and a stage-discharge relation was developed for inflow and outflow structures in order to compute stormwater discharge. Manning S-4050 automatic water samplers were used to collect samples of storm runoff at inflow and outflow points. Recording rainfall stations were operated by the City of Austin. Both discrete and discharge-weighted composite water samples were analysed for specific conductance, fecal-coliform and fecal-streptococci bacteria, suspended solids, dissolved solids, volatile dissolved solids, biochemical oxygen demand, chemical oxygen demand, total organic carbon, total nitrogen, total phosphorus, dissolved cadmium, dissolved lead, dissolved iron, and dissolved zinc.

---

[1] The use of trade names in this report is for identification purposes only and does not constitute endorsement by the U.S. Geological Survey or the City of Austin.

A typical storm event, September 19-20, 1982, shown in Figure 4

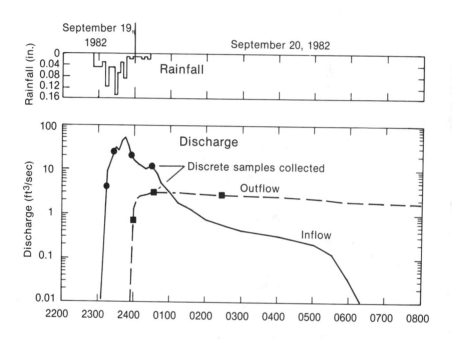

Figure 4.--Rainfall, discharge, and time of discrete sample collection at inflow and outflow stations at the Barton Creek Square Shopping Center for September 19-20, 1982.

illustrates the general nature of rainfall-runoff characteristics for pond 1. Gaged inflow peak was 53.2 cfs whereas gaged outflow peak was 3.06 cfs. In all, data were collected on 22 storms during the 1982-84 period. Using discharge-weighted concentrations and gaged water discharge for both inflow and outflow, pond 1 removal efficiencies of suspended and dissolved material were calculated as a percentage of total inflow load or total number of bacteria. Removal efficiencies were computed by dividing the difference between inflow loads and outflow and outflow loads by the total inflow load and multiplying by 100. Removal efficiencies are reported only for storms in which all flow leaves the pond through the sand filter.

Peak concentrations and loads of most constituents (and total densities of bacteria) at BCSM were substantially larger in the inflow than in the outflow. For example, figure 5 shows discharge-weighted concentrations of suspended solids and dissolved solids in the inflow and outflow at BCSM. Table 1 shows the inflow and outflow loads and removal efficiency for these same constituents. The report by Welborn and Veenhuis (1987) gives similar results for additional urban water-quality constituents.

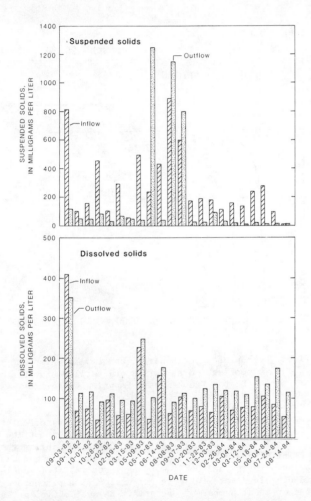

Figure 5.--Discharge-weighted concentrations of suspended solids dissolved solids, and volatile dissolved solids in the inflow and outflow of Pond 1 at Barton Creek Square Shopping Center

Table 1.--Loads for suspended solids and dissolved solids and removal efficiencies at Barton Creek Square Shopping Center [Runoff: 1, from shopping mall; 2, from shopping mall and intervening areas; 3, includes flow through drop outlet.]

| Run-Off | Year | Date of Storm | Suspended Solids | | | Dissolved Solids | | |
|---|---|---|---|---|---|---|---|---|
| | | | Inflow Load (lb) | Outflow Load (lb) | Removal Efficiency (%) | Inflow Load (lb) | Outflow Load (lb) | Removal Efficiency (%) |
| 1 | 1982 | Sept. 3 | 1,270 | 105 | 91.7 | 639 | 316 | 50.5 |
| 1 | | Sept. 19-20 | 805 | 350 | 56.5 | 540 | 792 | -46.7 |
| 1 | | Oct. 7 | 547 | 158 | 71.1 | 252 | 404 | -60.3 |
| 2 | | Oct. 28 | 2,800 | 417 | 85.1 | 287 | 456 | -58.8 |
| 2 | | Nov. 2 | 1,030 | 228 | 77.9 | 942 | 799 | 15.2 |
| 2 | 1983 | Feb. 9 | 2,050 | 308 | 85.0 | 403 | 435 | -7.9 |
| 1 | | Mar. 15 | 302 | 168 | 44.4 | 347 | 328 | 5.5 |
| 1 | | May 9 | 390 | 24 | 93.8 | 180 | 149 | 17.2 |
| 3 | | May 10-11 | 20,800 | 20,700 | 0.5 | 2,190 | 2,330 | -6.4 |
| 1 | | June 14 | 1,360 | 115 | 91.5 | 495 | 512 | -3.4 |
| 3 | | Aug. 8 | 22,900 | 21,400 | 6.6 | 1,580 | 1,710 | -8.2 |
| 3 | | Sept. 7-10 | 13,200 | 13,100 | 0.8 | 2,310 | 1,870 | 19.0 |
| 2 | | Oct. 20 | 1,330 | 81.1 | 93.9 | 530 | 293 | 44.7 |
| 1 | | Nov. 22-23 | 1,220 | 75.7 | 93.8 | 512 | 364 | 28.9 |
| 2 | | Dec. 3 | 1,350 | 568 | 57.9 | 478 | 821 | -71.8 |
| 2 | 1984 | Feb. 26 | 737 | 94.9 | 87.1 | 629 | 348 | 44.5 |
| 1 | | Mar. 4 | 625 | 57.8 | 90.7 | 281 | 344 | -22.4 |
| 2 | | Mar. 12 | 699 | 47.1 | 93.3 | 398 | 523 | -31.4 |
| 1 | | May 18 | 693 | 39.9 | 94.2 | 228 | 294 | -28.9 |
| 2 | | June 4-5 | 2,360 | 71.1 | 97.0 | 892 | 878 | 1.6 |
| 1 | | July 24 | 738 | 65.1 | 91.2 | 635 | 759 | -19.5 |
| 1 | | Aug. 14 | 76.3 | 82.4 | -8.0 | 388 | 789 | -103 |
| Average of 1 and 2 | | | | | 78.3 | | | -12.9 |

Average removal efficiencies of the pond and (or) the filter system for suspended solids, biochemical oxygen demand, total phosphorus, total organic carbon, chemical oxygen demand, and dissolved zinc were between 60 and 80 percent. However, the average dissolved solids load was about 13 percent larger in the outflow than in the inflow. Average loads of total nitrite plus nitrate nitrogen were about 110 percent larger in the outflow than in the inflow. Apparently, the increase in loads of these constituents is due to oxidation and mineralization of previously deposited material and subsequent leaching from the bed of the pond or from the filter system. As noted above, removal efficiencies are reported only for storms in which all flow leaves the pond through the sand filter. During larger storms a portion of the water passes through the pond unfiltered, thus potentially reducing removal efficiencies on an annual basis. Some resuspension of previously trapped pollutants may occur, thus further reducing removal of pollutants associated with sediment particles.

Austin Stormwater Monitoring Program

The present City of Austin SMP is designed to satisfy the data base needs which were not met by previous monitoring programs. The objectives of SMP are:

(1) To establish the relationship in Austin between percent impervious cover and total load of fifteen water-quality constituents in stormwater runoff from specific land uses.

(2) To determine the efficiencies of various structural stormwater controls in removing fifteen water-quality constituents from stormwater.

The SMP includes fourteen monitoring stations at eight locations and is expected to grow to eleven additional stations at three new locations (Figure 6 and Table 2). Some older stations, with sufficient data, will be discontinued. Since 1984, the city has monitored about 300 storm events at all sites and collected and analysed about 1400 water-quality samples. The city's monitoring system is completely automated and controlled through a central computer by use of telemetry transfer of data through dial-up telephone lines.

Each monitoring site has a programmable flow-metering system called "Quadrascan" (American Digital Systems). The battery-operated system provides on-site hardware that measures stream discharge, temporarily stores data, and controls the operation of water-quality samplers. The Quadrascan System includes ultrasonic depth and velocity measuring sensors which are used in conjunction with various kinds of flow-measurement flumes or weirs depending on gaging conditions. Rainfall measurement is obtained by weighing-type raingages and more recently, by tipping-bucket rain gages.

Monitored data are stored and backed up on the city's central computer. Diagnostic information, such a battery voltage, can be collected at the central computer. If data are being transmitted during a storm event, the operator can also direct a given water-quality sampler to take a sample. Monitoring results can be printed or plotted in various ways. When the data base allows, state-of-the-art stormwater management models will be investigated as a means of simulating the processes of stormwater runoff, pollutant accumulation and washoff and structural control. It is anticipated that pollutant loading rate and removal efficiency projections can be obtained from use of a stormwater model.

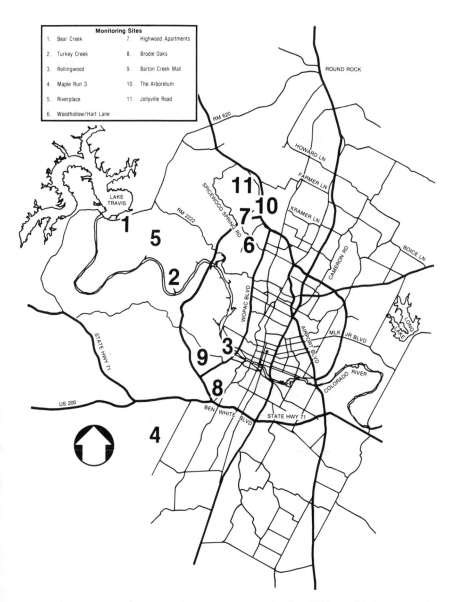

Figure 6. --Monitoring Site Map, City of Austin Stormwater Monitoring Program, Source is City of Austin

Table 2.—Stormwater Quality Monitoring Site Descriptions

| Site | Description | Control Measure | Area (acre) | Impervious Cover (%) |
|---|---|---|---|---|
| (1) Bear Creek | Undeveloped (control watershed) | None | 301.0 | 3 |
| (2) Turkey Creek | Some residential, mostly undeveloped | None | 1335.0 | 5 |
| (3) Rollingwood | Low density residential | None | 62.8 | 21 |
| (4) Maple Run 3 | Medium density residential | Detention/filtration | 27.8 | 36 |
| (5) Woodhollow/Hart Lane | Medium density multifamily residential | Detention | 371.0 | 39 |
| (6) Highwood Apartments | Medium density multifamily residential | Detention/filtration | 3.0 | 50 |
| (7) Brodie Oaks | Moderately developed, multiuse residential/commercial | Detention/filtration | 49.0 | 65 |
| (8) Barton Creek Square Mall | Highly developed commercial | Detention/filtration | 47.0 | 93 |
| (9) Barton Ridge Plaza | Moderately developed commercial | Sedimentation/filtration | * | * |
| (10) Jollyville Road | Highway right-of-way | Detention/filtration | * | * |
| (11) Urban core | High density urban | None | * | * |

*Under construction—values to be determined

Relation of Monitoring and Design

The city has a draft of design guidelines for water-quality control basins (City of Austin, 1988). Soon, these guidelines will be enacted as an administrative rule. Together with an enacted comprehensive watershed ordinance, the design guidelines should exert a strong positive influence in protecting the city's watersheds from the adverse effects of urban development.

The city's design guidelines provide assistance in designing water-quality control structures and outlines requirements for maintenance. The ordinances also make use of non-structural controls such as density restrictions, buffer zones, overland flow and limitations on impervious cover. Temporary erosion controls must be installed on all construction projects in accordance with specifications of the city's erosion control manual.

As stated, one of the primary objectives of the city's SMP is to improve the effectiveness of the various stormwater controls. Although much can be learned from work in other areas of the United States, the city believes it is necessary to "fine-tune" methods to the local characteristics of the Austin area. It is believed that differences in climate, soils, topography, and vegetative cover can have significant impacts on stormwater volume, pollutant loads, and the effectiveness of control technologies. Hence, only detailed monitoring can quantitatively determine the effectiveness of various control methods in Austin (or in other areas of the United States), as well as the effectiveness of watershed ordinances.

One of the most significant changes in the City's design requirements that has resulted from its monitoring efforts is the decision to place water quality control basins off-line from the main stormwater flow. The ordinance requires that ponds isolate and treat the first half inch of runoff from the drainage area. This design eliminates the resuspension and pass-through problems encountered at the BCSM site.

As noted above, the negative removal efficiency for nitrate + nitrate nitrogen for pond 1 at BCSM is most likely due to a chemical conversion taking place in the sand media used in the filtration system. As a result of this monitoring, new pond designs are required to have an impervious liner when they are located over the recharge zone of the local Edwards Aquifer. This is to prevent injection of nitrates into the aquifer. During the monitoring at pond 1 it was discovered that maintenance of the controls is at least as important to overall effectiveness of water-quality ponds as is their design. In February 1984, the top 10 inches of sand at BCSM was replaced because of clogging by sediment. Because clogged sand filters allow successive storm events to pass through the system relatively untreated, the design guidelines now require that filtration basins be preceded by a sediment trap or, in some cases, a full sedimentation basin in order to keep the bulk of settleable solids from reaching the filter surface. This type of design, coupled with semi-annual removal of accumulated sediment from the sand surface, should significantly delay need for sand media replacement. One of the new monitoring projects now being built involves a pair of filtration basins and will be used to experiment with filtration media. One filtration basin will act as a control while various filter media will be used experimentally in the other paired filtration basin.

## Conclusions

The City of Austin is committed to a comprehensive stormwater monitoring program, with one major objective being to validate and refine stormwater control basin design. Based on a cooperative monitoring program with the U.S. Geological Survey at a detention-filtration system, the city is learning the connection between stormwater control design, maintenance, and monitoring.

## References

1. City of Austin, "Stormwater Quality Modeling for Austin Creeks," Department of Public Works, 1984, 10 p. + appendix.

2. City of Austin, "Design Guidelines for Water Quality Control Basins Department of Environmental Protection, Water Quality Division, 32 p. + appendix.

3. City of Austin, "Design Guidelines for Water Quality Control Basins 1988, 32 p. + appendix.

4. City of Austin, Department of Public Works, Watershed Management Division, "INTERIM WATER QUALITY REPORT--Hydrologic and Water Quality Data for Barton Creek Square Mall and Alta Vista PUD, Fall, 1984, 6 p. + figures and tables.

5. City of Austin, Watershed Management Division, Department of Public Works, "City of Austin Stormwater Monitoring Program, 1986, 21 p. + figures.

6. Schueler, Thomas R., "Controlling Urban Runoff: A Practical Manual for Planning and Designing Urban BMPs," Metropolitan Washington Council of Governments, 1987, 275 p.

7. Slade, Raymond M., Ruiz, Linda, and Slagle, Diana, "Simulation of the Flow System of Barton Springs and Associated Edwards Aquifer in the Austin Area, Texas," U.S. Geological Survey Water-Resources Investigations Report 85-4299, 49 p.

8. Welborn, Clarence, T., and Veenhuis, Jack E., "Effects of Runoff Controls on the Quantity and Quality of Urban Runoff at Two Locations in Austin, Texas," U.S. Geological Survey Water-Resources Investigations Report 87-4004, 1987, 101 p.

## Discussion of Mr. Veenhuis' Paper

Question:

What do these filter systems cost?

Answer:

Not certain of the unit costs, but they can get fairly expensive.

Question:

What have been the monitoring costs:

Answer:

$700,000 in all, of which $250,000 were in sample analyses costs.

Question:

Have you had to do any retrofitting, as a result of changes in pervious versus impervious areas?

Answer:

Not yet. Not anticipating any, since these facilities are sized based on design levels of both pervious and impervious areas.

Question:

Where did the design criteria for these systems come from?

Answer:

They were based on the designs of older ones, as well as on particle-size analyses and experience with solids penetration into previous beds. That experience showed penetration of 13" to 15", so we made these beds 18" deep.

Question:

Why is Austin doing this?

Answer:

To protect the Barton Springs amenity specifically. and the Edwards aquifer in general.

Question:

There is a lot of concrete involved in some of these facilities. How are small-site developers providing this sort of thing?

Answer:

By using more earthern berms, avoiding concrete as much as possible. One developer actually put in a 2-chamber pond, with some concrete, for a small development.

Question:

Wasn't all this done downstream of Barton Springs?

Answer:

No; upstream.

Question:

Who's responsible for maintenance?

Answer:

That's divided. For residential areas, the City provides maintenance. For multi-family and commercial areas, the owners are responsible, although we admit that enforcement is a problem.

Question:

Let's get back to the costs. What do these things cost?

Answer:

I remember that a smallish one (for 6.5 acres) cost $99,000. A larger one (for 9.0 acres) cost $165,000.

# MULTIPLE TREATMENT SYSTEM FOR PHOSPHORUS REMOVAL

By James T. Wulliman,[1] M.ASCE, Mark Maxwell,[2] M.ASCE, William E. Wenk[3] and Ben Urbonas,[4] M.ASCE

Abstract. Detention ponds, wetlands, and infiltration basins have each been shown to reduce constituent loads carried by urban stormwater, including phosphorus (US Environmental Protection Agency 1986 and Martin and Smoot 1986). Design recommendations are presented for a system using a combination of these three treatment mechanisms to achieve at least a fifty percent reduction in total phosphorus load. Information is presented relating phosphorus load and control strategies to the extent of development in the upstream basin. Techniques for assessing overall phosphorus removal and guidelines for designing detention ponds and wetland/infiltration areas for water quality treatment are presented. Also addressed are considerations involved in integrating such a project into a natural prairie landscape and creating diversity in the wetland zones.

## Introduction

Cherry Creek Reservoir is a major flood control and recreational resource in the Denver Metropolitan area. Concern over the potential for accelerated entrophication of the lake has led to investigations into constituent loading from urban runoff. The rapidly developing Shop Creek Basin was identified as a key source of phosphorus entering the lake, which is the critical nutrient with respect to in-lake algae growth (Denver Regional Council of Governments 1982). At the same time, the Shop Creek channel was experiencing severe erosion as runoff from upstream developing areas increased. This created a substantial maintenance problem

[1] Project Manager, Muller Engineering Company, 7000 West Fourteenth Avenue, Lakewood, Colorado 80215

[2] Project Manager, Black and Veatch, 1400 South Potomac Street, Suite 200, Aurora, Colorado 80012

[3] President, William Wenk Associates, 1900 Wazee Street, Suite 360, Denver, Colorado 80202

[4] Chief, Major Drainageway Planning, Urban Drainage and Flood Control District, 2480 West 26th Avenue, Suite 156-B, Denver, Colorado 80211

dealing with the sedimentation that occurred at the lake. In 1987, the City of Aurora contracted with Muller Engineering Company, with Black and Veatch as water quality consultants and William Wenk Associates as landscape architects, to design a system of improvements to Shop Creek. The two-fold objective of this project was to stabilize the Shop Creek channel and to reduce the phosphorus load it conveyed to Cherry Creek Lake by at least 50 percent.

Three principal efforts were employed to arrive at the final configuration of improvements on Shop Creek. First, the system of control facilities that had the greatest potential for effectiveness, given the constraints of the project, was identified. Second, the potential treatment efficiency was estimated. Third, the effectiveness of each component of the control system was optimized. During each effort, design issues were raised, analyzed, and resolved. These issues are highlighted in this paper and their analysis and resolution is discussed in an effort to transfer information to designers of similar systems.

It should be recognized that this paper is authored subsequent to the design process and prior to construction and operation. The actual performance of the system is yet to be determined. A follow-up paper is planned to transfer information regarding the actual treatment efficiencies achieved, as well as other operational issues.

## Analysis of Phosphorus Loading Data

The 560-acre Shop Creek Basin, which is comprised of mixed residential and commercial uses, is currently approaching full development, with a projected imperviousness of 39 percent. Figures 1 and 2 summarize rainfall, runoff and phosphorus loading data for the basin during the period from 1982 to 1987. Several trends are evident.

A dramatic improvement in the quality of both base flows and storm flows has occurred as development in the basin has tapered off and disturbed lands have been stabilized by a vegetative or man-made cover. Figure 3 illustrates a correlation between storm flow phosphorus content and the extent of land disturbed in the upstream basin, based on an analysis of aerial photographs of the Shop Creek Basin. This information implies that runoff from an urbanizing basin normally experiences an increase in phosphorus concentration as development occurs and then a decrease and leveling off as disturbed lands are stabilized. This trend is illustrated in Figure 4 and attests to the importance of instituting effective erosion control measures during construction.

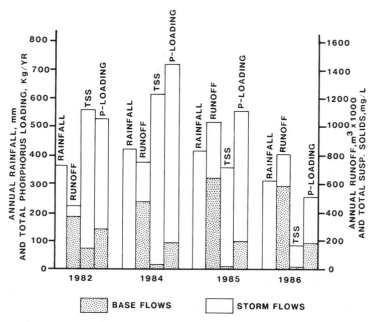

Figure 1. Rainfall, Runoff, and Phosphorus Loading Data

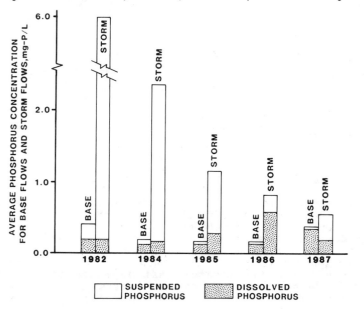

Figure 2. Phosphorus Concentration Data

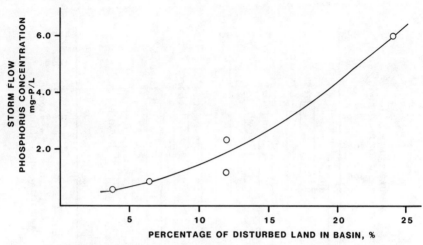

Figure 3. Phosphorus Concentration vs. Disturbed Land

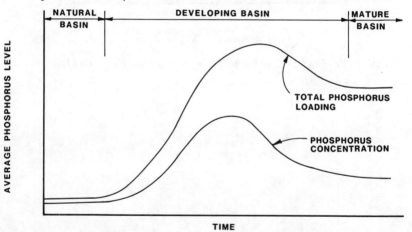

Figure 4. Phosphorus Trend Graph

Total phosphorus loading from a basin is directly dependent upon the quantity of runoff delivered by the basin. Because runoff increases with development, increased storm flow in a fully developed basin can offset improvements in water quality from land stabilization, resulting in increased total phosphorus loading. This phenomenon is also depicted in Figure 4 and illustrates that improvement in water quality as a basin matures does not necessarily negate the need for non-point source treatment facilities.

As shown in Figure 1, the majority of total phosphorus loading in the Shop Creek basin is contributed during storm events (an average of 80 percent in the analyzed years). Thus it is critical that storm flows, not just base flows, be treated in order to achieve the goal of at least 50 percent removal of all non-point source phosphorus. This is likely to be the case in other urbanizing basins as well.

## Identification of Best Management Practices for Phosphorus Control

Measures used to control phosphorus in urban runoff include the following (US Environmental Protection Agency 1986 and Martin and Smoot 1986)

1. Reduction in basin sediment yield through source control practices
2. Channel stabilization
3. Settling
4. Infiltration
5. Vegetative filtering
6. Biological (plant) uptake

The first two measures are watershed controls and the last four are operations performed on runoff. Although potentially very effective if instituted at the beginning of the development phase, erosion control measures will not achieve substantial phosphorus reductions in a largely developed basin such as Shop Creek's. Rather, facilities are required that can remove phosphorus from runoff, particularly during storm events.

For the Shop Creek project, recommended treatment facilities consist of a detention pond with a permanent pool for settling and biological uptake, followed by a series of combined wetland/infiltration areas (Muller Engineering Company 1988). This combination best satisfies the project objectives of developing a non-mechanical, low maintenance system of improvements that could be integrated into the natural landscape of the Cherry Creek State Recreation Area. Figure 5 presents a schematic plan and profile of the Shop Creek facilities. Treatment measures such as land application systems or non-vegetated rapid infiltration basins were not selected because they were perceived to require higher maintenance, be less aesthetic, or be less effective in treating storm runoff.

## Siting Considerations

In the Shop Creek project, a determination had to be made whether to locate the treatment system at a single site at the downstream end of the basin or to create multiple smaller facilities upstream in the basin. The former approach was chosen as the most effective in a similar

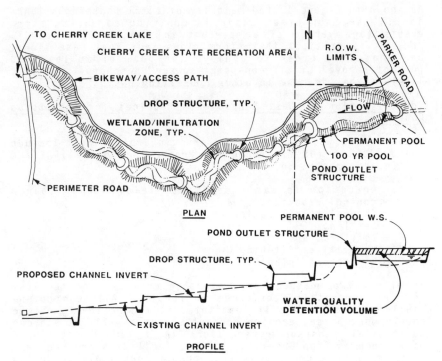

Figure 5. Recommended Shop Creek Facilities

manner as it is generally concluded that large, institutionalized detention facilities for runoff quantity control are preferable to multiple on-site facilities in terms of land use, maintenance and effectiveness (Hartigan and Quaswbarth 1985). In addition, facilities located at the outfall point of a basin can be effective in treating runoff prior to entry into receiving waters, regardless of the type of upstream controls.

## Prediction of Phosphorus Removal Efficiency

The phosphorus removal efficiency of any specific facility is difficult to predict in advance. In order to determine if the recommended combination of detention, wetland and infiltration would achieve the design goal of at least 50 percent removal of non-point phosphorus in Shop Creek, phosphorus mass balance equations and a computer spreadsheet were developed. A series of six cases were modeled using assumptions of flow distribution and treatment efficiency for each component that ranged from conservative to optimistic. The flow schematic for the proposed plan is shown in Figure 6, and the treatment assumptions are listed in Table 1. Mass balance equations utilized in the computer program follow Figure 6.

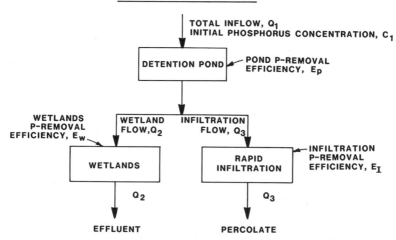

Figure 6. Treatment Flow Schematic

1. PHOSPHORUS LOAD

Base Load = $Q_{1b} * C_{1b}$ = 136 kg/year

Storm Load = $Q_{1s} * C_{1s}$ = 454 kg/year

Total Load = $Q_{1b} * C_{1b} + Q_{1s} * C_{1s}$ = 590 kg/year

2. PHOSPHORUS REMOVED

Base = $E_p * 136 + (1-E_p) * 136 * E_w * (Q_2/Q_1)_b + (1-E_p) * 136 * E_I * (Q_3/Q_1)_b$

Storm = $E_p * 454 + (1-E_p) * 454 * E_w * (Q_2/Q_1)_s + (1-E_p) * 454 * E_I * (Q_3/Q_1)_s$

Total = $E_p * 590 + (1-E_p) * 136 * (E_w * (Q_2/Q_1)_b + E_I * (Q_3/Q_1)_b) + (1-E_p) * 454 * (E_w * (Q_2/Q_1)_s + E_I * (Q_3/Q_1)_s)$

3. TOTAL PERCENTAGE REMOVED = $\dfrac{\text{Total Removed} * 100}{590}$

A summary of the results is presented in Table 2, indicating removal efficiency estimates ranging from 52 to 87 percent, depending upon the assumptions used. As this analysis was based on a wet year, the results provide reasonable assurance that the removal goal will be achieved on a long term

Table 1. Treatment Assumptions

| Case | Flow Source | $Q_1$ m³ x1000 | $Q_2/Q_1$ | $Q_3/Q_1$ | $C_1$ (mg-P/l) | $E_P$ | $E_W$ | $E_I$ |
|---|---|---|---|---|---|---|---|---|
| 1 | Base  | 864 | 0.50 | 0.50 | 0.15 | 0.30 | 0.25 | 0.80 |
|   | Storm | 617 | 1.00 | 0.00 | 0.74 | 0.30 | 0.25 | 0.80 |
| 2 | Base  | 864 | 0.25 | 0.75 | 0.15 | 0.30 | 0.25 | 0.90 |
|   | Storm | 617 | 0.75 | 0.25 | 0.74 | 0.30 | 0.25 | 0.90 |
| 3 | Base  | 864 | 0.25 | 0.75 | 0.15 | 0.40 | 0.25 | 0.90 |
|   | Storm | 617 | 0.75 | 0.25 | 0.74 | 0.40 | 0.25 | 0.90 |
| 4 | Base  | 864 | 0.25 | 0.75 | 0.15 | 0.40 | 0.30 | 0.95 |
|   | Storm | 617 | 0.75 | 0.25 | 0.74 | 0.40 | 0.30 | 0.95 |
| 5 | Base  | 864 | 0.25 | 0.75 | 0.15 | 0.40 | 0.40 | 0.95 |
|   | Storm | 617 | 0.50 | 0.50 | 0.74 | 0.40 | 0.40 | 0.95 |
| 6 | Base  | 864 | 0.00 | 1.00 | 0.15 | 0.40 | 0.50 | 0.95 |
|   | Storm | 617 | 0.50 | 0.50 | 0.74 | 0.40 | 0.50 | 0.95 |

Table 2. Summary of Estimated Phosphorus Removal

| Case | Analysis Conditions | Load (kg/year) | Removed (kg/year) | (%) |
|---|---|---|---|---|
| 1 | Most Conservative | 590 | 307 | 52 |
| 2 |  | 590 | 378 | 64 |
| 3 |  | 590 | 407 | 69 |
| 4 |  | 590 | 425 | 72 |
| 5 |  | 590 | 484 | 82 |
| 6 | Most Optimistic | 590 | 513 | 87 |

average basis. The analysis can be repeated with other flow and treatment values in order to assess the sensitivity of the results to the input assumptions. Principles of this analysis may be applied by other designers to water quality systems utilizing multiple treatment mechanisms.

The analysis tends to illustrate the value in utilizing multiple treatment mechanisms for phosphorus removal. Because the actual treatment efficiency of any individual facility may be better or worse than predicted, designing several processes in combination increases the likelihood the system will function satisfactorily as a whole.

## Optimization of Pond Design

A number of issues require resolution in optimizing the efficiency of a phosphorus control pond. As much as possible, standardized recommendations proposed for the Denver area (Urbonas and Ruzzo 1986) have been followed in the Shop Creek design. These recommendations include provision of a water quality detention volume equal to 12.7 millimeters of runoff from the impervious surfaces in the watershed, establishing a total drain time of approximately 40 hours, and inclusion of a secondary treatment system such as a sand filter. The drain time recommendations for surcharge storage above the permanent pool of a wet pond were subsequently revised to be 12 to 16 hours.

Detention Time. In order to assess the impact of detention time, the removal efficiency of suspended phosphorus was calculated and plotted versus median detention time. This analysis is based on the recommended water quality detention volume, which is assumed to be normally dry and lie above a permanent pool equal to half of this volume. The following removal efficiency equation for sedimentation in dynamic conditions was used (US Environmental Protection Agency 1986):

$$E = 1 - (1 + 1/n * V_s/(Q/A))^{-n}$$

where:

$E$ = fraction of initial solids removed ($E * 100$ = percent removal)
$V_s$ = settling velocity of particles

$Q/A$ = rate of applied flow divided by surface area of basin (an "overflow velocity," often designated the "overflow rate")
$n$ = a parameter which provides a measure of the degree of turbulence or short-circuiting, which tends to reduce removal efficiency

Table 3 presents settling velocities typical of suspended solids found in urban runoff, based on analyses made during the National Urban Runoff Program (U.S. Environmental Protection Agency 1986).

Table 3. Settling Velocities of Urban Runoff Solids

| Size Fraction | % of Particle Mass in Urban Runoff | Aver. Settl. Veloc.(m/hr) | Aver. Equiv. Diameter (mm) |
|---|---|---|---|
| 1 | 0-20% | 0.0091 | 0.014 |
| 2 | 20-40% | 0.091 | 0.042 |
| 3 | 40-60% | 0.46 | 0.11 |
| 4 | 60-80% | 2.1 | 0.28 |
| 5 | 80-100% | 20. | 2.5 |

The fourth column has been added to indicate the representative size of the particles. Of these size fractions, the four with finer particles have been shown to have an affinity for phosphorus adsorption (Sartor and Boyd 1972). In this analysis these four finer fractions have been assumed to make up 100 percent (25 percent each) of the particles carrying phosphorus in urban runoff. Because of the finer overall particle distribution, a pond's estimated removal efficiency for suspended phosphorus would be somewhat lower than it would be for all suspended sediment.

The removal efficiency equation was set up in a spreadsheet format to analyze a spectrum of input assumptions, including pond stage and outlet orifice size. The equation was applied with n equal to 3 (moderate performance) and Q equal to outflow at the orifice. Even though pond inflow may be greater than outflow, increased circulation in the pond as it fills would allow continuous settling to occur. Therefore, the assumption of Q as outflow is reasonable.

The results of the analysis are plotted in Figure 7. It is evident that the point of diminishing returns in suspended phosphorus removal efficiency is reached at a median detention time of approximately 12 hours, although 90 percent of the performance is achieved at a median time of approximately six hours. For maximum efficiency, a 12-hour median detention time is recommended as a possible standard for phosphorus control ponds.

This median detention time equates to a total drain time of approximately 36 hours assuming the control orifice is located at the permanent pool surface. The discharge

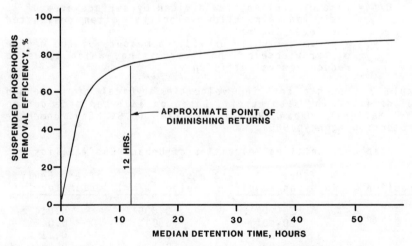

Figure 7. Removal Efficiency vs. Detention Time

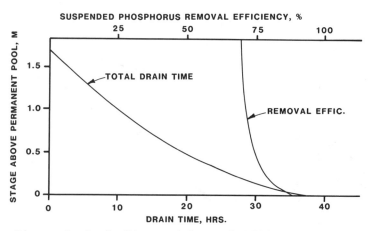

Figure 8. Drain Time and Removal Effic. vs. Stage

characteristics of an orifice maximize the detention of small, frequent inflows into the pond as desired for "first flush" treatment. Suspended phosphorus removal efficiency and drain time are related to pond stage for the proposed Shop Creek Pond in Figure 8.

Detention Volume. The previous removal efficiency analysis is based on a storm event just large enough to fill the recommended water quality detention volume (equal to 12.7 millimeters of runoff from the upstream impervious area). For the Shop Creek Basin, which has a projected imperviousness of 39 percent, this detention volume (11,200 cubic meters) is filled during a storm event with a two-hour rainfall depth of 19 millimeters. This storm has a return period of 1.5 years. It is expected that suspended phosphorus removal efficiency would be greater than shown in the previous analysis for storms smaller than 19 millimeters of rainfall and less for larger storms.

Storm size is correlated to cummulative percent of runoff in Figure 9. The rainfall and event data of Figure 9 are based on a 20-year analysis of rainfall in the Denver area (Urbonas and Ruzzo 1986). The cumulative percent of runoff was determined by combining the cumulative percent of rainfall with the rainfall/runoff relationship for the Shop Creek Basin depicted in Figure 10. Figure 9 implies that approximately 70 percent of the total volume of storm runoff is produced by storms which would surcharge the recommended water quality detention volume (greater than 19 millimeters of rainfall).

Suspended phosphorus removal efficiencies were calculated for three storms exceeding 19 millimeters of rainfall (the 2-year, 10-year and 100-year events), as well as for two

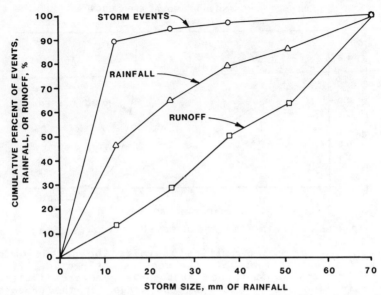

Figure 9. Cumulative Distribution of Rainfall and Runoff

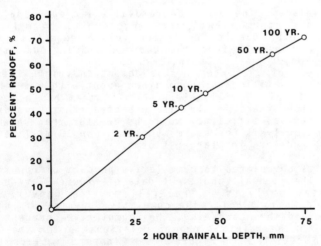

Figure 10. Rainfall-Runoff Relationship

smaller storms, assuming a median detention time of 12 hours for the water quality detention volume. These values are correlated to cummulative percent of storm runoff in Figure 11. The average phosphorus removal efficiency of the pond over the spectrum of runoff events is approximately 50

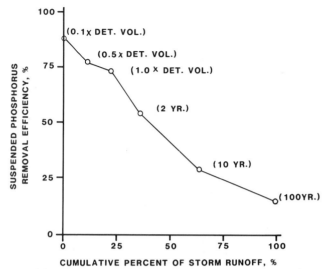

Figure 11. Removal Efficiency vs. Cumulative Runoff

percent. This overall efficiency appears to be a reasonable goal for suspended phosphorus removal in a water quality pond, and offers validation for the recommended water quality volume equal to 12.7 millimeters of runoff from the upstream impervious area.

It is important to note that the water quality volume functions best when designed to be normally dry between storm events. The suspended phosphorus removal efficiency in a storm of 19 millimeters of rainfall is calculated to be 73 percent assuming a median detention time of 12 hours. The same storm in a full pond yields a calculated removal efficiency of only about 35 percent, due to less attenuation and settling.

Permanent Pool Volume. The volume and depth of the permanent pool should be those necessary to produce low flow-through velocities during base flows and small storms and promote biological uptake of phosphorus. Safety considerations favor keeping the depth along the outside of the permanent pool to approximately 1.0 to 1.2 meters, even if the depth of the pool is greater at its center. It is recommended that the depth be sufficient, at least 1.0 meter, and circulation provided, to limit macrophyte growth and discourage the spread of such diseases as botulism among aquatic birds. The Shop Creek permanent pool volume is one half the design water quality detention volume.

Pond Layout. It is recommended that baffles and/or length to width ratios greater than two to three be utilized to minimize short circuiting (Kropp 1982). It is preferable that pond side slopes be 4 to 1 or flatter. It is recommended that adequate spillway capacity be provided to minimize the risk of embankment failure. If the elevation difference from the lowest point of the spillway to the invert of the outlet works at the centerline of the embankment is greater than ten feet, the pond would fall under the regulatory control of the State Engineer in Colorado.

Outlet Configuration. In order to maximize the area of the control orifice and minimize the potential for clogging, a single opening is preferred at the elevation of the permanent pool surface. Some debris, such as filamentous algae or plastic bags, tend neither to float nor settle to the bottom of the pond. Instead, these materials can move toward the orifice and therefore a hood, or shield, to skim off floating debris may not be sufficient to prevent clogging of the orifice. Instead, it is recommended that the orifice be protected by a large trash rack with the openings at least 4 times smaller than the orifice area. Pre-fabricated steel bar grating lends itself for this use. In order to minimize clogging, the ratio of trash rack area to orifice area is recommended be at least 50, and as much as 100 to 200. This sizing may be based on the anticipated quantity and type of debris in the pond, as well as the desired frequency of trash rack cleaning. It is recommended that the trash rack be structurally designed assuming the worst case of 100 percent clogging and no head behind the rack, and that access to the control orifice be provided, in order to clear the orifice should it ever become clogged. A gated drain incorporated into the outlet structure allowing release of the permanent pool volume would facilitate sediment removal operations.

## Optimization of Wetland/Infiltration Areas

Sedimentation and infiltration are key treatment process in these zones which, in size, are approximately two percent of the upstream impervious area in the Shop Creek Basin. Although wetland vegetation growth in areas designated for infiltration may inhibit percolation to some degree, it adds phosphorus removal via vegetative filtering and biological uptake and creates natural habitat desired in the project.

Infiltration Considerations. Five drop structures in the Shop Creek project, each seven feet high and provided with underdrains, function to stabilize the channel, slowing velocities, and to promote the infiltration of runoff. The cation exchange capacity of the underlying sandy soils is sufficient to provide a satisfactory level of phosphorus uptake. Although not planned at Shop Creek, the underlying soils could be replaced with soils possessing higher cation exchange capacity in order to improve phosphorus uptake and immobilization. It is important that the water quality pond be located upstream in the project in order to trap sediment and minimize plugging in the infiltration areas.

Wetland Considerations. A key issue in the design of the wetland areas is to minimize flow-through velocities in storms. The Shop Creek wetland areas maintain flow-through velocities in the 100-year storm of approximately 0.5 to 1.0 meters per second. Notched sills 0.52 meters feet high are provided at the drop structures. These allow low flows to pass, but detain storm flows and prevent flow acceleration upstream of the drop.

In the direction perpendicular to flow, the wetland areas are laid out in two zones, depicted in Figure 12. The meandering, central zone varies in width between 3 and 10 meters. It is to carry normal low flows and contain varieties of plants suited to continual submergence. The adjacent areas, varying in width between 12 and 35 meters, are to be somewhat higher than the low flow channel and are to be planted with vegetation suited for a depth to water table of up to 0.5 meters. This diverse growth of native willows, cattails, and other plantings will be inundated

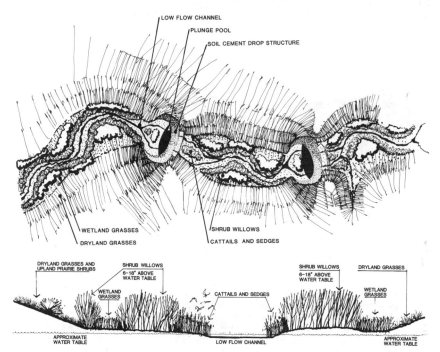

Figure 12. Typical Wetland/Infiltration Zones

during storm events and be designed to minimize wash-out from the wetlands of the phosphorus-rich organic humus.

## Integrating System into the Plains Landscape

To integrate the Shop Creek improvements into the existing landscape of the Cherry Creek State Recreation Area, the channel alignment, cross-sections, and drop structures are designed to be visually compatible with the materials and landforms typical of the prairie setting. The visual structure of the landscape in the vicinity of the channel is comprised of a series of gently sloping horizontal lines created by the low rolling hills. The existing stream meander, when viewed at eye level, repeats and reinforces the horizontal lines of the low hills. To avoid the imposition of a incongruent element such as a straight channel, the existing meander of the stream is to be retained and side slopes are to be graded to blend with the surrounding rolling hills. All channel side slopes are to be re-vegetated in upland prairie species currently found on the site. Channel drop structures are to be integrated into the landscape by using soil cement. Using on-site soil as the primary structural material visually blends the structures with the existing streambanks. The form of the drop structures, shown in Figure 13, repeat the gently sloping diagonal lines of the surrounding landscape.

Figure 13. Typical Soil Cement Drop Structure

## Conclusions

Based on the analysis and design of facilities for phosphorus removal at Shop Creek, the following conclusions are offered:

1. Phosphorus loading data imply that runoff from an urbanizing basin normally experiences an increase in phosphorus concentration as development occurs and then a decrease and leveling off as disturbed lands are stabilized, and attest to the importance of instituting effective erosion control measures during construction.

2. Because runoff increases with development, increased storm flow in a fully developed basin can offset improvements in water quality from land stabilization, resulting in increased total phosphorus loading.

3. For maximum effectiveness, it is recommended that large, regional water quality facitilies be planned, preferably just upstream of critical receiving waters, instead of multiple smaller facilities located throughout the watershed.

4. Because the actual efficiency of any individual treatment mechanism may be better or worse than predicted, designing several processes in combination increases the likelihood the system will function satisfactority as a whole. The Shop Creek treatment facilities consist of a detention pond with a permanent pool for setting and biological uptake, followed by a series of wetland/infiltration areas.

5. A spreadsheet sensitivity analysis based on a range of flow distribution and treatment efficiency assumptions for each treatment process provides reasonable assurance that the 50 percent phosphorus removal goal will be achieved on a long term average basis. Principles of this analysis may be applied by other designers to systems utilizing multiple treatment mechanisms.

6. The standardized recommendation of water quality detention volume proposed for phosphorus control ponds in the Denver area (Urbonas and Ruzzo 1986) is verified. This volume, equal to 12.7 millimeters of runoff from the upstream impervious area, yields an average suspended phosphorus removal efficiency of approximately 50 percent over the expected spectrum of runoff events. The water quality detention volume functions best when designed to be normally dry between storm events and located above a permanent pool.

7. For maximum efficiency, a 12-hour median detention time (36-hour total drain time) is recommended as a possible standard for phosphorus control ponds, although 90 percent of the performance is achieved in half the time.

8. The permanent pool depth should be sufficient, at least 1.0 meter, and circulation provided, to limit macrophyte growth and discourage the spread of diseases among aquatic birds. The Shop Creek permanent pool volume is one half the design water quality detention volume.

9. In order to maximize the area of the pond control orifice and minimize the potential for clogging, a single opening is recommended at the elevation of the permanent pool surface, protected by a trash rack at least 50 times greater in area than the orifice. A

gated drain allowing release of the permanent pool volume would facilitate sediment removal operations.

10. Five large drop structures in the Shop Creek project, each provided with notched sills and underdrains, function to stabilize the channel, reduce flow velocities, and promote the infiltration of runoff over wide, relatively flat areas of sandy soil.

11. A diverse system of wetland vegetation in the infiltration zones adds phosphorus removal via filtering and biological uptake and creates natural habitat desired in the Shop Creek project. Low flow-through velocities are maintained to minimize wash-out from the wetlands of phosphorus-rich organic humus.

12. The Shop Creek improvements are designed to be visually compatible with the materials, landforms and vegetation typical of the existing prairie landscape.

## Appendix. References

Denver Regional Council of Governments (1982) "Denver Urban Runoff Evaluation Program, File report."

Hartigan, J.P. and Quaswbarth, T.F. (1985) "Urban Nonpoint Pollution Management for Water Supply Protection: Regional vs. Onsite BMP Plans," Proceedings of Twelfth International Symposium on Urban Hydrology, Hydraulics, and Sediment Control, University of Kentucky.

Kropp, R.H. (1982) "Water Quality Enhancement Design Techniques," Proceedings of Conference on Stormwater Detention Facilities, American Society of Civil Engineers.

Martin, E.H. and Smoot, J.L. (1986) "Constituent - Load Changes in Urban Stormwater Runoff Routed Through a Detention Pond-Wetlands System in Central Florida" Report No. 85-4310, U.S. Geological Survey.

Muller Engineering Company (1988) "Shop Creek Drainage Outfall System, Preliminary and Final Design Report", with Black and Veatch and William Wenk Associates, prepared for City of Aurora, Colorado.

Sartor, J.D. and Boyd, G.B. (1972) "Water Pollution Aspects of Street Surface Contaminants," Report No. EPA-R2-72-081, U.S. Environmental Protection Agency.

Urbonas, B.R., and Ruzzo, W.P. (1986) "Standardization of Detention Pond Design For Phosphorus Removal," Urban Runoff Pollution, edited by Torno, H. Marsalek, J., and Desbordes, M., Springer-Verlag, Berlin.

## Discussion of Mr. Wulliman's Paper

Question:

Why was a phosphorous removal goal of 50% chosen?

Answer:

Not sure; Colorado Water Quality Control came up with that number.

## LOAD-DETENTION EFFICIENCIES IN A DRY-POND BASIN

By Larry M. Pope[1] and Larry G. Hess[2]

**Abstract:** Dry-pond detention basins commonly are used to mitigate peak stormwater discharges from urban areas; however, the effects that these basins may have on the quality of urban runoff are not clearly understood. The purpose of this paper is to describe the results of a cooperative study between the U.S. Geological Survey and the Kansas Department of Health and Environment to evaluate load-detention efficiencies for 11 water-quality constituents.

Inflow and outflow to a dry-pond detention basin in Topeka, Kansas, were monitored for 19 storms during a 14-month period. Samples of runoff were collected automatically at two inflow and one outflow locations. Inflow and outflow constituent loads were computed with subsequent computation of load-detention efficiencies.

Three constituents (dissolved solids, ammonia plus organic nitrogen, and total organic carbon) had negative (larger loads out than in) median detention efficiencies (-78.5 percent, -9.0 percent, and -3.0 percent, respectively). Median detention efficiencies for the other constituents were: suspended solids (2.5 percent), chemical oxygen demand (15.5 percent), nitrite plus nitrate nitrogen (20.0 percent), ammonia nitrogen (69.0 percent), total phosphorus (18.5 percent), dissolved phosphorus (0.0 percent), total lead (66.0 percent), and total zinc (65.0 percent).

## Introduction

During the past several years, the construction of dry-pond detention basins has become a common practice in mitigating peak stormwater discharged from urban areas in eastern Kansas. Many communities are now requiring that stormwater-detention facilities be incorporated in all new urban development. The effectiveness of dry-pond detention basins in regulating peak stormwater discharges from urban areas is well documented and can be predicted based on design characteristics and expected flow rates and volumes. The effect that these detention basins may have on the quality of urban runoff is less understood and in need of further study.

A previous study of the effects of urbanization on the quality of runoff in local receiving streams in Topeka, Kansas, concluded that some water-quality constituents, particularly total lead and total zinc, have

---

[1] Hydrologist, U.S. Geological Survey, 1950 Constant Ave. - Campus West, Lawrence, KS 66046, (913-864-4321).
[2] Environmental Engineer, Kansas Department of Health and Environment, Topeka, KS 66620.

a significant direct relation to the degree of urbanization (Pope and Bevans, 1987). Concentrations of these trace elements were determined to be related to the percentage of residential plus commercial land use within a basin and to street density. Other studies have documented the distribution of trace elements and other water-quality constituents in other urban environments and have quantified the occurrence of these constituents in street sweepings and stormwater solids (Wilber and Hunter, 1979; Miller, et. al., 1977).

With the occurrence of trace elements and other constituents, such as nutrients and organic carbon and nitrogen, prevalent in the urban environment, a method of limiting the transport of these constituents to local receiving streams would be beneficial in protecting the water quality of these streams.

Given the utilization of dry-pond detention basins in decreasing peak stormwater discharges from urban areas, it is important to investigate the effectiveness of these basins in intercepting the transport of urban-related constituents to local streams. Therefore, the purpose of this paper is to describe the results of a 14-month cooperative study between the U.S. Geological Survey and the Kansas Department of Health and Environment to evaluate the effects on load transport of selected water-quality constituents as a result of detention storage in a dry-pond detention basin in eastern Kansas.

## Site Description

Dry-pond detention basins in eastern Kansas have a variety of shapes and sizes, but typically are small, grass-lined, and receive inflow at one or many points. Inflow may come from several sources--direct runoff from streets and parking lots, overland flow, or as discharge from storm sewers. The dry ponds receive runoff from small urban basins, which are typically less than 50 acres. The outflow structures of the ponds may consist of drop boxes, riser boxes, or precast concrete pipe that serve as controlling structures to mitigate peak discharges. Discharge from the ponds is either into storm-sewer systems or natural drainage channels depending on location and local ordinances.

The dry pond that was studied in Topeka, Kansas, receives runoff from a 12.3-acre single-family residential area consisting of 25 houses either partly or completely within the drainage basin. A private school and church are also partly in the basin. Total impervious area equals 6.03 acres (49 percent of the basin) of which 4.29 acres (71 percent of the total impervious area) is classified as effective-impervious area (direct connection to a storm-sewer system). Location of the dry pond in relation to the drainage area and pervious and impervious areas are shown in Figure 1.

The dry pond is retangular in shape (130 feet by 290 feet), is grass-lined, and has a maximum storage depth of 4.5 feet. Inflow to the pond is through two 24-inch diameter concrete storm-sewer pipes. Outflow is through an open-top 3x4x3-foot concrete riser box with a 4-inch-diameter inlet opening and 12-inch-diameter outlet.

## Instrumentation

Measurement of flow into and out of the dry pond is accomplished with a combination of flumes and pressure transducers to sense hydraulic head. Inflow flumes consist of a modified Palmer-Bowlus design constructed in a 10-foot section of 21-inch-diameter polyvinyl chloride

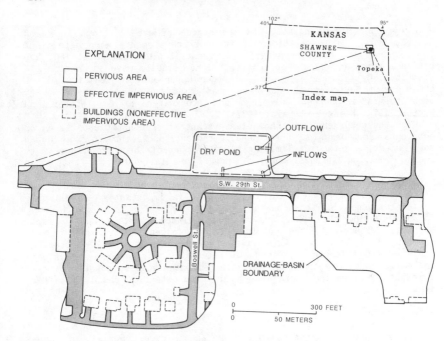

FIG. 1.--Location of Dry-Pond Detention Basin and Pervious and Impervious areas.

(PVC) storm-sewer pipe and inserted into the existing 24-inch-diameter concrete inflow pipes. Modification to the traditional Palmer-Bowlus design consisted of decreasing the slope of the sidewalls to 1:1 and raising the floor of the flume to compensate for the flatter sidewalls. These modifications were evaluated previously at a U.S. Geological Survey's urban-runoff test site in Jackson, Mississippi (Kilpatrick, et. al, 1985). The "pipe-insert method" of flume construction allowed the flumes to be calibrated, prior to installation, in a jet-tank facility at the hydraulic laboratory of the U.S. Geological Survey's Gulf-Coast Hydroscience Center located at the National Space Technology Laboratory, Mississippi.

Outflow through the riser box is measured with a 12-inch culvert-inlet (manhole) flume manufactured by Badger Corporation[1], Harvey, Louisiana. The manhole flume is a low-head loss, self-cleaning Parshall-type flume. The flume is secured in the riser box and sealed to prevent bypass flow. Hydraulic head measurements for the pipe-insert flumes and manhole flume are made with Schaevitz Model P-3061 pressure transducers connected to a pneumatic-bubbler system of dry nitrogen gas. Operating range of the transducer is 0 to 100 inches of hydraulic head with 0 to 5 volts direct-current output that is directly proportional to pressure difference. Hydraulic-head measurements are recorded with a

---

[1] The use of brand names in this paper is for identification purposes only and does not constitute endorsement by the U.S. Geological Survey or the Kansas Department of Health and Environment.

Campbell Scientific Instruments (Logan, Utah) CR-21 data logger. Rainfall is measured with a Weathertronics Model 6010-99 tipping-bucket rain gage (Qualimetrics, Inc., Sacramento, California) located at the dry pond (fig. 1) and recorded by the CR-21 data logger. Hydraulic-head and rainfall measurements are recorded at 5-minute intervals.

### Sample Collection and Analysis

Stormwater-runoff samples were collected at inflow and outflow locations with Manning 4051 (Manning Environmental Corporation, Santa Cruz, California) 3-liter, 24-discrete-sample automatic samplers. Sampler intake orifices were located in the turbulence downstream of the inflow and outflow flumes to insure that collected samples were from a well-mixed representation of the runoff. Sample-collection containers were housed in chest-type freezers that were modified to maintain a temperature of 4 °C to preserve samples until processing was completed. Discrete samples of runoff were collected throughout the runoff period of each monitored storm. Runoff-sample collection was controlled by a program of the CR-21 data logger and was based on intervals of time, thresholds of hydraulic head, and rate of change in hydraulic head; all values were user programmable. About 1,150 discrete samples were collected during the 19 storms monitored from September 1986 through November 1987.

To define the mean constituent concentrations necessary to compute inflow and outflow constituent loads for a storm, selected discrete samples of runoff were composited into discharge-weighted samples. Discrete samples were selected to adequately define variations in flow rate and constituent concentrations. The method of computing the scharge-weighted value of each discrete sample to be included in the composite sample was based on the mid-interval method of subdividing di(Porterfield, 1977).

For most monitored storms, one composite sample was made for each sampling location (two inflow and one outflow). An analysis of this sample provided mean water-quality consitituent concentrations for storm runoff at the sampling locations. However, because some storms consisted of several separate periods of rainfall (substorms), multiple composite samples were required at inflow locations. Analysis of these samples provided mean constituent concentrations for substorm periods. Total constituent loads for these storms were a summation of the loads computed for the substorms.

Chemical analyses of composite samples were made for all constituents except total ammonia plus organic nitrogen, total organic carbon, dissolved solids, and suspended solids, by the Kansas Department of Health and Environment, Division of Laboratories, Topeka, Kansas. Concentrations of total ammonia plus organic nitrogen and total organic carbon were determined by the U.S. Geological Survey, Denver, Colorado. Concentrations of dissolved and suspended solids were determined by the U.S. Geological Survey, Lawrence, Kansas. Concentrations of dissolved solids were determined for water samples that were filtered through 0.45-micrometer filters.

### Rainfall Characteristics

The 19 storms monitored during this study represented a variety of rainfall characteristics and antecedent conditions (Table 1). Total rainfall ranged from 0.04 inch for storm 15 to 2.15 inches for storm 1. Duration of rainfall ranged from 30 minutes for storm 2 to 2,255 minutes

Table 1.—Rainfall Characteristics and Antecedent Conditions for the Monitored Storms, 1986-87

| Storm no. (1) | Date (2) | Total rainfall (inches) (3) | Rainfall duration (minutes) (4) | Number of dry days preceding storm (5) | Rainfall in 3 days prior to storm (inches) (6) | Rainfall in 14 days prior to storm (inches) (7) | Number of days since last rainfall of indicated magnitude 0.25-0.50 inch (8) | 0.50-1.0 (9) | >1.0 inch (10) | Maximum rainfall intensity (inches per hour) for indicated period 5 minutes (11) | 15 minutes (12) | 1 hour (13) |
|---|---|---|---|---|---|---|---|---|---|---|---|---|
| 1 | 9/11/86 | 2.15 | 210 | 3.7 | 0 | 0.11 | 16 | -- | 33 | 2.64 | 2.24 | 1.57 |
| 2 | 9/15/86 | 0.21 | 30 | 6.1 | 0 | 2.27 | -- | -- | 4 | 1.32 | 0.72 | 0.21 |
| 3 | 9/18/86 | 0.06 | 90 | 3.1 | 0.22 | 2.49 | -- | -- | 7 | 0.12 | 0.12 | 0.06 |
| 4 | 9/22-23/86 | 1.47 | 1,410 | 4.3 | 0 | 2.47 | -- | -- | 11 | 3.84 | 2.40 | 0.88 |
| 5 | 9/26-27/86 | 0.70 | 380 | 2.9 | 0.59 | 1.77 | -- | 3 | 15 | 2.64 | 1.24 | 0.35 |
| 6 | 9/28/86 | 0.05 | 50 | 1.1 | 0.71 | 2.48 | -- | 2 | 17 | 0.24 | 0.12 | 0.05 |
| 7 | 10/2-3/86 | 1.58 | 2,255 | 1.9 | 1.37 | 3.70 | 2 | -- | 3 | 1.68 | 0.96 | 0.33 |
| 8 | 10/09/86 | 0.09 | 40 | 5.8 | 0 | 3.74 | -- | -- | 7 | 0.24 | 0.16 | 0.09 |
| 9 | 10/11/86 | 0.89 | 885 | 1.6 | 0.14 | 3.06 | -- | -- | 9 | 0.72 | 0.56 | 0.30 |
| 10 | 10/13/86 | 0.11 | 85 | 2.3 | 0.89 | 3.99 | -- | 2 | 11 | 0.24 | 0.20 | 0.08 |
| 11 | 10/22/86 | 0.32 | 255 | 8.2 | 0 | 1.18 | -- | 11 | 20 | 0.24 | 0.16 | 0.13 |
| 12 | 10/25/86 | 0.23 | 400 | 2.7 | 0.33 | 1.37 | 3 | 14 | 23 | 0.12 | 0.12 | 0.09 |
| 13 | 10/25/86 | 0.25 | 410 | 0.24 | 0.56 | 1.93 | 0.2 | 14 | 23 | 0.24 | 0.20 | 0.09 |
| 14 | 11/4-5/86 | 0.22 | 235 | 0.07 | 0.46 | 1.28 | 1 | 24 | 33 | 0.12 | 0.12 | 0.10 |
| 15 | 11/10/86 | 0.04 | 40 | 2.6 | 0.07 | .75 | 6 | 30 | 39 | 0.12 | 0.08 | 0.04 |
| 16 | 8/18/87 | 1.05 | 185 | 4.8 | 0 | 1.88 | -- | 5 | 6 | 03.24 | 2.36 | 0.82 |
| 17 | 9/9-10/87 | 1.13 | 235 | 3.1 | 0.06 | 1.84 | -- | -- | 14 | 02.52 | 1.60 | 0.68 |
| 18 | 11/15/87 | 0.96 | 290 | 20.5 | 0 | 0 | 21 | -- | 30 | 0.60 | 0.44 | 0.37 |
| 19 | 11/16/87 | .28 | 230 | 1.1 | .96 | .96 | -- | 1 | 31 | .24 | .20 | .13 |

for storm 7. Most of the storms monitored during this study occurred during an unusually wet period of late summer and early fall 1986. Of these storms, only storm 11 had an antecedent dry period longer than a week. Most storms had antecedent dry periods of 1 to 4 days. Only storm 18, which occurred during mid-fall 1987, had an extended antecedent dry period (20.5 days). Generally, rainfall volume in eastern Kansas is a direct function of rainfall intensity. Although intense rainfall may not result in large rainfall volume (storms 2 and 5), for the most part, intense rainfall typically is associated with thunderstorm systems that produce large volumes of rainfall, such as storms 1, 4, 7, 16, and 17.

### Flow Volumes and Load Computations

A comparison of inflow and outflow volumes is shown in Figure 2. Total inflow volume to the dry pond is a combination of two sources: (1) rain that falls directly into the pond, and (2) inflow discharged from the two storm-sewer outlets. Volume attributed to direct rainfall was computed based on measured rainfall per storm and known dimensions of the pond. Of the 19 storms monitored during this study, 2 (storms 7 and 9) had outflow that exceeded inflow. Outflow for storm 7 was substantially larger (24 percent) than inflow. Although not directly observed, it is believed that this additional outflow was the result of

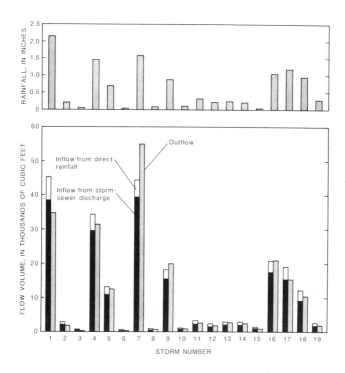

FIG. 2.--Rainfall and Inflow and Outflow Volumes for Monitored Storms

ungaged overland runoff from a residential area upslope from the detention basin or subsurface flow into the pond that may have occurred during the unusually lengthy duration of rainfall (2,255 minutes, Table 1) or both. Additionally, antecedent conditions for storm 7 were the wettest of any of the monitored storms; rainfall in the 3 days prior to the storm totalled 1.37 inches (Table 1). These factors probably caused the substantial runoff. Given this extraneous inflow and the adverse effect it would have in comparing inflow and outflow load computations, storm 7 was not used for analysis of load-detention efficiencies. Although outflow for storm 9 exceeded total inflow by 9.6 percent, about one-half of this difference could be the result of measurement error, which is considered to be plus or minus 5 percent. Therefore, data from storm 9 was included in subsequent computations of load-detention efficiencies.

In contrast to storm 7, storm 1 had an unusually large decrease in outflow volume (34,614 cubic feet) compared to inflow volume (45,235 cubic feet). The decrease in flow can be attributed to the extremely dry antecedent moisture conditions (Table 1). These dry conditions, with associated dry-cracking of bed soils, could make the pond susceptible to infiltration of a large volume of water. Additional factors affecting this loss would include evaporation, transpiration, and measurement error. Because the decrease in outflow volume for storm

1 appears to be a function of dry-pond conditions, the data from this storm was included in computation of constituent load-detention efficiencies. Average runoff-detention times for the 19 monitored storms ranged from about 6 to 16 hours depending on rainfall characteristics (volumes and intensities) and antecedent moisture conditions, which affect runoff rates and storage capacity of the bed soil of the pond.

Loads of selected water-quality constituents were computed from storm or sub-storm mean constituent concentrations determined by analysis of discharge-weighted composite samples of storm runoff, volume of flow associated with each composite sample of storm runoff, and an appropriate unit conversion coefficient. Total inflow loads are a summation of loads computed at both storm-sewer inflow locations.

## Load-Detention Efficiencies

Load-detention efficiencies for each selected water-quality constituent were computed from inflow and outflow loads by the following equation and expressed as a percentage:

$$\text{Load-detention efficiency} = \frac{(\text{Load In} - \text{Load Out})}{(\text{Load In})} \times 100 \quad . \quad (1)$$

Positive percentages of load-detention efficiency indicate the degree to which a constituent load is being decreased through detention storage. Negative values of load-detention efficiency indicate the degree to which loads of a constituent leaving the dry pond are greater than the loads entering the pond. Individual load-detention efficiencies for 11 water-quality constituents are presented in Table 2. Most of the constituents presented in Table 2 have a considerable range of detention efficiencies. With the exception of dissolved solids, total nitrite plus nitrate nitrogen, and total ammonia nitrogen, all constituents had individual load-detention efficiencies from substantially less than zero (negative detentions) to substantially more than zero (positive detentions). No positive detention efficiencies were computed for dissolved solids. All storms, except storms 5 and 11, had positive detention efficiencies for nitrite plus nitrate nitrogen. No negative detention efficiencies were computed for ammonia nitrogen. Load-detention efficiencies for storm 7 were not computed for reasons previously described. Other missing values in Table 2 were the result of lost samples or cases where mean concentrations for one or more storms or substorms were less than analytical detection limits.

Median values of detention efficiencies are shown in Figure 3. Of the 11 constituents shown, 3, dissolved solids, total ammonia plus organic nitrogen, and total organic carbon, had negative detention efficiencies (more load out than in). The -78.5-percent median detention efficiency computed for dissolved solids indicates that loads of dissolved solids are nearly 80- percent larger in the outflow than in the inflow. This increase in outflow load of dissolved solids is probably the result of salts leaching from bed soils and decaying organic material. Also, part of the increased outflow dissolved-solids load may have originated as deicing salts applied to streets during winter and subsequently transported into the dry pond via small-volume snowmelt runoff. Median detention efficiencies of total ammonia plus organic nitrogen(-9.0 percent) and total organic carbon (-3.0 percent) may not differ significantly from zero (no detention effect on loads) to warrant a definite statement concerning detention efficiency of either constituent other than that dry-pond detention storage seems to have little effect on loads of total ammonia plus organic nitrogen and total organic carbon. However, this effect concerning these organic constituents may be applicable only to the type of pond examined in this study--a grass-lined pond in which organic material accumulates through repeated mowings and deposition of clippings. Increases in organic constituents could reflect the deposition of inflow organic material being less than the suspension or leaching of indigneous organic material. This same process may be occurring for suspended solids, which had an unexpectedly small median detention efficiency (2.5 percent). The decrease of inflow loads of suspended solids as a result of deposition may be offset by suspension of indigenous material, possibly of organic origin.

Median detention efficiencies for chemical oxygen demand (COD), total nitrite plus nitrate nitrogen, total ammonia nitrogen, and total phosphorus are all significant, positive detention efficiencies (Fig. 3). Median detention efficiency for loads of oxidizable material, expressed as chemical oxygen demand, was 15.5 percent. Of the 18 storms for which detention efficiencies were computed, only 4 (storms 9, 11, 12, and 19) had negative efficiencies (loads out greater than loads in) for chemical oxygen demand. A 20.0-percent median detention efficiency was computed for nitrite plus nitrate nitrogen, with only two storms (storms 5 and 11) having negative detention efficiencies. The load decrease of nitrite plus nitrate nitrogen may be the result of denitrification and, possibly more importantly, loss of volume by infiltration into bed soils. Loads of total ammonia nitrogen were decreased at a median rate of 69.0 percent. Loads of ammonia nitrogen may be decreased through ionic adsorption on mineral surfaces with

Table 2.--Load-Detention Efficiencies and Statistical Summary
[Values are given in percent]

| Constituent (1) | Storm number (2) | | | | | | | | | | | | | | | | | | | Statistical summary | | |
|---|---|---|---|---|---|---|---|---|---|---|---|---|---|---|---|---|---|---|---|---|---|---|
| | 1 | 2 | 3 | 4 | 5 | 6 | 7 | 8 | 9 | 10 | 11 | 12 | 13 | 14 | 15 | 16 | 17 | 18 | 19 | Median (3) | Maximum (4) | Minimum (5) |
| Chemical oxygen demand (COD) | 20 | 27 | 45 | 37 | 23 | 22 | -- | 9 | -3 | 5 | -11 | -3 | 38 | 5 | 28 | 39 | 11 | 3 | -57 | 15.5 | 45 | -57 |
| Solids, dissolved (DS) | -8 | -66 | -208 | -48 | -90 | -54 | -- | -438 | -142 | -257 | -197 | -230 | -143 | -142 | -67 | -41 | -26 | -25 | -61 | -78.5 | -8 | -438 |
| Solids, suspended (SS) | 27 | 34 | -10 | 24 | 42 | -152 | -- | -43 | -33 | -115 | -42 | -105 | 45 | 40 | 25 | 35 | 9 | -4 | -134 | 2.5 | 45 | -152 |
| Nitrogen, $NO_2 + NO_3$, total, as $N(NO_2 + NO_3)$ | 29 | 29 | 16 | 30 | -1 | 2 | -- | 20 | 7 | 20 | -3 | 30 | 33 | 28 | 23 | 10 | 18 | -- | -- | 20.0 | 33 | -3 |
| Nitrogen, ammonia, total, as $N(NH_4)$ | 69 | 84 | 57 | 69 | 67 | 64 | -- | 86 | 44 | 80 | 74 | 10 | 31 | -- | 71 | 87 | 57 | -- | -- | 69.0 | 87 | 10 |
| Nitrogen, $NH_4$ + organic, total, as $N-(NH_4+ORG)$ | -26 | -16 | -99 | 35 | 18 | 14 | -- | 23 | -79 | -16 | 43 | -154 | -- | -86 | -101 | 3 | 60 | -2 | -- | -9.0 | 60 | -154 |
| Phosphorus, total, as P (P-TOT) | 11 | 49 | 50 | 16 | 17 | 4 | -- | 24 | -10 | 24 | 1 | 28 | 59 | 20 | 30 | 4 | 1 | -4 | 44 | 18.5 | 59 | -10 |
| Phosphorus, dissolved, as P (P-DIS) | -16 | 27 | 45 | 10 | -6 | -5 | -- | 16 | -55 | -- | -48 | 28 | 43 | 16 | 31 | 0 | -46 | -33 | -12 | 0.0 | 45 | -55 |
| Lead, total (LEAD) | 98 | 29 | 81 | 72 | 93 | 88 | -- | 66 | 70 | 51 | 66 | 66 | 66 | 41 | 58 | 50 | -783 | 92 | -126 | 66.0 | 98 | -783 |
| Zinc, total (ZINC) | 61 | 80 | 81 | 86 | 67 | -6 | -- | 78 | 79 | 33 | -27 | 35 | 67 | 48 | 46 | 69 | 63 | 74 | -99 | 65.0 | 86 | -99 |
| Carbon, organic, total (TOC) | -3 | 33 | 33 | 11 | -24 | -15 | -- | 12 | -17 | -4 | -8 | -7 | -- | 8 | 34 | -14 | 25 | 30 | -62 | -3.0 | 34 | -62 |

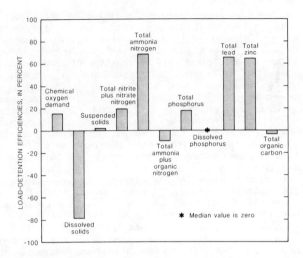

FIG. 3.--Median Values of Load-Detention Efficiencies for Selected Water-Quality Constituents

subsequent deposition during ponding, chemical oxidation to nitrate, and infiltration into bed soils. The decrease in the median load of total phosphorus was computed to be 18.5 percent. Only two storms (storms 9 and 18) had a negative detention efficiency. Much of the decrease in total phosphorus probably is through deposition of suspended phosphorus with minor additional decreases as a loss of dissolved phosphorus by infiltration into bed soils. However, on a long-term basis, the dry pond seemed to have little or no effect in decreasing inflow loads of dissolved phosphorus as indicated by the 0.0 percent median load-detention efficiency shown in Figure 3.

Median load-detention efficiencies for total lead and total zinc were markedly positive and virtually the same (66 percent and 65 percent, respectively). Because little of the total load of these two trace elements normally is transported in the dissolved state, it is assumed that decreases in load were the result of deposition of the suspended form of the elements during detention ponding. Detention efficiencies for dissolved lead and zinc could not be computed because of numerous "nondetected" analytical results for the discharge-weighted composite samples.

## Summary

Inflow and outflow to a dry-pond detention basin in Topeka, Kansas, were monitored for 19 storms during a 14-month period. Samples of runoff were collected automatically at two inflow and one outflow locations. Discharge-weighted composite samples of runoff were made from series of discrete samples. Chemical analyses of these composite samples provided storm or substorm mean constituent concentrations for 11 water-quality constituents. Inflow and outflow constituent loads were computed with subsequent computation of load-detention efficiencies. Three of the constituents examined (dissolved solids, ammonia plus organic nitrogen, and total organic carbon) had negative (larger loads out than in) median detention efficiencies (-78.5 percent, -9.0 percent, and -3.0 percent, respectively). Median detention efficiencies for the remaining constituents were: suspended solids (2.5 percent), chemical oxygen demand (15.5 percent), nitrite plus nitrate nitrogen (20.0 percent), ammonia nitrogen (69.0 percent), total phosphorus (18.5 percent), dissolved phosphorus (0.0 percent), total lead (66.0 percent), and total zinc (65.0 percent).

Appendix 1.--References

Kilpatrick, F.A., Kaehrle, W.R., Hardee, J., Cordes, E.H., and Landers, M.N. (1985). "Development and testing of highway storm-sewer flow measurement and recording system." U.S. Geol. Survey Water-Resources Inv. Report 85-4111.

Miller, T.L., Rinella, J.F., McKenzie, S.W., and Paramenter, Jerry. (1977). "Analysis of street sweepings, Portland, Oregon." U.S. Geol. Survey open-file report.

Pope, L.M., and Bevans, H.E. (1987). "Relation of urban land-use and dry-weather, storm, and snowmelt flow characteristics to stream-water quality, Shunganunga Creek basin, Topeka, Kansas." U.S. Geol. Survey Water-Supply Paper 2283.

Porterfield, G. (1977). "Computation of fluvial-sediment discharge." U.S. Geol. Survey Techniques of Water-Resources Inv., Book 3, Chap. C3.

Wilber, W.G., and Hunter, J.V. (1979). "Distribution of metals in street sweepings, stormwater solids, and aquatic sediments." Journal Water Pollution Control Federation, 51(12), 2810-2822.

# SIMULATED WATER-QUALITY CHANGES IN DETENTION BASINS
## Phillip J. Zarriello [1], AFF. ASCE

## ABSTRACT

Detention basins have been proposed as a method of decreasing peak discharge and contaminant levels in storm runoff in Irondequoit Creek watershed, tributary to Lake Ontario near Rochester, N.Y. A deterministic model was used to predict sediment retention by basins during three or four moderate to large stormflows at each of four sites in two types of basins–a temporary-storage basin, which retains storm runoff but allows normal runoff to pass unimpeded, and a maximum-storage basin, which maintains a permanent pool of water. The simulated outflow hydrographs from the basins differed little from the inflow hydrographs because each basin's storage capacity was small in relation to its drainage area. The decrease in peak flows ranged from 0 and 46 percent and, along with retention time, generally was greater in moderate stormflows than in large stormflows and in basins with a larger storage capacity in relation to its inflow. The predicted decrease in suspended-sediment load varied with particle-size distribution and detention time, both of which are a function of basin size and rate of inflow. The predicted annual decrease of suspended-sediment load, based on the mean particle-size distribution, ranged from 28 to 32 percent in temporary-storage basins and from 33 to 60 percent in maximum-storage basins; this range varied with the proportion of clay to sand. The rate of sediment accumulation within the basins indicates that maximum-storage basins would fill twice as fast as temporary-storage basins. The retention of particulate phosphorus, lead, and zinc was about 80 percent greater in maximum-storage basins than the temporary-storage basins because maximum-storage basins provide greater sediment retention. Average total annual load retention at each site indicated a reduction of 22 to 59 percent total phosphorus, 20 to 47 percent total lead, and 16 to 38 percent total zinc. Actual water-quality improvement may be less than predicted, however, because (1) the sediments may be resuspended mechanically, (2) chemical constituents may be released from the bottom sediments into the dissolved phase under anaerobic conditions, and (3) metals may remobilize in the presence of high concentrations of chloride from road salt.

## INTRODUCTION

Studies completed during the National Urban Runoff Program (NURP) indicate that stormwater-detention basins are one of the more promising means of diminishing nonpoint-source contaminant concentrations in storm runoff by decreasing the velocity of stormflow and allowing suspended sediment and associated constituents to settle and accumulate within the basin (U.S. Environmental Protection Agency, 1983). Six possible locations were identified for detention basins in the Irondequoit Creek watershed near Rochester, N.Y. (fig. 1), one of the NURP study sites, (O'Brien and Gere, 1983). The effectiveness of a detention basin to improve water quality will depend on fixed basin characteristics such as storage capacity and outlet design, as well as variable characteristics, such as stormflow volume, particle-size distribution, and the relations among dissolved and particulate constituent concentrations. To obtain information on how these variables influence detention-basin performance,

[1]Hydrologist, U.S. Geological Survey, 521 West Seneca St., Ithaca, N.Y.

the U.S. Geological Survey, in cooperation with the Monroe County Department of Engineering, began a study to simulate movement of stormflow through detention basins at four locations in the Irondequoit Creek watershed and determine the effect on peak flow and water quality. Two types of basins were simulated for each site--a temporary-storage basin, which impounds water during high flows, and a maximum-storage basin, which maintains a permanent pool of water. The simulations were derived from a water-quality model developed and used for the Irondequoit NURP program (Kappel and others, 1986) and from data collected during that study, August 1980 through August 1981 (Zarriello and others, 1985).

This paper describes the simulated stormflow attenuation and suspended-sediment removal by both types of basins at the four sites. It also examines (1) the effect of particle-size distribution on sediment removal in both types of basins; (2) the relation of suspended-sediment loads to adsorbed phosphorus, lead, and zinc loads; and (3) the effect that these basins could have on the transport of suspended constituents to Irondequoit Bay, to which Irondequoit Creek is the only large tributary.

## STUDY SITES

Four sites from which streamflow and water-quality data were available from the NURP study were selected for simulation. Three (Thornell, Linden, and Blossom) lie along Irondequoit Creek; the fourth (Allen) is on Allen Creek near its junction with Irondequoit Creek (fig. 1). The site and basin characteristics are summarized in table 1. The Thornell site is the furthest upstream; its 136-km$^2$ drainage area is main-

Table 1.--Detention-basin characteristics

| Site | Contributing drainage area (km$^2$) | Maximum pool altitude above sea level (m) | At maximum pool altitude | | |
|---|---|---|---|---|---|
| | | | Storage volume runoff (mm) | Equivalent rainfall[1] (mm) | Surface area (km$^2$) |
| Thornell | 136 | 125 | 7.1 | 138 | 0.55 |
| Linden | 262 | 113 | 2.3 | 25.4 | 0.24 |
| Allen | 78 | 91 | 9.1 | 77.2 | 0.17 |
| Blossom | 370 | 81 | 9.6 | 112 | 0.52 |

[1] Equivalent rainfall is based on runoff coefficients reported by Kappel and others, 1986

ly agricultural, rural, and undeveloped land. The Linden site, 8.7 km downstream, encompasses an additional 126 km$^2$ of more densely populated and commercially developed land. The Allen site, 0.31 km upstream of Irondequoit Creek, contains 78 km$^2$ in the most highly urbanized drainage area, which contributes higher chemical and sediment loads per unit area than the other major subbasins (Kappel and others, 1986). The Blossom site includes an additional 30 km$^2$ of moderately developed area downstream of the Allen and Linden sites and includes an active gravel quarry with an estimated storage capacity of 7.4 mm of runoff or about 87 mm of rain. If made available, the quarry could form a detention basin for Irondequoit Creek and was used as one of the configurations in the simulation.

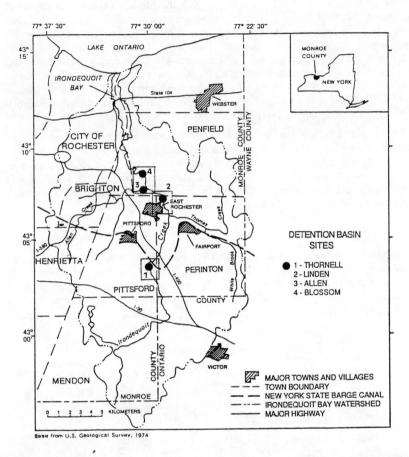

Figure 1.-- Principal geographical features of Irondequoit Creek basin and location of simulated basins

## METHODS

The simulation of flow-attenuation basins consisted of two parts--a flow model and a sedimentation model. The flow model uses measured streamflow data as inflow to the basin and generates an outflow hydrograph according to the configuration of the basin and its outlet. The sedimentation model then uses the inflow and simulated outflow data with the suspended-sediment concentration and particle-size distribution in the inflow to predict concentrations of suspended sediment in the outflow. The trap efficiency of the basin, expressed as a percentage, is calculated as unity minus the ratio of sediment load out to sediment load in.

The basin-outflow hydrographs were generated from the reservoir routing program of Jennings (1977), which is based on the modified-Puls method developed by the U.S. Soil Conservation Service (1972). This method does not account for evapotranspiration, changes in bank storage, or changes in ground-water storage, which are considered insignificant in relation to the volume of runoff during a storm.

The storage-to-outflow relations differed among the basins, depending mainly on the type of control structure. The temporary-storage control (fig. 2A) is designed to pass low flows unimpeded but to cause ponding during high flows. Criteria for ponding in temporary-storage basins were derived from a streamflow-duration analysis. A step-backwater analysis (Shearman, 1976) was used to derive approach elevations, assuming critical flow through the control, and the approach elevations were matched to the established ponding criteria by varying the control geometry. The maximum-storage type of basin (fig. 2B) was simulated with a fixed modified broad-creasted wier spillway (Hulsing, 1967) that maintains a permanent pool. Dimensions of the control geometry and therefore the storage-to-outflow relations, differed from site to site.

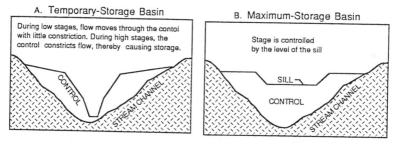

Figure 2.-- Front view of typical control configuration used for the simulations of temporary- and maximum-storage basins.

The transport and deposition rates of suspended sediment were obtained from the detention-storage part of the Multi-Event Urban Runoff Quality Model (DR$_3$M-QUAL) by Alley and Smith (1982). Particle entrapment is based on the "plug-flow" concept, in which discrete volumes or "plugs" of water are routed through a basin, and the settling of particulate matter within the plug is calculated according to Stoke's Law. The model calculates the time required to route a plug of water through a basin (detention time) in relation to the time required for particulate material of a given size to settle through the average depth of that plug within the basin. Particles that settle to the bottom of the basin within the time required for the plug to move through the basin are considered trapped.

The accuracy of the trap-efficiency predictions is uncertain because many simplifications and assumptions must be made for the simulation of the complex and variable sediment-transport mechanisms. The major assumptions related to flow in a basin are that (1) water flows in discrete plugs from a single inflow at one end of the basin to a single outflow at the opposite end; (2) flow within the plug is laminar; and (3) no mixing occurs between plugs. The model does not account for resuspension or movement of settled particles along the bed, changes in particle-size distribution during a stormflow, or chemical reactions that would remobilize constituents into a dissolved phase. Because the model does not account for chemical reactions that might occur, retention of phosphorus, lead, and zinc were estimated from the predicted sediment retention and the relation of phosphorus, lead, and zinc concentrations to the concentration of particulate matter reported in field studies (Brown and others, 1981; Rausch and Schreiber, 1981; Hey and Schaefer, 1983; Whipple and Hunter, 1981).

Sediment particles retained in a detention basin were converted to a unit volume through empirical relations of mass density determined for normally pooled and dry reservoirs by Lara and Pemberton (1963) to estimate the rate at which basins would fill with sediment.

## STREAMFLOW AND WATER-QUALITY CHARACTERISTICS USED IN BASIN SIMULATIONS

The storms selected for simulation represent differing seasonal periods and magnitudes of discharge. Runoff trends based on 24 years of record at the Allen Creek site and 40 years of meteorological record at the National Weather Service station near Rochester indicate that runoff during the water year used in this analysis (1981) was near normal. A log Pearson analysis of 1-, 3-, and 7-day maximum daily runoff indicated that the storms used for simulation have a recurrence frequency of less than 1 year except for the February storm, which had a recurrence interval of 1.9 to 8.8 years. The February storm therefore would be indicative of the lower limit of basin performance, while the other storms would represent a more typical range.

The size distribution of suspended particles varied during individual storms, from storm to storm, and from site to site. The discharge-weighted mean particle-size distribution (fig. 3) ranged from 10 to 20 percent sand, 47 to 56 percent silt, and 30 to 43 percent clay. The mean particle-size distributions were used for the simulations, and the sensitivity of the sedimentation rates over the range of the measured particle-size distributions at each site was tested. The distribution of particle size with discharge suggests increase in sand content with increased flow and would result in slightly underestimated model predictions.

Snowmelt and spring runoff produced 50 to 70 percent of the total annual suspended sediment, phosphorus, lead, and zinc loads, stormflows during the remainder of the year produced another 20 percent of the total annual loads (Kappel and others, 1986). The proportion of particulate to total phosphorus, lead, and zinc (the part that could settle) was estimated to be about 50, 70, and 60 percent, respectively. The particulate-constituent concentration are probably not evenly distributed among size classes, however, because clays have a greater ion-exchange capacity and because the larger surface area per unit mass of the finer particles provides a larger substrate area to which constituents can adhere (White, 1981; Carter and others, 1974; Jenne, 1968; Gibbs, 1973; Horowitz, 1984).

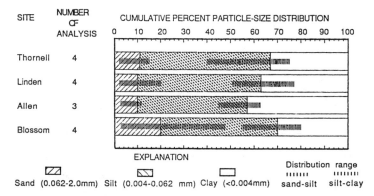

Figure 3.-- Particle-size distribution of suspended sedimnet obtained from August 1980 through August 1981. Valuse arer discharge-weighted averages and range of measured particle-size fractions.

## RESULTS AND DISCUSSION

### Flow Attenuation and Recession

The delay and reduction of peak discharge and the length of recession are a function of (1) the ratio of basin storage capacity to contributing drainage area, (2) the magnitude of the storm, and (3) the configuration of the control. The reduction of the peak discharge ranged from 1 to 39 percent in the temporary-storage basins and from 0 to 46 percent in maximum-storage basins; the lag time between inflow and outflow peaks ranged from 1 to 10 hours. Larger stormflows (lower probability of occurrence), such as that of February 1981, were attenuated the least; those of lesser magnitude showed a greater attenuation of peak flow and longer recessions.

### Sediment-Trap Efficiency

The predicted outflow concentrations of suspended sediment were proportional to the flow attenuation; the maximum-storage basins (fig. 4) and basins with a large storage capacity in relation to drainage area produced the greatest decrease in concentration. The predicted average trap efficiency is summarized in table 2 for mean particle-size distribution. Trap efficiency of both types of basins fluctuated with the particle-size distribution of suspended sediments. Within the measured range of sand and clays and silts content, a greater proportion of sand increased trap efficiency by 20 to 30 percent, whereas a greater proportion of clay decreased trap efficiency by 20 to 30 percent. The predicted trap efficiency of each type of basin is indicated in figure 5, which shows the regression line for the average particle-size distribution and an envelope curves for the range of particle-sizes. The amount of variation in the trap efficiency that is explained by discharge per unit area is 44 percent for temporary-storage basins and 68 percent for maximum-storage basins. The greater

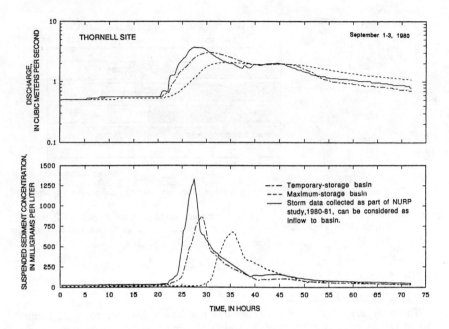

Figure 4.-- Measured and simulated strom discharge and suspended-sediment concentrations at Thornell site.

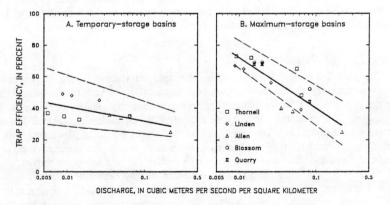

Figure 5.-- Relation of strream discharge to percent suspended-sediment retention for mean particle-size distribution and envelope cureves for measured range in particle sizes in temporary- and maximum sttorage basins.

variation in temporary-storage basins probably results from the wider range of storage capacity-to-outflow relations.

Table 2.--Suspended-sediment trap efficiency at temporary-and maximum-storage basins during selected simulated stormflows.
(Values in percent. Site locations shown in fig. 1.)

| Date of Site | [2]High storm | Temporary-storage basin | | | Maximum-storage basin | | |
|---|---|---|---|---|---|---|---|
| | | [3]High [1]Mean | High clay | High sand | Mean | clay | sand |
| Thornell | 9-1-80 | 33 | 21 | 58 | 72 | 66 | 84 |
| | 2-20-81 | 35 | 23 | 58 | 48 | 38 | 69 |
| | 7-10-81 | 37 | 24 | 60 | 65 | 70 | 86 |
| | 8-4-81 | 35 | 23 | 60 | 73 | 67 | 78 |
| Linden | 9-1-80 | 45 | 31 | 60 | 56 | 44 | 68 |
| | 2-20-81 | 35 | 23 | 49 | 39 | 28 | 52 |
| | 7-20-81 | 48 | 34 | 63 | 65 | 58 | 75 |
| | 8-4-81 | 49 | 36 | 64 | 67 | 59 | 79 |
| Allen | 2-20-81 | 25 | 16 | 34 | 25 | 16 | 33 |
| | 7-20-81 | 34 | 24 | 43 | 38 | 20 | 46 |
| | 8-4-81 | 36 | 26 | 46 | 40 | 32 | 48 |
| Blossom | 2-20-81 | 44 | 27 | 60 | 52 | 35 | 66 |
| | 7-20-81 | 68 | 58 | 78 | 69 | 64 | 84 |
| | 8-4-81 | 68 | 58 | 78 | 69 | 64 | 84 |

[1] Discharge-weighted mean.
[2] Minimum sand and maximum clay distribution.
[3] Maximum sand and minimum clay distribution.

## Annual Suspended-Sediment Removal

Annual retention of sediment (table 3) was derived from the discharge-to-retention relation developed from individual storm predictions and applied to seasonal loads calculated by Kappel and others (1986) for similar flow conditions. The estimated annual suspended-sediment retention in maximum-storage basins was 73 percent greater at the Thornell site, 40 percent greater at the Linden site, and 18 percent greater at the Allen site than temporary-storage basins but only 12 percent greater than at the quarry. The estimated retentions are weighted heavily toward the snowmelt and spring runoff period, when most of the annual load occurs, and therefore reflect the period when the basins are least effective and differences between the basins diminish.

Calculations of the volume of the annual load of sediment retained indicate that maximum-storage basins fill approximately twice as fast as temporary-storage basins. This rate will depend on the particle-size distributions of incoming sediments.and the degree of compaction or consolidation of sediments retained in the basin. Even though sediments that accumulate in temporary-storage basins have a higher density as a result of repeated exposure and drying, sediment retention averages about 50 percent more in maximum-storage basins.

Table 3.--Estimated annual retention of suspended-sediment loads.
[Numbers in parentheses represent range of trap efficiency as a percent based on retention estimated for full range of particle-size distributions.]

| Site | Annual load[1] (Mg) | Temporary-storage basin Estimated retention (percent) | Load retained (Mg) | Maximum-storage basin Estimated retention (percent) | Load retained (Mg) |
|---|---|---|---|---|---|
| Thornell | 3,400 | 32 (21-53) | 1,100 | 56 (52-80 | 1,900 |
| Linden | 5,800 | 34 (23-46) | 2,000 | 48 (38-61) | 2,800 |
| Allen | 2,200 | 28 (19-37) | 610 | 33 (24-41) | 720 |
| Blossom[2] | 15,000 | 53 (39-67) | 8,000 | 60 (46-73) | 9,000 |

[1] Annual suspended sediment loads reported by Kappel and others (1986).
[2] Temporary-storage site refers to the quarry.

The predicted annual trap efficiencies of maximum-storage basins were compared with those obtained from the Brune curve (1953) and a similar curve for small reservoirs (less than 39-km$^2$ drainage area) developed by Heinemann (1981), which express trap efficiency as a function of the ratio of storage capacity to inflow. As shown in figure 6, the study results agree closely with the trap efficiencies indicated by Brune's and Heinemann's curves. The underprediction for the Blossom site relative to the curves may be caused by the greater depth of the quarry through which particles must fall to be considered trapped.

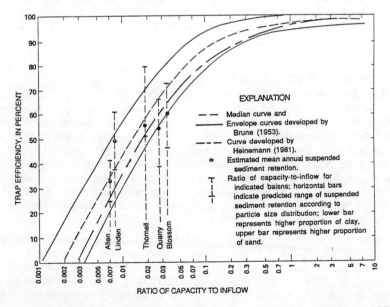

Figure 6.-- Simulated trap efficiency curves of maximum-storage basins compared to curve of Brune (1953) and Heinemann (1981) for normally pooled reservoirs.

## Removal of Phosphorus, Lead, and Zinc

The decrease in total annual load of these chemical constituents at each site ranged from 22 to 59 percent for total phosphorus, 20 to 47 percent for total lead, and 16 to 38 percent for total zinc. These estimates are based on the particulate proportion of each constituent and are weighted heavily toward the removal efficiency estimated for the basin during the snowmelt and spring runoff period, when most of the load was transported. On average, the maximum-storage basins provided 80 percent greater removal efficiency for each of these constituents than temporary-storage basins because of their greater sediment removal efficiency.

Remobilization of these constituents may occur through resuspension, chemical reactions, and biological activity. Phosphate may be released from trapped sediments under anerobic conditions by reduction of ferric complexes (Fillos and Swanson, 1975). Metals will remobilize under low pH conditions, but this is unlikely to occur in the Irondequoit basin. A greater factor may be the large amounts of road deicing salts used in the Irondequoit Creek basin (Kappel and others, 1986, Diment and others, 1974), which mobilize metals by forming a soluble chloro complex (Hey and Schaefer, 1983). These processes may considerably reduce the total removal of these constituents, but this was not examined.

## Combined Effects of the Four Detention Basins

The overall effect of detention basins on loads entering Irondequoit Bay from Iron addition to stormflow characteristics and other factors. Upstream flow-attenuation basins may decrease the performance of downstream basins by about 30 percent by increasing the proportion of finer suspended-sediment particles (Churchill, 1948), but this may be offset by the changes in flow characteristics. To estimate the decrease in load entering Irondequoit Bay from Irondequoit Creek, a mass-balance approach was used to calculate the combined effects of detention basins. Loads entering downstream basins were adjusted by the decreased load from upstream basins plus the load derived from the intervening area between basins. Trap efficiency was based on an increased proportion of finer particles. The estimated load reduction to Irondequoit Bay for suspended sediment, total phosphorus, lead, and zinc, respectively, is 43, 40, 30, and 26 percent for temporary-storage basins and 52, 48, 38, and 33 percent for maximum-storage basins.

## Design Considerations

The time required for a "plug" of water to move through a basin will largely determine the amount of particulate material that will settle. The relation between detention time and pool elevation, both of which are functions of discharge, is shown in figure 7 for the Thornell site for the storm hydrograph shown in figure 4. The figures show that as discharge increases, detention time decreases in maximum-storage basins, whereas it increases in temporary-storage basins. Because temporary-storage basins have an increased potential storage capacity, the outlet design should maximize detention time at peak flow and thereby increase trap efficiency during the time when most of the load is transported.

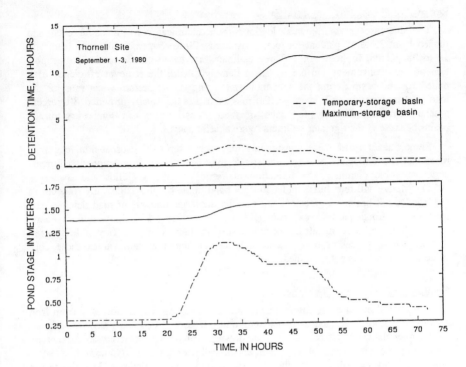

Figure 7.--Detention time and pool elevation as a storm is routed through the Thornell site.

## REFERENCES CITED

Alley, W. M., and Smith, P. E., Multi-Event Urban Runoff Quality Model: U.S. Geol.Surv. Open-File Report 82-764, 1982, 169 p.

Brown, M. J., Bondurant, J. A., and Brockway, C., Ponding Surface Drainage Water for Sediment and Phosphorus Removal: Transactions of the American Society of Agricultural Engineers, 1981, p. 1478-1481.

Brune, G. M., Trap efficiency of reservoirs: Transactions of the American Geophysical Union, v. 34, no. 3, 1953, p. 407-408.

Carter, D. L., Brown, M. J., Robbins, C. W., and Bondurant, J., Phosphorus Associated with Sediments in Irrigation and Drainage Waters for Two Large Tracts in Southern Idaho: Journal of Environmental Quality, v. 3, no. 3, 1974, p. 287-291.

Churchill, M. A., Discussion of "Analysis and Use of Reservoir Sedimentation Data," by L. C. Gottschalk; in Proceedings of the Federal Interagency Sedimentation Conference, Denver, 1947: U.S. Bureau of Reclamation, 1948, p. 139-140.

Diment, W. H., Bubeck, R. C., and Deck, B. L., Effects of Deicing Salts on the Waters of the Irondequoit Bay Drainage Basin, Monroe County, New York, in Proceedings of the Fourth Symposium on Salt: Cleveland, Ohio, Northern Ohio Geological Society, v. 1, 1974, p. 391-405.

Fillos, John, and Swanson, W. R., The Release Rate of Nutrients from River and Lake Sediments: Journal of the Water Pollution Control Federation, v. 47, no. 5, 1975, p. 1032-1042.

Gibbs, R. J., Mechanisms of Trace Metal Transportation in Rivers: Science, v. 180, 1973, p. 71-73.

Heinemann, H. G., A Sediment Trap Efficiency Curve for Small Reservoirs: Water Resources Bulletin, v. 17, no. 5, 1981, p. 825-830.

Hey, D. L., and Schaefer, G. C., National Urban Runoff Program; An Evaluation of the Water Quality Effects of Detention Storage and Source Control Final Report: Northeastern Illinois Planning Commission, 1983, 276 p.

Horowitz, A. J., A Primer on Trace Metal-Sediment Chemistry: U.S. Geol. Sur. Open-File Report 84-709, 1984, 82 p.

Hulsing, Harry, Measurement of Peak Discharge at Dams by Indirect Method: U.S. Geol. Sur.Techniques of Water Review Investigations Report, Book 3, Chapter A5, 1967, 29 p.

Jenne, E. A., Controls of Mn, Fe, Co, Ni, Cu, and Zn Concentrations in Soils and Water--the Significance of Hydrous Mn and Fe Oxides: Advances in Chemistry Series, v. 73, 1968, p. 337-387.

Jennings, M. E., Downstream-Upstream Reservoir Routing (program A697): U.S. Geol. Sur. unpublished report, computer contributions ,1977, 42 p.

Kappel, W. M., Yager, R. M., and Zarriello, P. J., Quantity and Quality of Storm Runoff in the Irondequoit Creek Basin near Rochester, New York; Part 2 - Quality of Storm Runoff and Atmospheric Deposition, Rainfall-Runoff-Quality Modeling and Potential of Wetlands for Sediment and Nutrient Retention: U.S. Geol. Sur. Water-Resources Investigations Report 85-4113, 1986, 93 p.

Lara, J. M., and Pemberton, E. L., Initial Dry Weight of Deposited Sediments; in Proceedings, Federal Interagency Sediment Conference: U.S. Department of Agriculture, miscellaneous publication 970, 1963, p. 818-845.

O'Brien and Gere, Nationwide Urban Runoff Program, Irondequoit Creek Basin Study Final Report: Syracuse, N.Y., O'Brien and Gere, 1983, 164 p.

Rausch, D. L., and Schreiber, J. D., Sediment and Nutrient Trap Efficiency of a Small Flood-Detention Reservoir: Journal of Environmental Quality, v. 10, no. 3, 1981, p. 288-293.

Shearman, J. D., Computer Applications for Step-Backwater and Floodway Analysis: U.S. Geol. Sur. Open-File Report 76-499, 1976, 103 p.

U.S. Environmental Protection Agency, Results of the Nationwide Urban Runoff Program: Executive Summary EPA PB84-185545, 1983, 24 p.

U.S. Soil Conservation Service, National Engineering Handbook, sec. 4, Hydrology, chap. 17, Flood Routing: U.S. Department of Agriculture, 1972, p. 17-1-17-93.

Whipple, William, Jr., Hunter, J. V., Settleability of Urban Runoff Pollution: Journal of the Water Pollution Control Federation, v. 53, no. 12, 1981, p. 1726-1731.

White, E. M., Possible Clay Concentration Effects on Soluble Phosphate Contents of Runoff: Environmental Sciences and Technology, v. 15, no. 1, 1981, p. 103-106.

Zarriello, P. J., Harding, W. E., Yager, R. M., and Kappel, W. M., Quantity and Quality of Storm Runoff in the Irondequoit Creek Basin near Rochester, New York, part 1, Data Collection Network and Methods, Quality-Assurance Program, and Description of Available Data: U.S. Geol. Sur. Open-File Report 84-610, 1985, 44 p.

# WATER QUALITY STUDY ON URBAN WET DETENTION PONDS

Jy S.Wu, Bob Holman and John Dorney

**ABSTRACT:** This paper summarizes results of a monitoring program conducted on three urban wet detention ponds located within the Piedmont region of North Carolina, in the city of Charlotte. Data collected from five storm events were employed to develop a relationship between detention pond performance and pond surface/watershed area ratios. In comparison with data from the NURP study, runoff quality of the study area is generally better and runoff sediment can be characterized by a finer particle size distribution. The attenuation of peak discharge appears unsatisfactory due to short circuiting of local drainage entering the detention ponds from surrounding areas. Despite the fact that these detention ponds were not built for water quality control, the observed improvement in water quality justifies the promising use of wet ponds for urban runoff pollution abatement.

The project was funded by The Water Resources Research Institute of The University of North Carolina.

## INTRODUCTION

Stormwater detention ponds, including both dry and wet ponds, are traditionally designed to reduce peak discharge in order to minimize downstream flooding. The secondary use of detention ponds for water quality improvement can not be ignored. A properly designed detention pond can serve not only for flood control, but also for retention of sediment and pollutants associated with settleable particulates. Knowledge is needed to better understand the design and maintenance strategies of such dual-function detention ponds including the removal mechanisms of pollutants occurring within the pond, and the hydraulic factors affecting the performance (Daily, 1985; DEM, 1985; Schueler, 1988; Wu and Ahlert, 1985 & 1986).

---

Respectively, Associate Professor, Department of Civil Engineering, UNCC, Charlotte, NC 28223; and Coordinator of the Watershed Protection Program and Supervisor of Special Projects Group, Division of Environmental Management, NC Natural Resources and Community Development, Box 27687, Raleigh, NC 27611.

The National Urban Runoff Program (NURP) initiated a number of projects to study the performance of nine wet detention ponds (U.S. EPA, 1983). The surface area ratios of these ponds (defined as the ratio of pond surface area to drainage area or SA/DA) ranged from 0.01 to 2.85%.

This paper presents results of a study aimed at evaluating the performance of existing urban wet detention ponds encountering a wide range of surface area ratios. A comprehensive data collection program was initiated to characterize the hydrologic and water quality responses of wet ponds during storm events. Water quality parameters examined included total suspended solids (TSS), total metals of lead, zinc, copper and iron, total Kjeldahl nitrogen (TKN), and total phosphorus (TP). In addition, runoff samples were collected from several storms and during the various stages of a storm event for analyzing particle size distributions. The research provides a data base for developing a relationship between pollutant removal, rainfall characteristics, particle size distribution and surface area ratio.

## STUDY SITE AND METHODOLOGY

Three urban wet detention ponds, namely Lakeside, LS (4.88 acres), Waterford, WF, (1.79 acres) and Runaway Bay, RB, (3.25 acres), were included in this study. They are located in the Piedmont region of North Carolina, in the City of Charlotte. The entire watershed has a drainage area of 437 acres and comprises three subareas of Lakeside (65 acres), Waterford (302 acres) and Runaway Bay (70 acres). The detention ponds are located on the respective subareas. These ponds were initially designed and built for storm runoff control; none of them were for water quality control. The watershed layout and information pertinent to each detention pond are given in Fig. 1 and Table 1, respectively.

### Field Monitoring and Runoff Sampling

A total of five stream gaging stations (stations 1, 2, 3, 4, and 5) and one rain gage were installed according to U.S.G.S. standards. Rainfall increments and stream gage readings were recorded at 5-minute intervals on paper tapes. A MITRON paper tape reader was employed to transmit the records into the U.S.G.S. Prime computer system for subsequent data processing. Rating curves relating stream stage to flow were developed for each gaging station. For large storms, a few recorded stages might exceed the range of the stage and discharge measurements. In these cases, rating curves were extended using the U.S.G.S. conveyance-slope method (USGS, 1977).

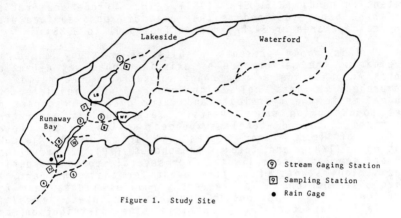

Figure 1. Study Site

- ⑨ Stream Gaging Station
- ⑨ Sampling Station
- ● Rain Gage

Table 1. Watershed and Detention Ponds Characteristics

|  | Lakeside | Waterford | Runaway Bay |
|---|---|---|---|
| Subarea Characteristics | | | |
| Land use | Single Family Apartment Condominium | Single Family Apartment Wooded | Apartment Wooded |
| Subarea, acres | | | |
| Major inflow | 30 | 302 | 367* |
| Local Drainage | 35 | - | 70 |
| Total | 65 | 302 | 437 |
| Detention Pond Characteristics | | | |
| Surface, acres | 4.88 | 1.79 | 3.25 |
| Volume, acre-ft | 38.84 | 5.08 | 12.27 |
| Mean Depth, ft | 7.97 | 2.84 | 3.78 |
| SA/DA, % | 7.51 | 0.58 | 2.27** |

\* Including outflows from Lakeside (65 acres), and Waterford (302 acres) subareas.
\*\* Overall including three ponds and the entire watershed.

It was not possible to monitor all local inflows from storm pipes. However, two representative pipes from the Runaway Bay subarea were chosen for runoff estimation using the TR-20 hydrologic model (SCS, 1983). Based on estimated values of runoff curve number and time-of-concentration, the model computed runoff hydrograph and volume from these pipes. The results were then extrapolated to the remainder pipes according to drainage area proportions, in order to derive the total local runoff. A simple calibration procedure involved fine-tuning of the curve number to achieve a reasonable flow balance between the inflows (total local runoff plus inflow at station 3) and the outflow from RB pond.

Stormwater samples were collected from outflows of each detention pond (stations 7, 8 and 11) using ISCO automatic samplers. The sampling frequency varied from half an hour to two hours depending on whether samples were taken during or after a storm event. In addition, runoff samples were collected manually at 10- to 20-minute intervals from the inflow of LS pond (station 6) and two storm pipes draining into RB pond (stations 9 and 10). Note, hydrologic simulation was performed for these two pipes.

Non-storm samples were collected once every two to three weeks at pond outlets to establish a background condition of water quality.

## Calculation of Removal Efficiency

The amount of pollutants transported was calculated by integrating the mass loading rate expressed as the product of flow and concentration. The removal or trapping efficiency of pollutants was then computed as the percent difference of the pollutant mass entering and leaving the detention pond.

The pollutant mass exported from areas surrounding the ponds was found by multiplying the areal runoff volume by an event mean concentration, defined as total constituent mass per unit runoff volume (mg/l) or a weighted average pollutant loading rate (lbs/acre/in runoff). The event mean concentration was derived from the runoff quality and flow information available at stations 9 and 10 (storm pipes). Table 2 presents the event mean concentrations for storm runoff from the Runaway Bay local drainage area, along with the NURP values.

## RESULTS AND DISCUSSIONS

A total of five storm events have been monitored and are presented in Table 3. Rainfall records obtained at the rain gage from 11/18/86 to 12/01/87 and the annual statistics for N.C. are also summarized in Table 3. The rainfall amounts of the monitored storms range from 0.64 inches to 3.6 inches with a mean of 1.76 inches.

### Runoff Characteristics

The event mean concentration (EMC) of pollutant constituents presented in Table 2 was derived from data collected on the Runaway Bay local drainage areas. The EMCs represent the averages of data from three storm events (storms 2, 3 and 4), and are representative of the runoff quality in the study area. In general, the EMCs obtained in this study are lower than those from the NURP study. TP and Zn are almost half and TKN is about 60% of the NURP values. The EMC of TSS is reported by NURP as one order of magnitude or more greater than those from secondary treatment plants. Taking a secondary effluent of 30 mg/l TSS, the NURP value for TSS would be 300 mg/l or greater. The average TSS obtained in this study is 135 mg/l. The EMCs for Cu and Pb could not be obtained because their concentrations in runoff samples were below the sensitivity limit of the flame atomization analytical technique (30 ug/l for Cu and 100 ug/l for Pb ). Maximum contaminant levels for Cu and Pb in drinking water are 1000 ug/l (federal secondary standard) and 50 ug/l (federal primary standard), respectively.

### Detention Pond Performance

The performance of detention ponds is influenced by particle size distributions of inflow sediment. The distribution may vary substantially from one area to another. Ten sets of storm samples were collected from a number of storms and at various stages of a storm event. The percent remaining corresponding to a settling velocity was analyzed. In order to compare the distributions, the data were averaged and presented in Fig. 2, along with the NURP result. In general, the particle sizes encountered in the study area are much finer than the national average, e.g. the 50 percentile is about 1/3 finer than the national observation.

The mean storm concentrations from WF pond outflow are generally higher than those from LS and RB ponds, particularly for TSS, Zn and Fe. The average of storm peak concentrations exceeds the background level, however, the

Table 2. Event Mean Concentrations of Storm Runoff

| Water Quality parameter | Runaway Bay Subarea | NURP Value* |
|---|---|---|
| TSS, mg/l | 135 | (1) |
| TKN, mg/l | 0.88 | 1.5 |
| TP, mg/l | 0.14 | 0.33 |
| Fe, mg/l | 6.1 | (2) |
| Cu, ug/l | (3) | 34 |
| Zn, ug/l | 70 | 160 |
| Pb, ug/l | (3) | 144 |

(1) One order of magnitude or more greater than those from secondary treatment plants
(2) Not reported
(3) Concentration below detection limit for flame atomic absorption spectrophotometric analysis
* U.S. EPA, 1983

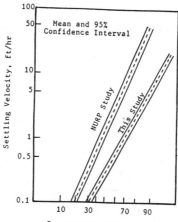

Figure 2. Probability Distribution of Settling Velocities

Table 3. Rainfall Statistics and Monitored Storm Events

| Storm No. | Starting Date | Volume (inches) | Duration (hrs) | Average Intensity (in/hr) | Time since last storm (hrs) |
|---|---|---|---|---|---|
| 1 | 01/01/87 | 1.39 | 17.83 | 0.078 | 184 |
| 2 | 02/26/87 | 3.60 | 66.67 | 0.054 | 89 |
| 3 | 03/25/87 | 1.73 | 146.83 | 0.012 | 137 |
| 4 | 04/15/87 | 1.48 | 80.00 | 0.019 | 383 |
| 5 | 06/04/87 | 0.64 | 2.83 | 0.219 | 9 |
| Mean |  | 1.76 | 62.83 | 0.076 | 160 |
| CV |  | 0.63 | 0.91 | 1.100 | 0.9 |
| Rain Gage Records (11/18/86-12/01/87) | | | | | |
| Mean |  | 0.50 | 6.61 | 0.222 | 112 |
| CV |  | 1.19 | 1.43 | 1.840 | 1.9 |
| Annual Values for N.C. * | | | | | |
| Mean |  | 0.36 | 5.90 | 0.066 | 77 |
| CV |  | 1.45 | 1.05 | 1.320 | 1.1 |

* U.S. EPA, 1986        CV = coefficient of variation

Table 4. Performance of LS and RB Ponds

| Storm No. | TSS | TP | TKN | Zn | Fe |
|---|---|---|---|---|---|
| LS Pond | | | | | |
| 1 | 82 | -20 | -58 | 72 | na |
| 2 | 94 | 10 | -7 | 82 | 78 |
| 3 | 95 | 82 | -9 | 69 | 78 |
| 4 | 85 | -55 | 4 | 71 | 71 |
| 5 | 100 | 100 | 100 | 100 | 100 |
| Avg. | 91 | 23 | 6 | 79 | 82 |
| RB Pond | | | | | |
| 1 | 56 | 55 | 5 | 46 | na |
| 2 | -7 | -10 | 27 | -29 | 2 |
| 3 | 62 | 62 | -4 | 85 | 46 |
| 4 | 74 | 19 | 37 | 67 | 55 |
| 5 | 87 | -4 | 35 | 40 | 79 |
| Avg. | 54 | 24 | 20 | 42 | 45 |

na = not available

Table 5. Relationship between Detention Pond and Surface Area Ratio

| SA/DA % | Pond System | TSS | TP | TKN | Zn | Fe |
|---|---|---|---|---|---|---|
| 0.75 | RB | 54 (26) | 24 (15) | 20 (10) | 42 (10) | 45 |
| 1.35 | WF+RB | 90 (95) | 26 (61) | 56 (42) | 53 (42) | 84 |
| 2.27 | LS+WF+RB | 58 (97) | 40 (62) | 41 (43) | 40 (43) | 49 |
| 7.51 | LS | 91 (99) | 23 (63) | 6 (44) | 79 (44) | 82 |

Numbers in paraenthesis representing long term averge removal (U.S. EPA, 1986)

mean storm concentrations are not significantly higher than the background level, with the exception of TSS.

Several remarks regarding pollutant removal can be made from Table 4. LS pond, because of its large area ratio, has a consistently excellent removal of TSS, Zn and Fe. However, its removal efficiencies for TP and TKN are inconsistent probably due to the variable input of waterfawl droppings into the pond (there are approximately 30 to 40 geese in this area). It has been reported that the daily excrement contributions from geese average at 2.7 lbs of dry excrement per 100 geese, and the percentages in dry weight of nitrogen and phosphorus in geese excrement are 4.5% and 1.0%, respectively (Green Lake study, Wisconsin-Personal Communication).

A removal efficiency of 100% is reported in storm 5 for all pollutant parameters for LS pond. This is because the subarea runoff produced by a small storm was totally retained in the pond. In fact, the runoff condition in storm 5 allows an independent evaluation of the removal efficiencies for LS pond, and for WF+RB ponds in combination.

The removal efficiencies for RB pond, on the average, are 54% for TSS, 20-24% for nutrients (TKN and TP), and 42-45% for metals (Zn and Fe). The influent of WF pond has not been sampled and monitored. It was assumed that the removal efficiency for WF pond is 40% for all constituents so that the overall removal efficiency for the entire watershed can be computed. Table 5 summaries the removal efficiencies for the various combinations of the pond system as a function of surface area ratio. The relationship is also presented in Figs. 3 and 4.

Included in Table 5 are the predicted long term average removal efficiencies (EPA, 1986) as well as results obtained in this study. In general, the predicted removals are less than the field results at SA/DA ratio of 0.75. A good agreement between the predicted and actual removals is obtained for TSS at SA/DA ratios of 1.35 and 7.51, except at SA/DA of 2.27. A low removal efficiency for the entire pond system (SA/DA of 2.27) is due to the unsatisfactory performance of WF pond whose removal was estimated at 40%.

The observed removal for TP is about half of the predicted values for SA/DA greater than 1. The actual and predicted removals for TKN and Zn agree well at SA/DA of 1.35 and 2.27. The TKN removal by LS pond (SA/DA of 7.51%) is extremely poor due to excessive nitrogen input from waterfawl droppings. LS pond also performs well in removing metals (79% removal for Zn and 81% removal for Fe).

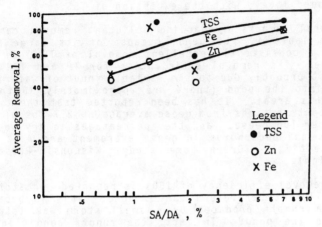

Figure 3. Effects of Surface Area Ratio on Pollutants Removal in Ponds

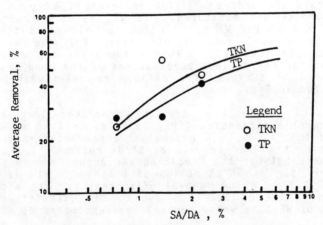

Figure 4. Effects of Surface Area Ratio on Pollutants Removal in Ponds

## CONCLUSIONS

Stormwater quality data and removal efficiency of pollutants have been presented characterizing the performance of urban wet ponds in the Piedmont region of North Carolina. The effect of surface area ratios on the performance of detention ponds was examined. Particle sizes in storm runoff were found to be much finer than the national average. The hydrologic performance of detention ponds included in this study is generally poor in terms of peak discharge reduction. This is due to the drainage of local runoff along the sides of the detention ponds. If local inflows can be diverted away from the pond outlet, the hydrologic and pollutant removal efficiencies of these detention ponds would be greatly enhanced. Despite the fact that the detention ponds were not initially designed and constructed for water quality control, an improvement of water quality has been observed to justify the use of wet detention ponds for urban runoff pollution abatement.

## LITERATURE CITED

1. Dally, L.K. 1983. Operation of Detention Facilities for Urban Stream Quality Enhancement. M.S. Thesis, University of Washington.
2. DEM. 1985. Toxic Substances in Surface Waters of the Falls and the Neuse Lake Watershed. Report no. 85-08. Division of Environmental Management, N.C. Department of Natural Resources and Community Development.
3. Schueler, T.R., 1987. Controlling Urban Runoff: A Practical Manual for Planning and Designing Urban BMPs. U.S. Metropolitan Washington Council of Governments.
4. SCS. 1983. Computer Program for Project Formulation: Hydrology. Technical Release 20, US Soil Conservation Service.
5. U.S. EPA, 1983. Results of Nationwide Urban Runoff Program-Executive Summary. US Environmental Protection Agency.
6. U.S. EPA, 1986. Methodology for Analysis of Detention Basins for Control of Urban Runoff Quality. EPA 440/5/87-001, US Environmental Protection Agency.
7. USGS. 1977. National Handbook of Recommended Methods for Water-Data Acquisition. US Geological Survey.
8. Wu, J.S. and R.C. Ahlert, 1985. A Trajectory Model for Analyzing Sediment Trapping Efficiencies in Storm Water Detention Basins. In: Proceedings Conference on Stormwater and Water Quality Management Modeling. W.James (ed.) CHI Report R149, McMaster University, pp.257-263.
9. Wu, J.S. and R.C. Ahlert, 1986. Modeling Methodology for Dual-Function Stormwater Detention Basins. PB87-159711, National Technical Information Service.

DESIGN AND CONSTRUCTION
OF
INFILTRATION TRENCHES

Bruce W. Harrington[*]

This paper discusses the use of infiltration trenches as a stormwater management practice. Feasibility, design and construction techniques are presented to improve the performance and prolong the useful lifespan of infiltration trenches. Practical site design applications are given to illustrate the type of landuse developments that are most suitable for infiltration trenches. The results of a statewide survey of infiltration practices in Maryland is also provided.

## Introduction

The State of Maryland requires infiltration practices to be considered first among all other stormwater management practices to provide urban runoff control from proposed developments. This requirement was developed as a result of the increasing awareness of water quality degradation from urban runoff. Infiltration practices have the ability to remove water borne pollutants by capturing surface water runoff and filtering it through the ground. Infiltration practices can also reduce runoff volume, provide groundwater recharge, reduce thermal impacts on fisheries, and augment low flow stream conditions. In fact, infiltration is the only stormwater management practice that can closely satisfy the goal of maintaining the pre-development runoff characteristics. This is true because there is no other stormwater management practice that can reduce runoff volumes to the predevelopment level. The multiple benefits make these stormwater management measures suitable for satisfying the intent and goals of maintaining the predevelopment runoff characteristics. However, infiltration practices do have their shortcomings.

---

[*]Water Resources Engineer, Md. Water Resources Administration, Dam Safety Division, Tawes State Office Building, Annapolis, Maryland 21401.

Infiltration practices can only control runoff from small drainage areas, and their feasibility is highly dependant on the underlying soil conditions. If not designed properly, they have the tendency to clog rapidly due to sediment entry during and after construction. Careful construction is necessary to assure that infiltration practices will operate as designed. Another shortcoming is the possible risk of groundwater contamination.

A survey was of over 200 infiltration practices constructed from 1984 through 1986 was completed in Maryland. The results of the survey indicated that the most frequently used infiltration practices were infiltration trenches and 80 percent of the trenches were working as designed (Clement & Pensyl, 1987). It is the purpose of this paper to discuss design and construction techniques that will improve the success rate and prolong the lifespan of infiltration trenches.

## Infiltration Trenches

An infiltration trench is an excavated trench that has been backfilled with stone aggregate and forms an underground storage reservoir. Runoff is captured either by depressing the trench surface or placing a small berm at the low side of the trench surface. It generally consist of a long, narrow excavation ranging from 3 to 12 feet in depth and typically serves a drainage area less than 10 acres. The surface of the trench can consist of stone or a vegetated surface with inlets that distribute the runoff into an underground stone reservoir as shown in Figure 1.

Infiltration trenches can be designed either to capture the increased runoff from developed areas or to capture a smaller amount of runoff to provide water quality control. If a trench is designed to satisfy peak control it will also provide water quality control. Because of the economic burdens of providing total peak control with infiltration alone, many trenches built in Maryland today consist of water quality control only. Other conventional stormwater management practices, such as detention basins, can be used in conjunction with infiltration trenches to provide peak control. A detailed design procedure for trenches can be found in the Standards and Specifications for Infiltration Practices (Md WRA, 1984).

Water quality control consist of sizing the trench to capture the first flush of runoff which contains the bulk of pollutants. The minimum volume rule used in Maryland is to capture the runoff associated with the first 1/2 inch of runoff from the contributing impervious areas (Md WRA, 1986). In sizing the trench, the stone void space ranges from 30 to 40 percent depending on the stone size. The

Figure 1. Types of Infiltration Trenches

(a) trench with a stone surface

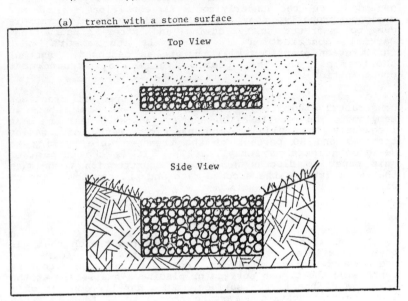

(b) trench with a vegetated surface and inlets

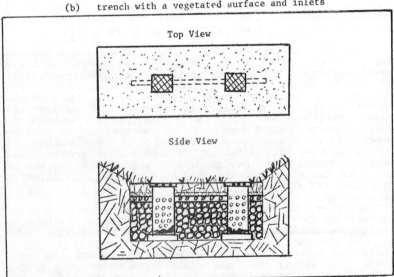

# INFILTRATION TRENCHES CONSTRUCTION

standard stone aggregate used in Maryland ranges from 1 to 3 inches in diameter and is designed with a void ratio of 40 percent. Other conventional stormwater management practices, such as detention basins, can be used in conjunction with infiltration trenches to provide peak control.

Since the first trenches were built in Maryland in the early 1970's, it was recognized that sediment clogging of the stone voids was occurring and that many trenches were not exfiltrating water. This problem is primarily the result of sediment contamination during construction. Another problem was the trenches could not be examined other than from the surface to determine if runoff was draining out the bottom into the underlying soil. In an attempt to overcome these difficulties, several new design and construction techniques were developed.

To determine if the infiltration trenches are operating as designed, the installation of observation wells are required in trenches to provide the inspector with a means of determining what was going on underground. The observation well consist of a vertical section of 4 to 6 inch diameter perforated PVC pipe which sits on a foot plate and has a secure removable locking cap as illustrated in Figure 2. The observation well allows the inspector to monitor the infiltration trenches after rainfall events.

A filter fabric is required around the walls and bottom of the trench to prevent the migration of fine soil particles from entering into the stone voids. A six inch layer of sand may also be used as a filter media at the bottom of the trench. Filter fabric is also required one foot below the surface of the trench to capture sediment and prevent it from clogging the bottom of the trench (stone surface trench). This may create more frequent maintenance cleanouts but will minimize the possibility of the entire trench from clogging with sediment. In the case of a vegetated surface, the top filter fabric will prevent the soil from migrating into the stone.

It is also required that the entire contributory area to the trench be stabilized before construction of the trench could begin. If the drainage area cannot be stabilized prior to the start of construction, sediment laden runoff must be diverted away from the trench construction area. Otherwise, the entire stone aggregate may have to be excavated and cleaned or replaced.

To promote longevity and minimize maintenance, runoff must be filtered prior to entering into the stone reservoir. A minimum 20 foot grass buffer strip is necessary to remove the larger sediment particles that will clog the stone voids in the trench. It is highly recommended to avoid concentrated flow which would not be filtered adequately

Figure 2. Infiltration Trench with Observation Well Detail

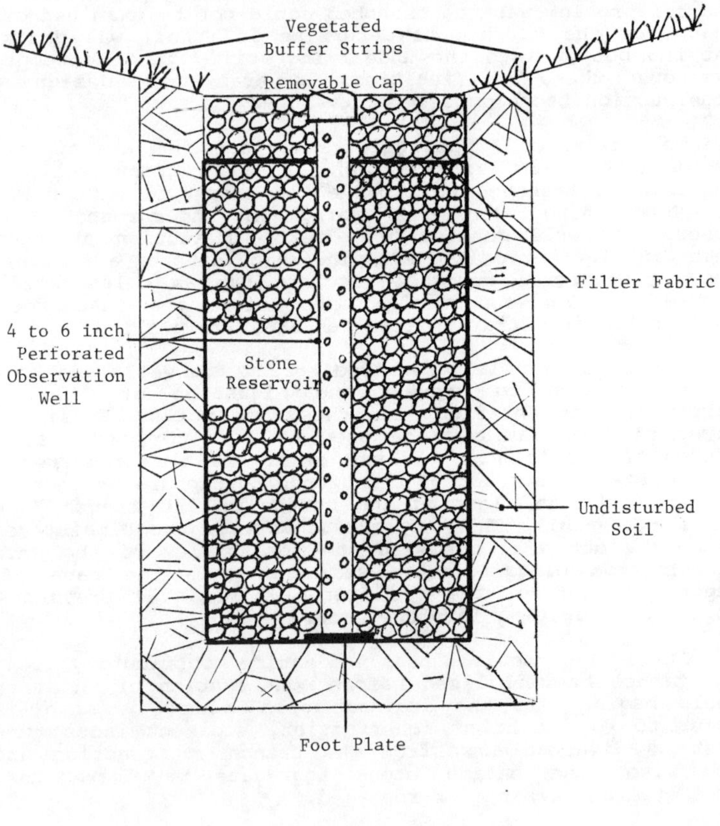

Source: Modified after Md. WRA, 1984

before entering into the trench. To avoid this condition, grade the site so that runoff is flowing in a sheet flow condition across the grass buffer strip. If runoff must be piped or channeled to the trench, install a level spreader to create a sheet flow condition over the grass buffer strip. A level spreader can simply be created by a small linear basin where concentrated water is spread out and discharged over a long weir.

Many of the infiltration trenches that were surveyed were designed and constructed without any runoff filtering and were piped directly to stone reservoirs beneath parking lots. This has been a common way in commercial areas to design trenches because of the limited open space available. Maintenance of trenches under parking lots would require removal of the overlying asphalt which is not practical. This type of trench design has a short lifespan and is not recommended.

### Feasibility and Design of Infiltration Trenches

A number of feasibility requirements can be tested to determine whether a an infiltration trench can be constructed on a specific site. These feasibility test include:

1. The minimum infiltration rate of the underlying soil layers at least 4 feet below the bottom of the trench must equal or exceed 0.50 inches/hour. Acceptable USDA soil texture classes include sand, loamy sand, sandy loam, and loam. This generally applies to soils with less than 20% of clay and less than 50% of silt. Table 1 lists the minimum infiltration rates for the acceptable soils and is based on laboratory examination of more than 5000 soil horizons (Rawls, Brakensiek & Saxton, 1982).

Table 1.  Allowable Soils for Infiltration Trenches

| Soil Texture | Minimum Infiltration Rate (f, in/hr) | SCS Soil Group | Maximum Trench Depth (in) |
|---|---|---|---|
| Sand | 8.27 | A | 1489 |
| Loamy Sand | 2.41 | A | 434 |
| Sandy loam | 1.02 | B | 183 |
| Loam | 0.52 | B | 93 |

2. The maximum allowable underground storage time cannot exceed 72 hours (3 days). This maximum storage time was established based on the time between average rainfall events being approximately 3 days. Therefore, an infiltration trench should be dry prior to the start of the next rainfall event. Based on the allowable storage time (T) of 72 hours, the minimum infiltration rates (f), and a stone void space (Vs) of 40 percent, the maximum allowable trench depths (d) listed in Table 1 are computed by: $d = (f*T)/Vs$.

3. The depth between the bottom of the trench and the seasonal high groundwater table or depth to bedrock must be greater than or equal to 4 feet. This information can be preliminarily determined from soil survey reports and verified by soil borings during a wet period.

4. Water supply wells must be located at least 100 feet from infiltration trenches to prevent any possible well contamination.

5. Infiltration trenches must be located at least 20 feet away from building foundations to avoid possible hydrostatic pressures on foundations or basements.

6. The ground slope downstream of the trench must not exceed 20%. Ground slopes steeper than 20% (5h:1v) increases the chance of downstream seepage and possible slope failure.

7. Infiltration trenches generally can serve drainage areas no greater than 10 acres.

All of the above feasibility conditions should be investigated and each of them are important in ensuring the proper functioning of the proposed infiltration trench. If a site investigation reveals that the feasibility conditions are not satisfied, an infiltration trench should not be pursued.

### Site Design of Trenches

Infiltration trenches may be designed for a wide variety of landuse conditions. The following figures illustrate the most frequent application of trenches on commercial, residential, and highway development areas.

## Commercial Trench Design

With respect to landuse, infiltration trenches are most frequently located in commercial areas where only limited open space is available. A trench is commonly constructed adjacent to the parking lot area as shown in Figure 3. It is important to convey runoff across the length of the trench to maximize the removal of larger sediment particles in the grass buffer strip. Slotted curbs will act as a level spreader to prevent concentrated flow. The trench should have an overflow to pass storms larger than the design storm downstream to a stabile watercourse. Many of the first trenches built in Maryland where constructed under parking lots without any runoff filtering. These types of trenches should be avoided.

**Figure 3.    Commercial Trench Design**

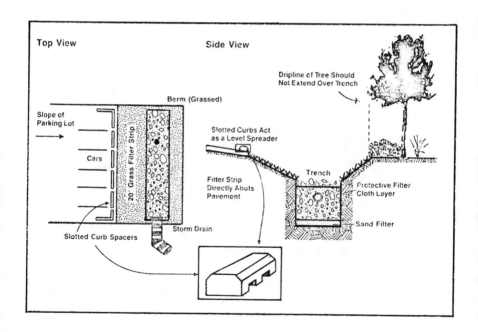

Source:    MWCOG, 1987

## Median Strip Design

Trenches are also frequently placed in median strips on highway projects or in parking lot islands as illustrated in Figure 4. The trench is constructed in a depressed area with a minimum of 20 feet of vegetated filter strip on each side. To maximize the underground storage volume for runoff, the bottom slope of the trench should be flat. An overflow pipe or swale is installed to pass excess runoff. The trench should not be constructed in fill material which normally has a low permeability rate and may create a seepage problem at the in-situ soil interface.

Figure 4.   Median Strip Design

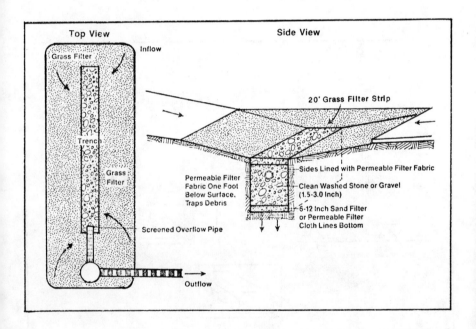

Source:   MWCOG, 1987

## Residential Trench design

In low density residential areas, trenches may be installed in roadside swales as shown in Figure 5. To capture runoff into the stone reservoir, place small check dams or railroad ties across the swale. The bottom slope of the swale should not exceed 5 percent. Slopes greater than 5 percent may erode the swales and contaminate the trench. To increase the storage volume, several trenches can be installed in the swale with check dams or railroad ties.

**Figure 5. Residential Trench Design**

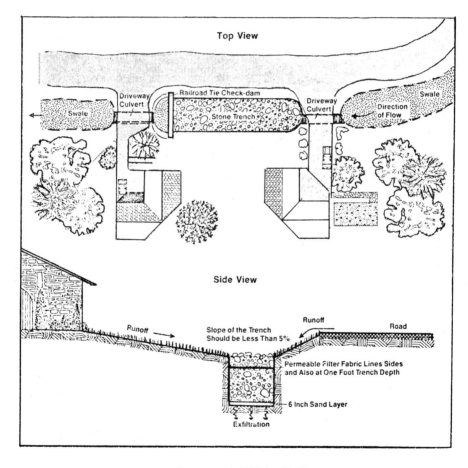

Source: MWCOG, 1987

Infiltration trenches installed at convenience stores or gas stations are often contaminated with oil and grit. To prevent this type of contamination, runoff should be filtered through water quality inlets prior to entering the trench. The water quality inlets in Maryland have three chambers; a sediment chamber, an oil/grit chamber and an exit chamber that conveys runoff to the trench area as shown in Figure 6.

Figure 6. Water Quality Inlet

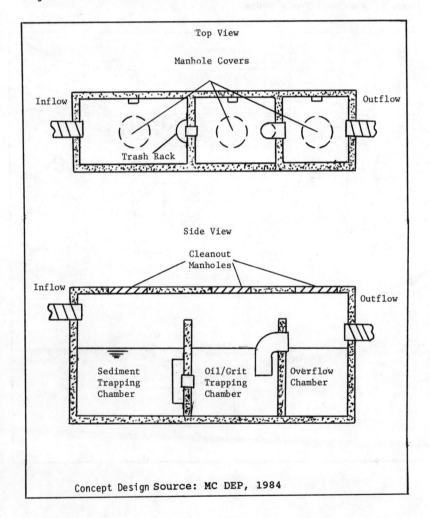

## Statewide Infiltration Survey

A survey was conducted in Maryland to examine all infiltration practices constructed between 1984 and 1986. The survey looked at infiltration basins, infiltration trenches, dry wells, porous pavement, and vegetated swales with check dams installed in 22 counties. Of the more than 200 infiltration practices surveyed, 67 percent of the practices were functioning as designed. The trenches had the highest use rate at 45 percent and the highest success rate of 80 percent. The high success rate of the trenches surveyed may be misleading because approximately 50 percent of the working trenches had no observation wells. Therefore, it is very possible that some of these trenches had standing water below the surface. Infiltration basins had the second highest use rate of 30 percent and the lowest success rate of 48 percent.

The vast majority of infiltration trenches were constructed on commercial sites. Over 85 percent of the trenches that were not functioning did not have vegetated buffer strips. Without buffer strips, there is no means of preventing sediment from clogging the trench. More than 50 percent of the non-working trenches did not include soil borings. Without soil borings, there is no exact way to know if the underlying soil is suitable for infiltration. Approximately 70 percent of the failed infiltration trenches showed signs of sediment entry either during or after construction.

### Trench Construction

Trench construction is very important to assure that the trench will operate as designed. Many of the trenches constructed in Montgomery County, Maryland in the early 1970's have failed. The reason for failure is primarily due inadequate soils investigations, a lack of sediment control, the absence of vegetated buffer strips, and poor construction techniques. It has been estimated that 50 percent of the trenches failed during construction due to the premature clogging of the stone void space. The following construction procedure should minimize any risk of premature sediment clogging.

Construction Procedure

1. Prior to any construction activity on the site, rope off the trench area to prevent heavy equipment from compacting the underlying soil

2. Before excavation begins, the entire area draining to the trench should be stabilized. If this is not possible, place a diversion berm around the perimeter of the trench to prevent sediment entry during construction.

3. Excavate the trench with a backhoe or a wheel or ladder type trencher (Hannon, 1980). Bulldozers or front-end loaders should not be used because the weight and the equipment blades can seal the soil surface. Place the excavated soil material 10 to 15 feet away from the trench to minimize migration of the soil back into the trench.

4. Inspect the bottom and side walls for removal of any protruding material such as tree roots that may puncture or tear the filter fabric. Contact the design engineer or supervisor if there is any evidence of impermeable clay soils, bedrock material, or groundwater seepage.

5. Following the trench excavation, line the bottom and sides of the trench with approved filter fabric to prevent the soil particles from migrating into the stone voids. The fabric should be laid with a sufficient length to cover the top of the stone aggregates. Overlap the filter fabric rolls by a minimum of 2 feet. An acceptable filter fabric will have an equivalent opening size that will prevent soil piping and a much higher permeability than the underlying soil. Table 2 gives a partial list of acceptable filter fabric brands (Prince Georges County, 1984). Because of some clogging experiences, a six inch deep layer of clean sand may be used on the bottom of the trench in replace of the filter fabric.

Table 2. Approved Filter Fabrics For Infiltration Trenches

AMOCO 4545

EXXON Geo-textiles No. 125D, 130D, & 150D

Mirafi 140-N

Supac 4NP, 5NP, & 8NP        Source: Prince George's County
                                     SWM Bulletin No. 4, 1986
Terratex SD

Typar 3401

NOTE: This is a partial list of acceptable filter fabrics for use in infiltration trenches in the Washington, D.C. area. The use or brand name does not constitute an endorsement for any particular product or company.

# INFILTRATION TRENCHES CONSTRUCTION

6. Install the observation well with a foot plate to prevent the pipe from sinking into the ground. The observation well should consist of a 4 to 6 inch PVC pipe with a secure but removable cap. The well will allow the inspector to determine if the trench is dewatering after storm events. A cross section of a observation well in a trench is shown in Figure 4.

7. Place 1-3 inch clean washed stone aggregates in the excavated trench with a backhoe or front-end loader in loose 12 inch lifts and lightly compact with plate compactors. Cover the stone with the filter fabric with at least a six inch overlap. Place at least one foot of stone or soil above the filter fabric. Other sizes of stone aggregates may be used if the void space has been accurately determined from laboratory test and is incorporated in the design. In the case of a stone surface, pea gravel may be substituted in replace of the top 1 foot of 1 to 3 inch stone for improved sediment filtering and cleanout. Trenches that have soil backfill with a vegetated surface require inlets to capture and transmit runoff to the stone reservoir. The inlets may simply consist of 2 to 3 foot diameter perforated pipes with top grates. Small perforated pipes connected to the inlets should be installed along the length and at the stone surface of the trench. To avoid premature clogging, use sod as the surface vegetation.

8. It is critical that the site area be totally stabilized with vegetation before the trench is put in service to capture runoff. Make sure the overflow from the trench is directed to a non-erosive outlet which leads to a stabilized watercourse.

## Conclusions

In Maryland, infiltration trenches are in widespread use to fulfill the intent of stormwater management of maintaining the predevelopment runoff characteristics as nearly as possible. To improve the success rate and prolong the lifespan of infiltration trenches, careful design and construction is required. Based on past experiences and the most recent field survey of infiltration practices, it is essential to prevent sediment entry from clogging the stone voids. Construct trenches only after the entire site is stabilized or divert sediment laden runoff away from the trench construction area. After construction, make sure that an effective vegetated buffer strip is provided to minimize the risk of sediment contamination. With careful design and construction, infiltration trenches can continue to operate as designed.

## References

Clement, P. and Pensyl K. (1987). Results of the State of Maryland Infiltration Practice Survey. Sediment and Stormwater Division. Maryland Department of the Environment. Annapolis, Maryland. 47 pp.

Hannon, J. B. (1980). Underground Disposal of Stormwater Runoff. California Dept. of Transportation and the Federal Highway Administration. U.S. Department of Transportation, Washington D.C. (FHWA-TS-80-218). 236pp.

Maryland Water Resources Administration (Md WRA, 1984). Standards and Specifications for Infiltration Practices. Maryland Department of Natural Resources. Annapolis, Maryland. 125pp.

Maryland Water Resources Administration (Md WRA, 1985). Inspector's Guidelines Manual for Stormwater Management Infiltration Practices. Maryland Department of Natural Resources. Annapolis, Maryland, 60pp.

Maryland Water Resources Administration (Md WRA, 1986). Minimum Water Quality Objectives and Planning Guidelines for Infiltration Practices. Maryland Department of Natural Resources. Annapolis, Maryland. 31pp.

Metropolitan Washington Council of Governments (MWCOG, 1987). Controlling Urban Runoff: A Practical Manual for Planning and Designing Urban BMP's. Department of Environmental Programs. Washington D.C. 268pp.

Montgomery County Dept. of Environmental Protection (MC DEP, 1984). Construction and Design Criteria for Water Quality Inlets. Stormwater Management Division. Rockville, MD. 5 pp.

Rawls, W. J., Brakensiek, D. L., and Saxton, K. E. (1982). Estimation of Soil Properties. Transaction of the American Society of Agriculture Engineers, Vol. 25, No. 5. pp 1316-1320.

## Discussion of Mr. Harrington's Paper

Question:

Given the failure rates shown for these infiltration trenches (33%), we don't accept those kinds of failure rates for other public works facilities--bridges, for example, do we?

Answer:

True; but these trenches work better than ponds.

Question:

How do we calculate the minimum allowable infiltration rate?

Answer:

On the basis of the bottom area only.

Question:

Are these being used now for water quality control, in addition to their earlier use for peak-flow shaving?

Answer:

Yes, more than ever.

Question:

Any quality results?

Answer:

None noted thus far. There are no quality-related design criteria other than the vegetative buffer strip, which is there to filter out solids.

Question:

Have you noticed any increased flows in sanitary sewers where these trenches are used?

Answer:

There are no such cases known to me.

Question:

In residential areas, do people park their cars on top of these trenches?

Answer:

I don't know.

Question:

Is there any particular concern at the state level about the 100-feet-from-a-well criterion?

Answer:

Yes, the state people take that seriously.

Question:

(Observation, really). These criteria for infiltration trenches are not completely new or untried. They are similar to old-septic tanks and leachfield criteria, for which there are at least 100 years of supporting data. One of the surest things we know about the systems is: They will eventually fail; they are to one degree or another temporary. On a second matter, in addition to a minimum infiltration rate criterion, there should perhaps be as well a maximum infiltration rate criterion. This relates in a way to, but may be more scientific thn, the 100-feet-from-a-well criterion, which is fairly arbitrary.

SWEDISH APPROACH TO INFILTRATION AND PERCOLATION DESIGN

Peter Stahre[1] and Ben Urbonas[2], M.ASCE

ABSTRACT

This paper describes design procedures of infiltration and percolations facilities, which in part were originally authored by Dr. Stahre for the Swedish Association of Water and Sewage Works (1983). Design of infiltration and percolation facilities is similar to detention facilities. The storage volume of an infiltration basin being the ponding volume on top of the infiltration surface and the volume of a percolation facility being the pore volume inside the underground percolation pit.

It has been observed that infiltration and percolation facilities work best for very small runoff basins such as individual lots. They have to be designed conservatively using low unit loading rates and need to "rest" between runoff events to let the pores in the receiving soils rejuvenate. When the loading rates are too high, pores tend to seal.

DESIGN FLOW

The volume of runoff reaching an infiltration or percolation facility depends on several factors. These include the tributary basin size, the degree of development in the basin (i.e., amount of impervious surface) and the characteristics of rainfall and snowmelt in the area. First, we will discuss the suggested techniques for estimating rainfall and snowmelt runoff for the design of infiltration and percolation facilities.

Stormwater Runoff

As infiltration and percolation facilities are mainly used

---

[1] Assistant Manager, City of Malmo Water and Sewage Works, Malmo, Sweden.

[2] Chief, Master Planning and South Platte River Programs, Urban Drainage and Flood Control District, Denver, Colorado.

for small runoff basins, runoff calculations can be based on the Rational Formula, namely:

$$Q = C*I_t*A \qquad (1)$$

in which, $Q$ = runoff rate for a T-year storm, in liters/second,
  $C$ = runoff coefficient, non-dimensional,
  $I_t$ = rainfall intensity for a T-year storm at a storm duration t, in liters/second/hectare,
  $A$ = area of the tributary watershed, in hectares.

By multiplying the average runoff rate, Q, by the design storm duration, t, we obtain the cumulative volume, namely

$$V = C*I_t*t*A*3600*10^{-3} \qquad (2)$$

in which, $V$ = total runoff at time t, in cubic meters,
  $t$ = storm duration in hours.

Thus, the runoff volume can be approximated using Intensity-Duration-Frequency curves. Given a design storm duration, the volume of rainfall (i.e., Block Rain = I*t) can be calculated. This is illustrated in Figure 1 for three storm durations.

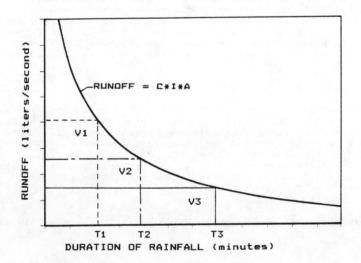

Figure 1. Derivation of Block Rain inflow hydrograph.

Block rain obtained using the I-D-F data represents the average intensity for the most intense parts of recorded rainstorms. The rain that falls before and after this period is not included in the I-D-F curve and consequently is not reflected in the block rain. Since the sizing of basin depends on runoff volume, it is necessary to somehow account for those parts of the rainstorm not included in the block rain. Based on studies by Sjoberg and Martensson (1982) it was recommended to increase block rainfall volume by 25 percent, which in Sweden improved the runoff predictions when using Rational Formula. As a result, it is suggested that the volume of runoff reaching an infiltration or a percolation facilities be estimated using Equation 3.

$$V = 1.25*C*I_t*t*A*3600*10^{-3} \tag{3}$$

Although the choice of the recurrence frequency and storm duration is often dictated by local criteria, it is recommended that the design for on-site disposal facilities be sized to handle no less than a 2-year storm. On the other hand, designing for stormwater quality, the design may be for smaller storm events and still be effective in removing most pollutants; however, sizing for smaller events will shorten the expected life of such facilities.

U.S. EPA suggests that water quality facilities be designed for the average rainstorm, which happens to be much smaller than the 2-year storm. In the Denver area for example, the average runoff producing storm has only 0.44 inches of rainfall, while the 2-year design storm has 1.15 inches of rainfall. Since runoff from these average rainstorms is relatively small, the infiltration and percolation surfaces designed to handle them will be loaded to capacity quite frequently.

Snow Melt

In some parts of the United States, snow melt can govern the sizing of the infiltration and percolation facilities. This is especially the case when the basins have a relatively small amount of impervious surface compared to the pervious surface. Under extreme conditions, actual snow melt intensities could reach as much as 0.15 inches of water per hour. However, typical runoff rates from melting snow appear to be less than these extremes. Unlike rainfall on unfrozen and unsaturated soils, melting snow will contribute runoff from the pervious surfaces. Although it is not possible to generalize how much runoff can be expected under the varying climatic conditions, it is suggested, where applicable, the design be checked using the following minimum snow melt rates:

MINIMUM SNOW MELT RATES FOR DESIGN

| Impervious Surfaces | | Pervious Surfaces | |
| --- | --- | --- | --- |
| English Units | Metric Units | English Units | Metric Units |
| 0.04 ft$^3$/s/ac | 2.8 l/s/ha | 0.02 ft$^3$/s/ac | 1.4 l/s/ha |

## SELECTING INFILTRATION BASIN SITES

There are several conditions that will rule out the site as an infiltration facility. If the following conditions are discovered or are likely, disposal of stormwater by infiltration is not recommended:

* Seasonal high groundwater is less than 2 feet below the infiltrating surface

* Bedrock or impervious soils are within 4 feet of the infiltrating surface

* The infiltrating surface is on top of fill

* The surface and underlaying soils are SCS Hydrologic Group D, or the saturated infiltration rate is less than 0.3 inches per hour.

### Sizing an Infiltration Basin

If the above conditions do not rule out the site, the authors suggest the site be then evaluated using the Swedish Association for Water and Sewer Works (1983) recommendations. This procedure is based on assigning points for various site conditions according to Table 1. When the sum of points for a site is less than 20 points, it is considered unsuitable for infiltration. On the other hand, a site with more than 30 points is considered to be an excellent candidate, while sites receiving 20 to 30 points are considered good candidates. Figure 2 summarizes this screening process. This evaluation system is intended for preliminary screening of potential sites and is not a substitute for sound engineering and site-specific investigation.

If the initial screening determines the site as acceptable, the needed infiltration surface area and storage volume must then be determined. Table 1 suggests that the infiltrating surface be no smaller than one-half of the impervious surfaces tributary to the infiltration basin. This establishes the lower limit. On the other hand, there is no defined upper limit.

Storage volume is a function of the infiltration rate which is difficult to accurately determine. It is not possible to generalize what infiltration rates should be used at various sites; however, Table 2 contains data illustrating the diversity even among similar soils. It is therefore recommended that infiltration tests be performed at each site and that the lowest measured rates be used for actual design.

TABLE 1. POINT SYSTEM FOR EVALUATING INFILTRATION SITES

1. Ration between tributary connected impervious area ($A_{IMP}$) and the infiltration area ($A_{INF}$):

    * $A_{INF} \geq 2\ A_{IMP}$      20 points
    * $A_{IMP} < A_{INF} < 2\ A_{IMP}$      10 points
    * $0.5\ A_{IMP} \leq A_{INF} \leq A_{IMP}$      5 points

    Pervious surfaces smaller than $0.5\ A_{IMP}$ should not be used for infiltration.

2. Nature of surface soil layer:

    * Course soils with low ratio of organic material      7 points
    * Normal humus soil      5 points
    * Fine grained soils with high ratio of organic material      0 points

3. Underlaying soils:

    * If the underlaying soils are courser than surface soil, assign the same number of points as for the surface soil layer assigned under item 2 above.

    * If the underlaying soils are finer grained than the surface soils, use the following points:

        * Gravel, sand of glacial till with gravel or sand      7 points
        * Silty sand or loam      5 points
        * Fine silt or clay      0 points

4. Slope (S) of the infiltration surface:

    * $S < 0.07$ ft/ft      5 points
    * $0.07 \leq S \leq 0.20$ ft/ft      3 points
    * $S > 0.20$ ft/ft      0 points

5. Vegetation cover:
   * Healthy natural vegetation cover:  5 points
   * Lawn is well established  3 points
   * Lawn is new  0 points
   * No vegetation, bare ground  -5 points
6. Degree of traffic on infiltration surface:
   * Little foot traffic  5 points
   * Average foot traffic (park, lawn)  3 points
   * Much foot traffic (playing fields)  0 points

TABLE 2. TYPICAL INFILTRATION RATES

| SCS Group and Type | Infiltration Rate (Inches per Hour) |
|---|---|
| A. Sand | 8.0 |
| A. Loamy Sand | 2.0 |
| B. Sandy Loam | 1.0 |
| B. Loam | 0.5 |
| C. Silt Loam | *0.25 |
| C. Sandy Clay Loam | 0.15 |
| D. Clay Loam and Silty Clay Loam | <0.09 |
| D. Clays | <0.05 |

* Minimum rate, soils with lesser rates should not be considered as candidates for infiltration facilities.

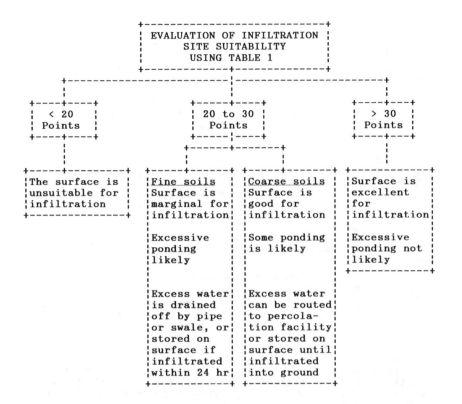

Figure 2. Evaluation of infiltration site suitability.

The storage volume can be calculated graphically using a known inflow envelope curve and infiltration rate. It is found by determining the largest difference between the inflow envelope function, V(t), and the cumulative infiltration function F(t) as illustrated in Figure 3. In this figure we see two distinct regions. When the storm duration is less than $t_b$, the runoff arrives at the site faster than it can be infiltrated into the ground. When the duration exceeds $t_b$, infiltration capacity is greater than the inflow rate and the stored water drains out through infiltration. It is possible to increase the infiltration area so that the runoff envelope curve V(t) never exceeds the cumulative infiltration line F(t) and ponding does not occur. The designer needs to vary the infiltration surface area until satisfactory results are obtained.

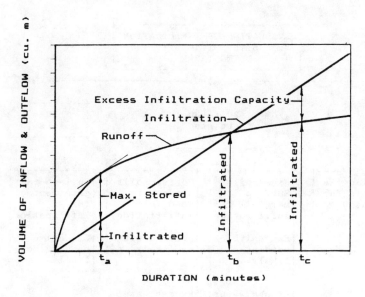

Figure 3. Sizing storage volume using runoff envelope.

PERCOLATION BASINS

The factors affecting the suitability of a site as a percolation facility are similar to those listed for infiltration facilities. If the following conditions are discovered or are likely at the site, disposal of stormwater by percolation is not recommended:

* Seasonal high groundwater is less than 2 feet below the bottom of percolation bed.

* Bedrock is within 4 feet of the bottom of the percolation bed.

* Percolation bed is located in fill.

* The adjacent and underlying soils are classified by SCS as Hydrologic Group C or D, or the field saturated hydraulic conductivity is less than $2 \times 10^{-5}$ m/s.

If the above conditions do not rule out the site, the Swedish Association for Water and Sewer Works (1983) provide sound recommendations for design. The procedure is based on metric units.

# INFILTRATION AND PERCOLATION DESIGN

## Darcy's Law

The rate at which water percolates into the ground can be estimated with the aid of Darcy's Law, namely:

$$U = k*I \qquad (4)$$

in which, $U$ = flow velocity in meters per second
$k$ = hydraulic conductivity in meters per second
$I$ = hydraulic gradient in meters per meter

In theory, Darcy's Law applies to groundwater flow in saturated soils; however, it is safe to assume that the soils will be saturated when the facility is operating. Also, because the bottom of the percolation field is above normal groundwater, it is possible to assume that the hydraulic gradient $I = 1.0$.

Assuming infiltration facilities, it is not possible to generalize what hydraulic conductivity should be used for the design of percolation facilities. Ranges in hydraulic conductivity are listed in Table 3 for various types of soils. As can be seen, the conductivity can vary by as much as four orders of magnitude for some soils groups, further reinforcing the need for site-specific testing.

When performing field hydraulic conductivity tests, use the lowest conductivity for design. In addition, remember that the soils will clog with time. It is recommended that the field conductivity values be reduced by a safety factor of 2 or 3 for design. Since a failed percolation basin is very expensive to rebuild it, the designer needs to be conservative and not leave behind a system that, after several years, will become a burden on the community.

## Effective Porosity of Percolation Media

The effective porosity of the fill media in a percolation pit or trench tells us how much of the total basin volume is available for the storage of water. The porosity of the media depends on the type of material used and Table 4 lists some of the representative values for some of the common used materials.

## Effective Percolating Area

The Swedish guidelines recommended that the bottom of a percolation pit or trench be considered as impervious. It has been observed that the bottom becomes clogged in time by the accumulation of sediments. This recommendation anticipates this and assumes that all water percolates into the ground only through the basin sides.

TABLE 3. HYDRAULIC CONDUCTIVITY OF SEVERAL SOIL TYPES.

| Soil Type | Hydraulic Conductivity (meters per second) |
|---|---|
| Gravel | $10^{-3} - 10^{-1}$ |
| Sand | $10^{-5} - 10^{-2}$ |
| Silt | $10^{-9} - 10^{-5}$ |
| Clay (saturated) | $< 10^{-9}$ |
| Till | $10^{-10} - 10^{-6}$ |


TABLE 4. EFFECTIVE POROSITY OF TYPICAL STONE MATERIALS.

| Material | Effective Porosity (percent) |
|---|---|
| Blasted rock | 30 |
| Uniform sized gravel | 40 |
| Graded gravel (3/4-inch minus) | 30 |
| Sand | 25 |
| Pit run gravel | 15 - 25 |

## Calculating Storage Volume

The water depth in the basin will vary during the filling and emptying process. Since the sides of a percolation trench will not be inundated fully during the entire storm, it is suggested that only one-half of the trench depth be used as the effective area for infiltrating rainstorms (i.e., $A_{perc}/2$). With a hydraulic gradient equal to 1.0, Darcy's Law gives the following expression for the outflow from the basin:

$$V_{out}(t) = k*1.0*(A_{perc}/2)*t*3600 \qquad (5)$$

in which, $V_{out}(t)$ = Volume of water percolated into the ground, in cubic meters
$\quad$ k $\quad$ = Hydraulic conductivity of soil, in meters per second
$\quad$ $A_{perc}$ = Total area of the sides of the percolation facility, in square meters
$\quad$ t $\quad$ = Percolation time, in hours.

The maximum volume of water stored, V, in the facility is the difference between $V_{in}(t)$ and $V_{out}(t)$, which can be expressed as follows:

$$V = \max [\ V_{in}(t) - V_{out}(t)] \qquad (6)$$

after substituting Equations 3 and 5 into Equation 6, it becomes,

$$V = \max [t*3600*10^{-3}*I_t*C*A*1.25 - k*(A_{perc}/2)*t*3600] \qquad (7)$$

in which, $I_t$ = rainfall intensity at time t, in liters/second/hectare.

The task now is to design a percolation basin with the least amount of storage volume sufficient to capture the runoff from the design storm. Dividing both sides of the above equation by C*A, one gets,

$$V/(C*A) = \max \{1.25*3.6*t*I_t - [A_{perc}/(C*A)]*(k/2)*3600*t\}$$

Making the following substitutions in the above expression,

$$D = V/C*A$$

$$E = (A_{perc}/2)*k*[1000/(C*A)]$$

one gets the following:

$$D = \max [\ 4.5*I_t*t - E*t*3.6\ ]$$

The parameter D represents the specific percolation storage volume, namely the storage volume expressed in cubic meters per hectare of connected impervious area. The parameter E represents the specific outflow from the basin expressed as liters per second per hectare of connected impervious area. As a first step an envelope curve, similar to Figure 3, of the specific inflow, $V_{in}$, is constructed. This is done by plotting ($4.5*I_t*t$) vs. t, where the values of $I_t$ are taken from the intensity-duration-frequency (I-D-F) curves for the site.

Various specific outflow rates, $V_{out}$, are superimposed on the same diagram as straight lines and the maximum specific storage volumes, $D_{max}$, for each specific outflow rate is obtained graphically as illustrated in Figure 4. These maximum values of D are then plotted against the values of E, giving us a diagram similar to Figure 5. The optimizing of a percolation facility size can now be performed using a trial and error procedure.

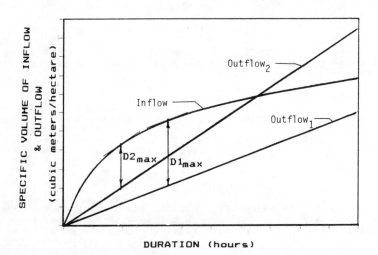

Figure 4. Graphic calculation of specific storage D.

EXAMPLES OF EVALUATION AND DESIGN

The simplest way to explain the above procedure is through examples. The first example illustrates how an infiltration site is evaluated for adequacy. The second example shows how a percolation facility is sized.

# INFILTRATION AND PERCOLATION DESIGN

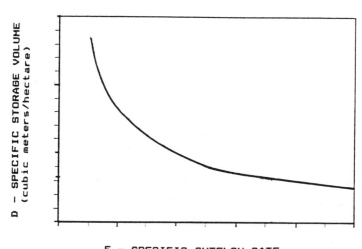

E − SPECIFIC OUTFLOW RATE
(liters/second/hectare)

Figure 5. Specific storage volume D as a function of E.

Example 1: Evaluation of an Infiltration Site

An infiltration site is to dispose runoff from with a tributary roof area of 100 square meters. It is a lawn with a surface area of 210 square meters and a 1.0 percent slope draining away from the house. The topsoil and the underlaying soils are composed mostly of course silt. Determine if the lawn is a good candidate as an infiltration facility.

Using the point system presented in Table 1, the results are as follows:

1. The ratio between the impervious surface and the infiltration area results in $A_{inf} = 2.1 * A_{imp}$

   This gives the site 20 points.

2. The top soil is course silt.

   This gives the site 5 points.

3. The underlaying soil is course silt.

   This gives the site 5 points.

4. The slope of the infiltration surface is < 0.07 ft/ft.

   This gives the site 5 points.

5. The infiltration surface is a new established lawn.

   This gives the site 0 points.

6. It is expected the lawn will have normal foot traffic.

   This gives the site 3 points.

7. The total is 38 points for this site. According to Figure 2, the site can be used for infiltration and excess runoff from this site is not likely.

## Example 2: Sizing of a Percolation Facility

A runoff from a roof with an area of 800 square meters and a lawn of another 800 square meters is directed to a percolation basin. Given the following, find the length of the percolation trench:

Tributary area: $A_{imp}$ = 800 m²; $A_{perv}$ = 800 m².

Percolation facility: Height of stone filling = 0.8 m;

Width of trench = 1.0 m; Porosity of stone filling = 0.4;

Minimum depth to seasonal groundwater = 2 m.

Hydraulic conductivity of soil = $4*10^{-5}$ m/s.

Design rainfall: 2-year storm.

The calculations are performed as follows:

1. Construct a design curve, Figure 6, using the I-D-F curve for the site.

2. Hydraulic conductivity of soils is $4*10^{-5}$ m/s. To compensate for possible non-homogeneity of the soils, a safety factor of two is used and the hydraulic conductivity for this design is equal to $2*10^{-5}$ m/s.

3. As a first try, assume that the length of the trench is 35 meters. The value for E is calculated as follows:

$$E = (A_{perc}/2)*k*[1000/(C*A)]$$

$$E = [2*(35+1.0)*0.8/2]*(2*10^{-5})*(1000/0.08)$$
$$= 7.2 \text{ liters per second hectare.}$$

4. Using Figure 6, D = 165 cubic meters per hectare.

5. Multiplying D by the tributary impervious area of 0.08 hectares, we get a storage volume of 13.2 cubic meters.

6. Since the effective porosity of the stone media is 0.4, the trench volume is $V_{trench} = (13.2/0.4) = 33$ m³.

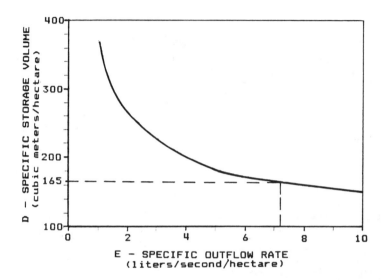

Figure 6. Design curve using a 2-year design storm I-D-F.

Using the trench length assumed in step 3, the total volume of the trench is 35*1*0.8 = 28 cubic meters. Comparing this to the 33 cubic meters calculated in step 6, we see that the assumed trench was too short. Calculations are repeated assuming 40 meters as the new trench length. We find,

Calculated: $V_{trench}$ = 30 cubic meters.

Assumed: $V_{trench}$ = 32 cubic meters.

The second assumption resulted in a trench that is longer than needed. Since the oversizing is less than 10 percent, the 40 meter length is accepted for final design.

As a final step, the design is tested to see if it is adequate under snow melt conditions. Using the recommended design snow melt rates, check to see if:

$$(A_{perc}*k*I*1000) > (S_{perv}*A_{perv} + S_{imperv}*A_{imperv})$$

We see that,

$$A_{perc}*k*1*1000 = [2*0.8*(40+1)]*2*10^{-5}*1000 = 1.312$$

and

$$S_{perv}*A_{perv} + S_{imperv}*A_{imperv} = 1.4*0.08 + 2.8*0.08 = 0.336$$

Since 1.312 > 0.336, the snowmelt condition does not govern and the original design is considered adequate. Note that in checking for snow melt, we used the full trench depth as the effective area and not one-half of it as was done for rainfall runoff. This is justified by the protracted and steady state nature of snow melt runoff. If it is intense enough, it will fill a trench and keep it inundated for extended period of time as new snow melt keeps feeding it.

## REFERENCES

Swedish Association of Water and Sewage Works, Local Disposal of Storm Water, Publication VAV P46, 1983. (In Swedish)

Sjoberg, A., Martensson, N., "Analysis of the Envelope Method for Dimensioning Percolation Facilities," Chalmers University of Technology, 1982. (In Swedish)

## Discussion of Mr. Stahre's Paper

Question:

Are you using your factor of 1.25 for all storms?

Answer:

Yes. (Comment from the floor: That may be an overestimation for very large storms).

Institutional Aspects of Stormwater
Quality Planning

Nancy U. Schultz, P.E. and Ronald L. Wycoff, P.E.*

ABSTRACT

Planning and implementation of programs to control the quality of stormwater runoff in urban areas are carried out within institutional frameworks that directly influence the effectiveness of the programs. Case studies in six metropolitan areas across the United States point out the differing motivations, responsible agencies, and control types that can be encountered in water quality planning for urban runoff and the way the framework within which that planning occurs can enhance or detract from successful implementation.

INTRODUCTION

Urban stormwater runoff (SWR), including combined sewer overflow (CSO), is a significant source of water pollution that must be considered in any comprehensive urban water quality improvement program. Identification of the water quality impacts of SWR and planning for its control involve a complex interplay of technical and site-specific local issues. Among the most significant is the institutional structure within which the planning and implementation must take place.

Water pollution control is clearly a function of government. However, in the United States, comprehensive water quality planning and subsequent urban SWR control usually involves multiple levels of government (local, state, and federal) and frequently the courts. The roles of authority and responsibility are often difficult to establish and the question of "who pays?" is likely to determine how much, if any, action is actually taken.

Six case studies of SWR quality control programs were selected to represent a variety of institutional approaches to the problem faced by urban areas. The study locations are Chicago, Seattle, Milwaukee, St. Louis, Santa Clara Valley in California, and Boston. The points of the case study histories pertinent to an understanding of the importance of institutional aspects of SWR

---
*Water Resources Engineers, CH2M HILL, Inc., Milwaukee, Wisconsin and Gainesville, Florida, respectively.

programs are the motivation for initiating the program, the agencies made responsible for the program and the controls implemented--or not--through the program.

## CASE STUDIES

### Chicago

The motivation for Chicago's first major SWR quality management project was compelling; the late 1800's had seen repeated outbreaks of cholera attributed to contamination of the City's water supply by wastes carried in the rivers and drainageways. The City constructed the Chicago Sanitary and Ship Canal to divert the Chicago River drainage basin from its historic outflow into Lake Michigan toward the Mississippi River Basin through the Des Plaines and Illinois rivers (Figure 1).

By the 1960's, the Metropolitan Sewerage District of Greater Chicago (MSDGC) had been granted authority and responsibility to maintain the sewerage and drainageways. Chicago had long since completed a program to intercept dry weather sanitary sewage and divert it to secondary treatment plants, and MSDGC was diligently pursuing a policy of prohibiting discharge of any wastewater into Lake Michigan. By this time, however, the combined sewage/stormwater system was severely under capacity, evidenced in frequent and widespread "basement backups" with combined sewage temporarily "stored" in residences and commercial buildings during intense rainstorms. The resulting complaints and obvious health hazards motivated pursuit of a second major SWR project, the Tunnel and Reservoir Plan (TARP).

When federal funding for water quality improvement projects became available through the 1972 Clean Water Act, MSDGC sought funding for construction of TARP. The U.S. Environmental Protection agency (USEPA) concluded that the tunnel portion of TARP would sufficiently reduce SWR loadings and approved grants for TARP Phase I. Construction began in the mid-1970s and is now nearly finished. Upon completion, SWR from all but the largest storms should be captured, stored, and treated, reducing stormwater discharges from 50 to 100 occurrences per year to about 10.

The U.S. Army Corps of Engineers may fund TARP Phase II, construction of additional tunnels and a large storage reservoir, under an urban flood control program. Although the funding is motivated by the project's ability to reduce flooding, it would result in virtually all stormwater being captured, stored, and subjected to at least secondary treatment before release to the waterways.

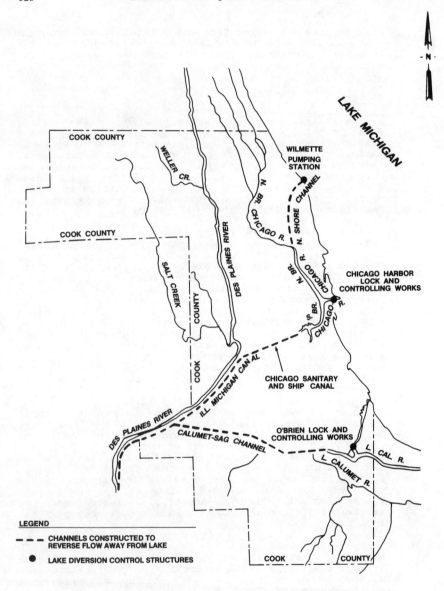

**FIGURE 1.**
Chicago Diversions.

## Seattle

Seattle was spurred into addressing SWR planning by well-founded local concerns that CSOs were restricting recreational uses and aesthetic enjoyment of the freshwater lakes and streams in the area. The City of Belleview, a suburb adjacent to Lake Washington, further recognized that separated stormwater from its area was contributing to water quality problems in the lake.

The regional utilities agency, METRO, initially addressed the problems by installing a realtime computer-aided telemetry and control network (CATAD) to use inline storage capacity in the interceptors and selectively route overflows to more tolerant saltwater systems (Figure 2). The City of Seattle undertook a program to separate or provide offline storage to prevent overflows into the freshwater systems. The City of Bellevue formed a Drainage Utility to raise funds, develop, implement, and manage a joint stormwater quantity/quality management plan.

By 1985, programs judged sufficient to meet the area's water quality objectives were being implemented by the various agencies. However, the State of Washington concluded that "the greatest reasonable" reductions in discharges of sewage contaminated overflows, even to saltwater systems, should be implemented, in part to alleviate toxics problems. METRO has ranked a series of relief, separation, and offline storage projects on the basis of largest reduction in overflow volume per dollar expended and will implement these projects as funds are available. The City of Seattle is similarly continuing pursuit of separation and offline storage projects. Belleview is aggressively pursuing quality conscious drainage development.

## Milwaukee

Milwaukee was first motivated to address SWR when the USEPA funded construction of the Humboldt Avenue detention tank, an offline storage device, as a demonstration project. The City was further compelled when the State of Illinois sued in federal court to require Milwaukee to curtail all discharges of untreated sewage into Lake Michigan and its tributaries. At the same time, the Wisconsin Department of Natural Resources (WDNR) sued in the Dane County (Wisconsin) court to require the Milwaukee Metropolitan Sewerage District (MMSD) to aggressively pursue improvement of their wastewater management facilities.

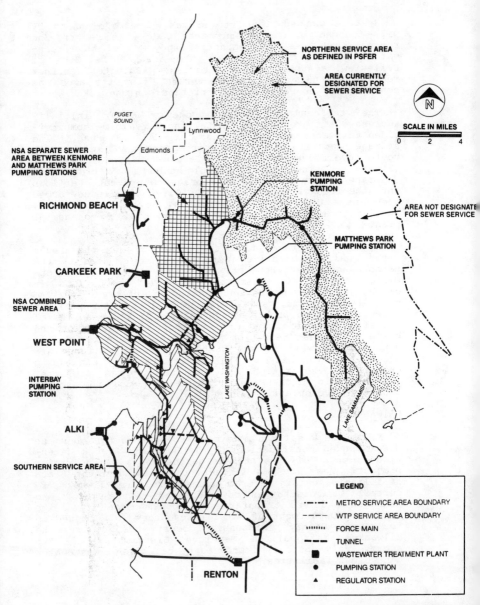

**FIGURE 2.**
Seattle Metro Collection System and Receiving Waters.

The U.S. Supreme Court eventually concluded that the District Court had no standing to require a greater level of pollution control than that mandated by federal law. The MMSD implemented a Water Pollution Abatement Program to meet the stipulated agreement with WDNR to eliminate all separate sewage overflows and reduce CSOs to the extent required to achieve receiving water quality standards.

Design of the components of the Water Pollution Abatement Program is nearly complete, and construction of the recommended facilities is about half done. Separate sewage overflows have been reduced through a combination of near surface relief sewers, 13 percent average reduction in the amount of stormwater (inflow) entering the sanitary sewers, realtime control to maximize use of existing sewer capacity, and construction of tunneled, oversize relief sewers to provide inline storage during extreme wet weather (Figure 3). CSOs are being reduced primarily by providing facilities to use the tunneled inline sewers for combined sewage storage during periods when the full volume is not required for sanitary sewage storage.

Construction of the separate sewage facilities is being funded with USEPA Clean Water grants, Wisconsin grants, and local bond issues and increased sewerage fees. Construction of the combined sewage facilities is funded through a special Wisconsin Fund, local bond issues, and increased sewerage fees. Planning, design, and construction have proceeded under the stipulated schedule despite the fact the local municipalities and the MMSD have not reached agreement on apportionment of the local share of the costs.

The designated areawide water quality management agency, the Southeastern Wisconsin Regional Planning Commission, has determined that control of sanitary and combined sewers alone will be insufficient to achieve receiving water quality standards. They have recommended both urban and rural non-point source controls to achieve the goals. Several of the affected watersheds have been given high priority control in the Wisconsin non-point source control program and are receiving limited matching funds from the state for construction of control facilities. To date, these funds have been applied primarily for studies and a limited number of agricultural non-point source control installations.

St. Louis

The St. Louis Metropolitan Sewerage District (MSD) first addressed SWR quality to comply with federal grant

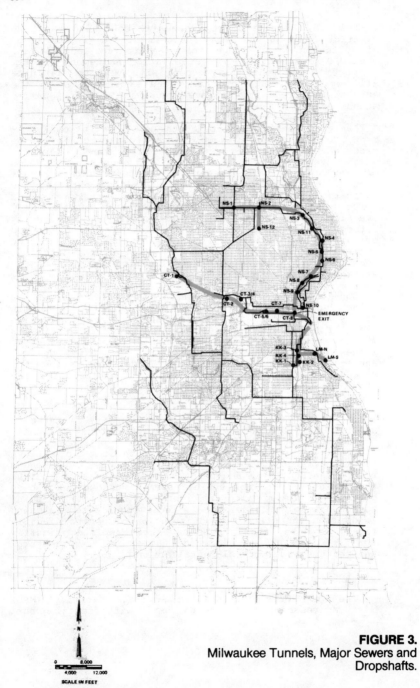

**FIGURE 3.**
Milwaukee Tunnels, Major Sewers and Dropshafts.

requirements for facility planning to upgrade the Bissel Point and Lemay wastewater treatment plants.

The Bissel Point facility plan recommended extensive rehabilitation and relief of the combined sewers along the Mississippi River. The Lemay facility plan concluded that pollutant loads from non-point sources and separate storm sewers far exceeded those from combined sewers, and that reduction of CSOs would not significantly improve receiving water quality of the River Des Peres. The plan recommended a limited program of combined sewer rehabilitation in areas experiencing flooding or where intercepting structures or flap gates required repair.

The St. Louis MSD has responsibility for both sewage and stormwater drainage in the service area. To date, it has not initiated evaluations of the feasibility of controlling pollutant loads from non-point or separate stormwater facilities. The St. Louis MSD has begun implementation of the combined sewer rehabilitation projects. Funding was to be provided through local sewer revenues and Federal grants. However, a recent court decision that the MSD has no authority to charge fees in areas not immediately benefited by the projects being funded has jeopardized funding for combined sewer projects.

## Santa Clara Valley

The southern portion of San Francisco Bay, the receiving water for drainage from the Santa Clara Valley, has had water quality limitations on recreational use and fishery since the area was developed late in the 19th century. Although wastewater treatment plants now provide tertiary treatment and sanitary sewers have no known overflows, the South Bay still does not achieve desired water quality standards.

In 1986, the California Regional Water Quality Control Board developed a basin plan for San Francisco Bay that required development of an action plan for evaluating and abating non-point source pollution affecting the South Bay. A task force of representatives from the Santa Clara Valley Water District and the affected municipalities was formed to pursue the non-point source aspects of the basin plan (see Figure 4).

As a first step, the task force commissioned a study to develop an action plan that met the requirements of the Regional Water Quality Control Board and defined a step-by-step, technically sound procedure to quantify the non-point source loads, identify sources, and select the

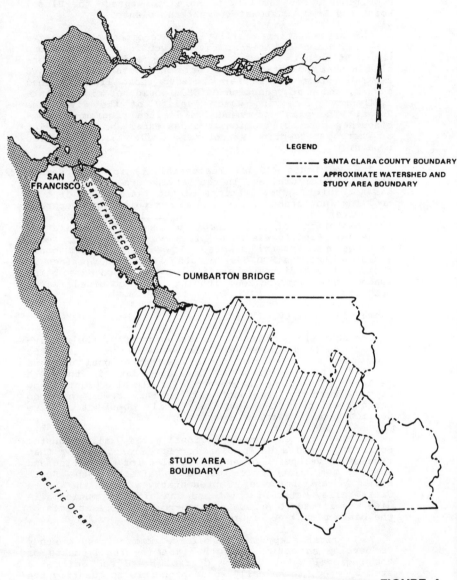

**FIGURE 4.**
Santa Clara Valley Non-Point Source Project Location.

most cost-effective means of achieving an as yet undefined desired level of non-point source control. The resultant action plan was adopted by the Board in 1987 and is currently being implemented.

The recent inception of this program makes it uncertain which, if any, non-point source controls will be implemented. However, the case study is informative for the following two characteristics:

1. It is perhaps the first such study undertaken as a result of full recognition that even the most stringent point source controls would be insufficient to achieve water quality standards.

2. The implementing agency is a cooperative task force of several entities potentially affected by non-point source controls rather than a single responsible agency funded through Clean Water grants.

Boston

CSO was recognized as a source of Boston Harbor pollution more than 50 years ago. In 1936, a comprehensive early study of water quality conditions in Boston Harbor and its tributary streams recommended treatment at the main interceptor outlets, as well as adequate control of stormwater overflows that prevent recreational water use along the waterfront of Boston Harbor and its estuaries and tributaries.

A proposed solution to Boston Harbor CSO pollution presented in 1967 involved the storage of CSO in deep tunnels. The plan was designed to meet significant flood control goals and therefore had extremely large components. In 1972, the U.S. Senate and House of Representatives passed resolutions requesting the U.S. Army Corps of Engineers to undertake a joint study with the Commonwealth of Massachusetts to recommend wastewater management improvements and alternatives for the Boston metropolitan area. The resulting study, published in 1976, considered three general approaches to CSO control: sewer separation, deep tunnel storage/treatment, and decentralized storage/treatment. Sewer separation was eliminated because of the high costs and the impractical nature of such construction in the central city. The deep tunnel approach, based on the 1967 plan, was also eliminated based on the high estimated construction cost. The recommended decentralized plan included an array of transport, storage, and treatment facilities located throughout the combined sewer service area. It had the lowest estimated cost of the three approaches and allowed for staged implementation with immediate opportunities for solving high priority problems.

Between 1978 and 1982, detailed CSO facilities planning was performed for Boston Harbor based on the recommended decentralized approach to CSO control. Several projects are in various stages of planning, design, or construction, or have been completed and are in operation. For the most part, however, the 1982 plan is largely unimplemented.

In response to federal court actions and proposed NPDES permit requirements, the Massachusetts Water Resources Authority (MWRA) initiated in April 1986 a review of past planning for CSO control in Boston Harbor. Implementation of many of the 1982 CSO Control Facilities Plan projects is questionable, either because the proposed site is no longer available or because the control technology proposed is unproven at the scale required. In addition, recent emphasis on both public and private waterfront improvements in the Boston area also negatively affects program implementation as proposed in the 1982 plan, with reduced construction site availability for CSO treatment facilities located at or near the overflow points and increased emphasis on the aesthetic characteristics of the receiving waters and waterfront property.

Based on the results of the planning review, the MWRA has recently undertaken a comprehensive update and expansion of the CSO facilities plan, subject to a Court-ordered time schedule. Figure 5 delineates the limits of the current planning area. The new CSO facilities plan will consider deep tunnel storage designed for pollution control only, along with an update of the 1982 decentralized storage/ treatment approach. Comprehensive receiving water quality modeling considering both separate SWR and CSO is included, as is development of short-term dry- weather overflow and wet-weather best management practices programs.

Since this facilities planning effort has just begun, the results and implementation are unknown. However, with an areawide agency such as the MWRA as the lead, and the federal court setting the overall Boston Harbor cleanup schedule, the probability of implementation of significant CSO controls is greater now that at any time in the past.

SUMMARY

Table 1 summarizes the motivation, responsibilities, implementation, and funding associated with each of the previously discussed case studies. A discussion of each

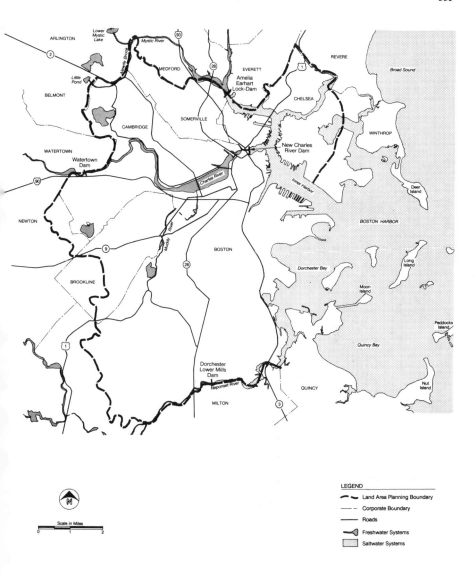

**FIGURE 5.**
Boston Harbor CSO Planning Area.

Table 1
SUMMARY CASE STUDIES

| Case Study | Motivation | Responsibilities | Implementation | Funding |
|---|---|---|---|---|
| Chicago | Health Hazards | Single Agency (MSDGC) | Collect, store, and treat (Tunnel and Reservoir Plan) | USEPA USCOE Local |
| Seattle | Aesthetics/Recreation | Two agencies: METRO City of Seattle | Inline Storage Offline Storage Sewer Separation | Local USEPA |
| Milwaukee | Court Orders | MMSD - combined sewage ??? - non-point source | Inline storage and treatment ??? | Local USEPA State |
| St. Louis | Facility Plan Requirements | Single Agency St. Louis MSD | Rehabilitation and Repair recommended | Uncertain |
| Santa Clara Valley | Non-attainment of water quality objectives | Multi-agency: SCVWD Cities County | Undetermined | Studies - Local Implementation - ? |
| Boston | Court Orders | Single Agency (MWRA) | Undetermined | Studies - Local and State |

of these factors as it affects the institutional character of a SWR program is provided below.

Motivation

The most effective motivation for implementing SWR quality control planning appears to be a local recognition or perception of a health hazard or water use impairment associated with SWR. Chicago and Seattle both proceeded with those SWR controls that the local community had determined would have positive impact on a desired receiving water use. The Santa Clara Valley plan, and to some extent the Milwaukee non-point source program, offer some evidence that demonstration that further consideration will be given to SWR and non-point source control when it is shown that point source controls alone are inadequate to protect water quality. Milwaukee and Boston demonstrate that, in the absence of adequate local motivation, court orders can achieve progress toward SWR control.

Responsibilities

The case studies demonstrate that a wide variety of institutional responsibility matrices can be effective, or ineffective, in pursuing SWR quality planning. State regulatory agencies and the USEPA must be actively involved in the planning and decision process. However, the key to success appears to be an areawide agency or umbrella organization, such as a wastewater or stormwater utility, which has responsibility, and authority across local geopolitical boundaries. It is also important for these agencies to have independent and continuous sources of funding. These funds provide the responsible agency the resources to develop the background data and studies necessary to understand SWR and CSO pollution and the impact on area receiving waters. This understanding is required as a prerequisite to rational pollution control planning.

Implementation

Of the case studies reviewed, the only SWR controls being aggressively implemented are CSO storage/treatment programs using pre-existing treatment facilities. Repair and rehabilitation of decaying combined sewers and appurtenances appears to be a palatable first step for communities initially addressing SWR quality.

It is regrettable that only the Belleview portion of the Seattle case study involves implementation of measures addressing separate stormwater facilities. This reflects the perceived relative water quality significance of combined sewage versus stormwater in urban areas. Most communities will concur that CSOs are undesirable, but few outside the technical profession acknowledge that urban stormwater can also cause adverse receiving water impacts.

Funding Source

Since most of the programs reviewed received federal, state, and local funds, it is difficult to attach any great significance to the funding sources. Although it is frequently assumed that the availability of federal and state funds encourages implementation of water quality improvement programs, available non-local funding alone does not provide sufficient motivation for SWR quality project implementation. Five of the six case studies had access to federal grant programs, yet only those motivated by local perceptions of water quality problems and/or court orders are proceeding. Despite state matching funds for non-point source controls in the Milwaukee watersheds, few such controls have been implemented. Conversely, Santa Clara Valley has initiated over $1,000,000 in planning work with local funding only. Without a local funding source, little progress will be made even if state and/or federal funds are available.

## Discussion of Ms. Schultz's and Mr. Shaver's Papers

Question:

Nancy, Earl, where should institutions be headed in this urban runoff field?

Answer (Mr. Shaver):

Start! Whether there is an EPA or not, get started. Everybody starts too late.

Answer (Ms. Schultz):

Educate people that there is a problem. Lots of our clients (public works agencies) don't seriously believe urban runoff is a problem (yet). How many public works department or consulting engineers care about urban runoff as a quality problem?

Comment:

No matter where the regulatory, institutinal motivation comes from (EPA or the state in some instances), actual implementation and the perceptions of value that prod implementation must eventually take place at the local level. In many places, there does not yet seem to be sufficient or very much grassroots, local support for runoff quality controls.

Comment:

Nancy's (and Ron Wycoff's) data **demonstrate** that local motivation is the key to runoff controls implementation.

# INSTITUTIONAL STORMWATER MANAGEMENT ISSUES

H. Earl Shaver*

Research, innovation, and structure performance are necessary components of an overall stormwater management program, but program success or failure rests on its institutional framework. Program elements such as problem definition, enabling legislation, regulations, and design criteria must clearly provide guidance and leadership to all impacted segments of the affected jurisdiction.

In addition to those clearly stated objectives, other issues such as manpower, training, program funding, and structure maintenance must be addressed. Only If each of these elements receives adequate support can a successful stormwater management program be implemented.

## Introduction

Research and monitoring to develop new stormwater practices or evaluate existing practices is probably the most exciting aspect of stormwater management. A programs' success or failure, though, is determined by its institutional framework and the support that the program receives from a monetary and manpower standpoint. To often, legislation is enacted to address stormwater problems and attempts to simplify a complex problem. Stormwater management control, from a quality or quantity standpoint is not an exact science and will require constant attention to be successful. Effective program implementation can be considered as a chain with all program elements acting as links in the chain. A weakness in any program element will seriously weaken the entire structure of the program. If a jurisdiction adopts a stormwater management program, it must understand the commitment that implementation entails.

---

*Department of the Environment, Stormwater Management Administration, Tawes State Office Building, Annapolis, Maryland 21401

## Problem Identification

One mistake that legislation will frequently make regarding stormwater management is that it can be too general or attempt to address too many problems at one time. The problem being addressed by the legislation must be clearly defined so that all impacted parties have a clear direction. If flooding is a problem, that should be stated. If water quality protection is the driving force for the legislation, then that should be clearly stated.

Habitat protection is another area which should directly be addressed. Stormwater runoff has significant environmental impact to receiving waters and guidance needs to be provided to determine the level of protection attempted. This issue will determine to a large extent whether a regional approach to stormwater management is possible.

The problems being addressed will determine the types of pactices which can be utilized. Flood control will, of necessity, require fairly large regional structures to provide measurable benefits. Water quality protection will require a variety of practices including extended detention, wetland creation, deeper water ponds, and the use of infiltration practices. Environmental protection may be the most difficult to provide due to the variety of impacts associated with urban development. The cumulative effects of channel instability, sediments, soluble and suspended pollutants, and lowered groundwater levels all contribute to species decline. Cold water fisheries such as trout may be seriously impacted by elevated temperatures of water released by a shallow pond.

A clear definition of the problem will provide a stable foundation upon which the implemented program can develop and expand.

## Enabling Legislation

As previously mentioned the legislation must clearly state the program objectives, but there are other issues which must be included for program success. The enabling legislation must provide for the promulgation of regulations by the reponsible local agency. These regulations will then provide guidance on specific design, construction, and maintenance.

When the area regulated has differing geomorpology or developmental pressures, the legislation must provide flexibility for the regulating agency to address their unique situations. The legislation may require watershed studies to identify specific problems and strategies for each watershed. It is imperative that funding by provided for watershed studies if that is a component of the overall program. A watershed wide approach would determine where on-site and off-site management will be required, and the soils information will determine the specific practices that should be employed.

The legislation must clearly state who needs to provide stormwater controls. Are there types of activities where stormwater controls would be an undue hardship? Agricultural activities are probably the most controversial when placing additional requirements but governmental projects also must be considered. Federal compliance may be difficult to obtain, and State and local governmental projects will require additional funding to meet local requirements.

Regardless of who is regulated by the program, that entity must be required to obtain a formal stormwater approval prior to initiation of construction. A convenient mechanism to ensure compliance would be to require stormwater approval prior to issuance of any building or grading permit. Also important in the permitting process is the requirement of a review fee. That fee should be structured to make the applicant support the plan review and inspection staff. It could be based on the limits of site disturbance or on the amount of impervious area created to be equitable.

Governmental agencies present unique problems in implementing a stormwater program. Federal agencies may refuse to recognize local permitting requirements, or that they contribute to an overall stormwater problem. Highway projects are more difficult to provide environmental control then any other type of construction. The planning and design of a highway project may take over seven years to complete and new stormwater criteria imposed at the end of the design phase will be difficult to comply with, especially if additional land must be acquired for the stormwater control structure. Also, highway projects are linear in nature and cross numerous watershed boundaries. They will require significant cuts and fills which makes a management strategy very difficult. Phased plans may be necessary on highway projects to provide for environmental protection.

Legislation must also provide enforcement penalties for non-compliance with stormwater requirements. These penalties should be extreme enough to act as a disincentive to individuals intent on circumventing the law. A civil citation section would povide an effective tool allowing inspectors to write a "ticket" at a non-conforming site and assessing an administrative penalty. Penalties will also require close support from a legal staff, who must be aggressive in their support of the stormwater program.

### Regulations

Regulations must clearly specify the necessary program elements and design requirements. They are specific in nature as opposed to the legislation which is more general. They are a legal document and subject to public scrutiny. The regulations must define the functions of the review agencies and provide a framework of minimum requirements which can be exceeded if a local component desires to be more restrictive.

Design criteria is the driving force behind the regulations. The design criteria clearly presents the strategy necessary to achieve program goals. An option to design criteria may be the establishment of performance criteria but most design consultants will need assistance in the design strategy necessary to meet the performance criteria. If that type of assistance must be provided then specific design criteria may be the prefered choice.

The strategy of control will depend on whether water quality or water quantity is the main program emphasis. Some level of water quantity control can be provided by a water quality control strategy, but water quality protection in a water quantity control program does not necessarily follow. If water quality control is necessary, the pollutants exiting a site cannot be the only concern. Increased volumes of runoff and increased peaks of discharge cause receiving channel instability so peak discharges and increased volumes of runoff must also be considered in the development of a water quality strategy. Design strategy must also reflect possible conflicts with other regulatory requirements. Placing a pond in non-tidal wetlands may alter the character of the wetland and violate Corps of Engineer requirements protecting non-tidal wetlands. The use of open, vegetative swales may violate a local public works requirement for curbs and gutter. Recognition of other local, state, and federal regulatory requirements will provide valuable insight into possible strategies.

Minimum submittal requirements must be specified with respect to major components. Exemptions and waivers, criteria and procedures, and maintenance responsibility must be clearly stated. Guidelines for detailed submittal requirements should also be developed with a checklist for submittal information. The guidelines may detail design procedures and accepted models, format for design reports, necessary notes to be placed on plans, and will provide detailed procedures for project review.

The regulations must address construction inspection. Stormwater management practices greatest cause of failure is improper construction. Infiltration practices must be protected from sediment entry or clogging will occur. Infiltration practices cannot be allowed to accept drainage until the contributary areas have been stabilized. Pond construction has certain critical times when inspections must be accomplished to ensure compliance with design plans. Core trenches, anti-seep collars, and water-tight connectors on pipes cannot be verified once construction is completed. The regulations may provide flexibility for inspection by either a local inspector or certification by the developers consultant of construction adequacy, but timely inspection must be accomplished. The regulations must detail the stages of construction that inspections will be made.

Maintenance requirements must also be clearly stated in the regulations. Provisions must be provided to specify inspection frequency and maintenance responsibility. Documentation of inspection and items checked shall be detailed on inspection reports. Procedures must be specified to ensure that deficiencies detailed by the inspections are rectified and enforcement procedures must be available in the event that repairs are not undertaken or are not done properly.

## Manpower

Implementation of any program can only be effective if adequate manpower is provided. Experience in Maryland has indicated that one engineer can review approximately two stormwater management plans per day. Some projects may take less time then that, but large highway projects may take a full day to review the detailed design report.

Inspectors, on the average, inspect three projects per day for stormwater management and sediment control. It takes about two years to adequately train an inspector so that he can adequately interface with a contractor in the field when problems occur.

Administrative support must be provided to the plan reviewers and inspectors and they must be supported in their efforts. Legal staff must be provided and educated to adequately support necessary enforcement actions. The whole issue of enforcement strategies and tools can be addressed by another paper, but those strategies are one of the very important elements of a stormwater control program.

## Education and Training

Any new program requires a significant effort in education to ensure proper implementation. All elements impacted by the program must have a clear understanding of program goals and specific responsibilities placed upon them. Workshops need to be regularly held for plan designers and reviewers to present proper design procedures for all engineered practices. Inspector workshops need to stress proper inspection techniques and important aspects of control practice construction.

Possibly the most important benefit of workshops is bringing people together to share experiences. It is important for these individuals to realize that they are not alone in their efforts. Also, communication facilitates program evolution by sharing ideas and efforts. Some of the most important changes to a design or construction standard have resulted from communication with consultants and inspectors.

An annual event, such as a conference, is an excellent mechanism to bring together all affected interests to foster communication. Interaction between design and construction personnel is extremely valuable in program effectiveness. Legislation and regulation alone will not provide a successful program. Cooperation of all affected parties is required and the most effective means to obtain cooperation is to open and maintain avenues of communication.

## Maintenance Infrastructure

Proper design and construction represents only two elements of a successful program. The long term functioning of the control structure, especially a water quality control structure, depends on the maintenance applied to that structure. Stormwater management required on residential developments must be publicly maintained. Private home owners will not understand the importance of maintenance nor the costs associated with that maintenance. Commercial properties can be expected to maintain a control structure if that structure is visible to the public. If the control structure is located behind a building, out of public view, maintenance will be a problem.

Adequate manpower must be provided for maintenance inspections and the manpower needs can be expected to increase each year as more structures are installed. Ideally, all structures should be publicly maintained and the structures should be designed to minimize maintenance. The major problem with the assumption of public maintenance is the cost. Completed structures should be placed on a computer listing with type of structure and inspection results entered. The program would notify the inspector when a site inspection is necessary so that a daily work load is established. The number of structures and their age will determine the number of inspectors necessary to handle the work load.

## Utility Concept

A funding mechanism needs to be established to fund all aspects of the stormwater program, but particularly maintenance. Local governments historically have paid for both construction and maintenance with general revenues from property taxes. Although these revenues are relatively stable, they cannot be relied upon as a secure source of funds as maintenance is generally a low priority item. One possibility is to finance maintenance by creating a stormwater uitility that levies "user charges."

Since the American Public Works Association (APWA, 1981) recommended user charges on the utility approach as the best method of financing stormwater management, many stormwater managers have written about their successful utilities in journals like Public Works. We recently surveyed 25 utilities in the United States and have compiled the most thorough data on utilities that is available to the public and local officials. Based on the survey, we have prepared a planning guide to utilities.

Basic concepts central to the operation of all utilities are that:

1. All properties contribute to or benefit from the stormwater system.
2. Users (ie, properties that generate stormwater) should pay in relation to the level of service provided or amount of service consumed.
3. Charge systems are designed to reflect the amount of runoff from properties and commonly are based on Rational Method Coefficient.

While many people benefit from the water quality improvements that result from stormwater management (particularly in areas with critical water resources like the Chesapeake Bay), it is emphasized that charges relate to the use of the stormwater system, not the benefits derived from it. In this way, charges can be distinquished from traditional benefit assessments.

The utility approach to financing a local program could solve many problems. In addition to funding maintenance, master watershed plannng, personnel costs, and retrofitting may also be included as funded elements. the calculation of charges and the method of billing are the key elements of a utility. The level of the charges depends on the amount of funds needed. The method of billing includes factors such as the billing mechanism, frequency of billing, and compatibility with existing data management systems. Most existing utilities have simply added a stormwater charge to existing utility bills.

## Conclusion

The implementation of a successful stormwater management program depends, to a large extent, on the institutional framework of the program. Those individuals responsible for the implementation of controls on site will attempt to reduce their expeditures as much as possible to remain economically competitive. The program must place equitable requirements on all impacted individuals and those requirements must be legally defensible as challenges to the authority can be expected.

Protection and use of our water resources may be the most important issue that confronts us as we enter the twenty-first century. Effective infrastructures need to be established to maximize program effectiveness.

It is important to recognize that stormwater controls can only mitigate adverse impacts of developmental activities to a certain extent. Land use is what creates these adverse impacts and should be carefully evaluated in conjunction with the effectiveness of control practices. Land use decisions must consider water resource impacts as a component of the overall approval process.

Urban Runoff Management - An industry Perspective
An Opinion

Neil G. Jaquet

Most of what I read and hear about urban runoff deals with the technical end of how urban water flows are generated, concentrated, degraded, detained, trickled, criticalized and eventually disposed. Representing a long-established major employer in Colorado, my perspective is not so much the technology of water quality or quantity--but how to get it managed in the least time, for the least cost and in a manner my management can understand and--hopefully fund.

## Industry Perspective of Runoff Quality and Quantity

I work for a company where water is one of the important ingredients in our product, and within an industry which brought images of water quality into the household long before the public knew or cared about water quality. Even in my childhood I remember the slogans--"From the land of sky blue waters", "It's the Water", "Brewed with water from gods country--some say it comes all the way from Canada", and the one which is still present on the cans and bottles of my employer--"Pure Mountain Spring Water".

I think any brewer is sensitive about he quality of the water going into the product--just the same as an engineer is concerned about the content quality of their reports or calculations. Coors has always been--especially now. It's hard not to pick up a copy of the Denver Post or Newsweek and not see an article reporting yet another incident of ground water contamination.

Coors presently produces about 17 million barrels of beer per year from a single plant about 15 miles west of Denver. To put that in engineering jargon, that's 1700 acre-feet, $5.3 \times 10^8$ 12 oz. cans, or $8.7 \ 10^7$ six packs. What makes us a little more sensitive abut water quality is that we operate our own water supply system. All of the other large breweries in the US receive their water supply from the local municipality. We've chosen to keep
--------------------
Manager, Water Resources Operations, Adolph Coors Company, Mail #RR856, Golden, CO 80401

the responsibility for the quality of our main ingredients in our own hands and have developed our own water supply system--consistent with a corporate philosophy of vertical integration. The supply system consists of diversion and distribution of over 5,000 acre-feet annually of ground water (referred to in the market place as "Pure Mountain Spring Water") and nearly 10,000 acre-feet annually of surface water--used for all non-brewing processes and cooling water.

All of our water is derived from the Clear Creek basin, an eastward-flowing, Colorado Front Range stream with about 40 square miles of tributary area above our plant. Clear Creek is a stream basin who's history strongly affects the runoff quality we see at our plant. It is a stream with a rich history of precious metal mining, on urbanized lower basin between Golden and Denver and an upper basin which is also urbanizing. We operate a system consisting of four off-channel reservoirs, supplied by several miles of supply ditches.

Water Resources

Being in the reservoir business and ditch business is where we've found ourselves coming into contact with issues of runoff quality and quantity. First, the quantity--we live in a semi-desert and our prime directive is to provide enough water so as production is not curtailed during the 100 year drought. That, my friends, means we're into the reservoir business. And, in the reservoir business we're concerned about two types of water quality events which can "muddy" our reservoirs and cause water shortages. Those are the catastrophic and insidious events.

The catastrophic events are those which have been dealt with in traditional engineering hydrology--the 100 year event or PMF. But, these are events which in general can be relatively easily planned for, and built around--although sometimes the engineering profession keeps changing its mind about how big an event really should be--a topic I'll address later.

I believe the events which are insidious are the more troublesome, more difficult to plan around, and difficult to convince management & government of it's potential for financial loss. It is also the most difficult to see clear benefits accruing to the "public" when runoff quality is at issue in private development or public works projects.

## Quality and Quantity Issue

We've become keenly aware of the role that runoff quality plans in our reservoirs and ditches. The basin upstream of Golden is subject to the random and unpredictable discharges of tailings and heavy metals from abandoned mine workings. These are a product of inadequate planning and regulation which occurred several generations before our time. But these random releases are generally the more catastrophic events. They are dealt with rather easily by simply ceasing to divert water from the stream. That's a good solution for water users who can switch to reservoirs when not diverting from the stream.

The chronic urban runoff issues which directly affect us concern the urbanization of the basins tributary to our reservoirs and ditches. Because our reservoirs and ditches are located in a relatively narrow valley--we're at the bottom of the hill--with everyone else wanting to add their increment of flow and degraded runoff to the water which flows onto and into our property.

In dealing with new developments desiring to discharge onto our property, we've had to assume the role of the diligent regulator--when it comes to water quality. It seems that the state of the governmental planning review is to be certain that the flows arrive at the fence line at the same rate of flow as historically. But, we've found ourselves differing with regulations in terms of total quantity delivered and its quality. The thoughts go like:

- the flow may be the same as during the pre-development period, but the total quantity of discharge is substantially greater. That's important when the discharge is into a ditch, which overflowed at some downstream point for 2 hours before development, now it will overflow 3 or 4 hours. Even though the rates the flow are the same as historic, its seems to me that damage is more severe the longer flood water inundates property.

- The nature of the runoff quality has now changed, and it will certainly affect our reservoirs as year after year of the two-year storm runoff enters our lakes. It will also affect us as the operator of an irrigation ditch when we are required to treat our "tail" water, or excess water, which enter public streams. It appears that water quality regulation will eventually require us to treat that "tail" water--at considerable expense.

We've found that the regulators typically aren't dealing with runoff quality as it enters our property. The issue being addressed is quantity. With two recent proposed developments we've had to say "no" to adjacent landowners who've assumed they could discharge their detained runoff onto our property. I found that the engineers both for the developer and the local government had a wide-eyed empty look when we make our feelings known about runoff quality. The level of ignorance was such that we offered our consultants to them to design a runoff detention system which could satisfy our concerns. Specifically we requested a detention pond of adequate size to fully contain the two year storm. The detained water will discharge by percolation. Storms in excess of the two year would overflow a grassy berm--to provide some level of treatment prior to discharge onto our property.

My experience in these situations was that the local governmental officials were not aware of our concerns and would certainly not have required runoff treatment if it were not for our insistence.

In both of these situations the treatment techniques were relatively inexpensive. That won't always be the case. I refer to the treatment of runoff discharging from older developments. Irrigation ditches in our area cross the topography of suburbs developed 20 to 30 years ago. I see the drift in the legislative and regulatory process toward regulation of "non-point" source runoff into receiving waters. These irrigation ditches, operated as private, break-even corporations, discharge their "tail" water into tributaries of the South Platte River. These discharges will be difficult and expensive to treat. Difficult because the ditches lie in urbanized areas with high land values and little available land. Expensive to treat because the flows are laden with un-detained and un-treated runoff from older parking areas, developments and street systems. If and when it will be the law to treat these waters, it will be necessary to provide treatment with public funds.

I believe the public should bear the expense of treatment of ditch tail water in urban areas because the public has viewed ditches as convenient open disposal trenches--a convenient means of disposing of grass clippings, and spent crank case oil. Governments have viewed them as an inexpensive place to drain a parking lot or street instead of building a proper storm water drainage system. It seems that the quality problems are being created by the public in their de facto use of the ditches as storm water conveyance structures--and the public should pick up the tab in cleaning up those waters when they reach the receiving streams.

We keep working so that the legislators, regulators, engineers and technical people keep some horse sense in mind when it comes to creating and enforcing regulations. That is, that we regulate runoff quality because low quality runoff has a tangible effect upon some downstream user--it lowers the quality of a reservoir or injures an irrigated crop. I am concerned that considerable public and private money will e spent to clean up runoff only to benefit a few dozen downstream carp--it doesn't seem that the benefits justify the expense.

Can we agree on this?

The engineers can't agree on how big the events really are and the type of structures which can accomplish the goal of runoff quality enhancement. Case in point--it seems that the magnitude of the 100-year storm keeps changing. Maybe it is coincidental that the 100-year and PMF gets larger just after a big storm in the Black Hills or along Colorado's Big Thompson River. Sometimes it goes the other way--a new paper is published and it reduces the intensity of the major storm event because of a "new" theory about meteorology--and suddenly 10 years of flood planning and large sums of private and public money have been spent on a storm who's magnitude drops by 30%, or even 50%. On Clear Creek for example, current studies are initially concluding that the 100-year event should be about half the size we've been planning and building for the past 10 years. For the period of 10 to 20 years ago, the magnitude of the 100-year event seemed to increase with the latest edition of the flood planning.

The engineering profession seems at times to be bent on losing its credibility within itself and with those people who pay the bills. One of the most important assets any profession has is its credibility with its constituents. Thats something I cherish highly with my management and board of directors. I'm constantly amazed about the quality of their memory. That is, if I'm not consistent from one month or year to another, I either will lose some credibility or had better have a bullet-proof reason why I've changed my mind.

Another case in point--the adaptation of requirements for trickle channels at the same time when treatment for quality of runoff is needed. It seems illogical to build channels to quickly convey the frequent storms to the detention ponds--and then build extra berms, install hardware or build larger capacity ponds to provide treatment. It seems to make more sense to leave soils on the channel floors, establish grass and encourage wetland development to provide treatment.

### Where do we go from here?

I'm positive about the future. I do see the regulators and administrators becoming more aware of quality concerns. Sometimes it may be, more by education--when we as private land owners, and corporations operating ditches keep saying "no" to the neighbors and governments who think it's OK to discharge untreated runoff onto or into our property. I'm encouraged by this conference--to see the breadth of knowledge being displayed and knowing that it will be dispersed throughout the country.

Some of the optimism is seen by development of pragmatic type of land stewardship. That is, land owners wanting to preserve their water quality assets in their reservoirs, streams or ditches and recognizing aspects of runoff which can degrade their property. At the same time, those owners don't want to throw away their money on low return investments--such as keeping trash fish alive in a receiving stream. Or, demanding that their neighbors install unnecessary structures for storm water detention and treatment facilities.

## Discussion of Mr. Jacquet's Paper

Comment:

I agree with your summarizing comment that we need a "substantial effort to document the tangible benefits to the public that justify the costs" of urban runoff quality controls.

Question:

With respet to riparian rights, how do you get to exercise controls on development of neighboring lands not under your (private industry) control?

Answer:

The local agency brings the plans of new developers to us to review, for which I give them a pat on the back. Without that foreknowledge, some developers would simply have thrown their (increased) drainage water over on our property before we had any chance to object.

Question:

Have you had any increased-damage volume problems from existing developments around you?

Answer:

No.

Question:

At what point do you consider the drainage from those neighboring developments a "degradation?"

Answer:

We want to control or contain the 2-year storm. After first-flush, we don't care as much.

Question:

Are the NPDES permits on storm water of concern to you, as a private industry?

Answer:

We notice that the cities around us, particularly downstream of us, are concerned about them. Rather than runoff, our concern is runon--to protect our rather famous water supply.

## SUMMARY OF INSTITUTIONAL ISSUES
By D. Earl Jones*

We have just heard from persons representing three different interests about institutional aspects of stormwater quality planning. These views were expressed by representatives of industry and a regulatory community as well as by consultants having a broad national overview of institutional water quality problems. As might be expected, the viewpoints differed, but their important messages involved mutually reinforced opinions.

The consultant, Ron Wycoff, viewed institutional problems in terms of program or process elements essential for effective program implementation. He provided several significant conclusions:

o   Motivation to create an effective program is essential. Absent local motivation but in the face of need, court-ordered action is a possible alternative.

o   Effectiveness requires formation of an area-wide agency or umbrella organization having authority to require and direct actions by each member political subdivision.

o   The umbrella organization must have an independent and continuous source of funding.

o   The professions must educate the public to recognize that stormwater runoff is just as serious a source of pollution as combined sewer overflows. Those who stimulate actions focused upon the latter must learn to direct equal effort toward the former.

o   Without local funding sources (reflecting local interest and concern), there will be little water quality progress even if non-local funding is available. Motivated local water quality advocates are absolutely essential.

These broad conclusions about essential institutional elements do not address detailed programmatic needs. They identify a minimum framework or skeleton upon which to build a water quality program.

The water quality regulator, Earl Shaver built upon the foregoing and added the following features:

o   Effective program implementation requires absolute local commitment to water quality. Problems must be defined clearly and enabling legislation, regulations and design criteria

---

*Consultant, Wright Water Engineers, Inc., 2490 West 26th Ave., Suite 100-A, Denver, Colorado 80211

# INSTITUTIONAL ISSUES SUMMARY

must provide clear guidance and leadership to all affected interests. Objectives and directions must be clearly understood and agreed upon.

o   One entity must guide implementation.

o   Anyone proposing construction must obtain water quality impact approvals.

o   There must be resources to assure continuing program compliance and authority to enforce compliance and levy significant penalties for non-compliance.

o   Regulations should define functions, responsibilities and minimum performance objectives to be attained.

o   Published design criteria, either in performance or specification terms, must be provided. Performance standards are the more desirable even though they may be somewhat more difficult to administer.

o   Regulations must specify minimum submittal requirements for each element of concern, must identify construction inspection requirements and timing, must provide for both short- and long-term maintenance, and must provide for documentation of approvals, special requirements and inspections.

o   The entire water quality and quantity management program must be provided sufficient qualified manpower, properly and regularly trained and given consistent and dependable administrative support.

Mr. Shaver expressed a strong opinion that system maintenance should be by the public, a point with which many of us agree in general. In many instances we can point toward better maintenance actions by private interests, especially in the too-common situation where public maintenance funding is inadequate. We cannot quarrel with Mr. Shaver's conclusion that protection and wise use of water resources may be the most important future issue facing us, and his observation that even the best water quality controls, rigorously implemented and enforced, can only partially mitigate adverse impacts of development activities. This latter observation, coming from a regulator, provides a ray of hope that such interests may ultimately recognize the value of practical quality objectives rather than requiring compliance with arbitrary theoretically ideal objectives that cannot be justified by actual benefits consistent with compliance costs.

Mr. Jaquet presented an industry view of runoff management. Many of you probably were surprised to hear him express greater concern for water quality than traditional regulating officials --- and with excellent supporting reasons. His industry, and many others as well, is extremely concerned about water quality as it affects product quality and the costs of specialized water treatments to assure quality consistency. A hidden cost for many industries is removal of fouling from steam condensers, an expense measured not only by the cost to clean a set of condenser tubes but also costs for condenser redundancy and possibly for additional down-time in the event of serious unexpected short-term fouling conditions.

We should engrave in our minds industry's frequent needs to maintain near-constant water quality over time, for waters having specific chemical contents and for waters that will produce only minimal fouling of condensers and other processing equipment. These concerns, vital to many industries, often will be more challenging than mere control of maximum contaminant levels. This identifies another set of water quality dimensions that may require public regulation and management.

Messrs. Shaver and Jaquet each called for bottom line application of common sense and practicality in establishing standards and procedures. Whereas the public objective has been to assure provision of some stipulated minimum quality level, the industry concern is to achieve broader positive control in the least time for the least cost, with a minimum of red tape. A common thread among the three presentations is the absolute need for cooperation among parties and for continuing communication among them to avoid adversarial situations, and strengthen mutual support. When everyone realizes this, we may see rapid progress for everyone.

Mr. Jaquet also discussed two particular water quality concerns, catastrophic and insidious events, neither of which is addressed fully by conventional regulatory approaches. An example of a catastrophic event of particular concern is the random release of concentrated toxics from a drainage area, such as might result from their sudden exposure by a rainfall-induced landslide in old mining or processing wastes, with their resulting downstream transport. Although this can be dealt with by ceasing withdrawals, triggered by continuous water quality monitoring, such a system is expensive, is not fail-safe and could result in costly protracted production interruptions. Additional effort obviously is needed to forestall explosion of such time bombs.

Mr. Jaquet views cumulative water quality effects of upstream development as insidious, because of their gradual development. Because of them, his industry finds itself concerned about both water quality and total discharges from upstream developments, whereas traditional public regulation has focused only upon post-development as compared to predevelopment peak discharges. An easily overlooked consequence of insidious impacts is that they can materially increase both effluent and process water treatment costs. These factors should at least be considered in public runoff management programs, and should be recognized to be as important as other aspects of runoff water regulation. Mr. Jaquet cites the unbelieving amazement of both project consultants and regulators when industry concern is expressed for both total discharge volume and water quality controls on upstream developments. In at least one Colorado case, both of these concerns have been held valid; such precedents are very strong tools for water quality proponents.

To me, the most important thrust evident from these presentations is emerging recognition that standards and regulations should realistically permit flexibility to respond sensibly to varying physical, biologic and economic conditions and needs, guided by common sense comparisons of benefits, costs, practicality and cost-sharing alternatives. I urge each of you to help further stimulate such directions.

Combined-Sewer-Overflow Control in West Germany
-History, Practice and Experience-

Hansjoerg Brombach *

## Abstract

Almost three quarters of the sewerage systems in West Germany are combined sewers. In the past hundred years, the combined sewer system has passed through several development stages. The latest development has been the wide spread utilization of "storm-overflow" tanks, which have effectively reduced the occurrences of overflow and the pollutant impacts to receiving waters. There are already more than 8000 such tanks with a total storage capacity of almost 5 million cubic meters. These tanks have been responsible for a marked improvement in the water quality of rivers and streams. The design criteria for the storm tanks are explained. New problems that accompany stormwater treatment, such as outflow throttling, cleaning the tanks, and monitoring are discussed. Successful technical innovations for dealing with these problems are described.

## 1. Historical Development of the Sewerage System

About 100 years ago, sewerage systems were systematically built in Germany. This program was in response to devastating outbreaks of typhoid, dysentry and malaria that continued into this century. Construction began in the major cities, where conditions were the worst. The first sewerage-system planning was entrusted to engineers from England because, at that time, there were no German specialists. These drains transported the wastewater directly into the nearest river without treatment. People even considered flushing public and household waste and garbage into the sewer system /1/, but this was found to be unwise and was prevented.

The English workmanship was a huge success. The epidemics disappeared but the rivers and streams became polluted. Shortly before World War I intolerable water pollution conditions in the congested industrial areas led people to build sewage plants. One man who especially contributed to the development of sewage treatment was Karl IMHOFF (1876 - 1965) /2/, who is also known in the USA as the inventor of the "Imhoff tank". By 1911, there were 70 sewage plants with Imhoff tanks in Germany.

Further development of sewerage systems was greatly hindered by the two world wars. The economic upswing following the last war was not significantly concerned with environmental protection. In 1957, for

---
*President, Umwelt- und Fluid-Technik GmbH, D-6990 Bad Mergentheim, Steinstrasse 7, Federal Republic of Germany

| No. | Federal State name | Total Area km² | Population 1000 people | Total length of combined and sanitary sewer lines km | Percentage of combined to total sewer lines % |
|---|---|---|---|---|---|
| 1 | Schleswig-Holstein | 15,676 | 2,567 | 7,466 | 18.7 |
| 2 | Hamburg | 753 | 1,812 | 3,023 | 42.0 |
| 3 | Niedersachsen | 47,407 | 7,125 | 26,542 | 14.3 |
| 4 | Bremen | 404 | 757 | 1,559 | 49.2 |
| 5 | Nordrhein-Westfalen | 34,044 | 17,207 | 51,904 | 71.7 |
| 6 | Hessen | 21,110 | 5,461 | 23,513 | 89.3 |
| 7 | Rheinland-Pfalz | 19,837 | 3,684 | 16,039 | 90.5 |
| 8 | Baden-Württemberg | 35,750 | 8,996 | 39,559 | 89.3 |
| 9 | Bayern | 70,440 | 10,644 | 40,497 | 84.7 |
| 10 | Saarland | 2,568 | 1,127 | 4,967 | 93.7 |
| 11 | West-Berlin | 4,800 | 2,130 | 3,439 | 36.6 |
|  | Federal Republic of Germany | 248,469 | 61,510 | 218,510 | 71.2 |
| For comparison: Missouri |  | 180,486 | 4,800 |  |  |

Table 1: Distribution of Population and drainage systems in the Federal Republic of Germany

example, only 50% of all West German households were connected to sewerage systems. The water quality of rivers and streams was so affected that, even today, bathing in rivers and streams is prohibited nearly everywhere.

Today, about 92% of the population is connected to sewerage systems and 86.5% to wastewater treatment plants (WWTP) /3/. The percentage of combined-sewage systems is 71.2%. In recent decades, the portion of separate-sewerage systems rose slightly but remains clearly in the minority with 28.8%. Sewer lines extend a total distance of 220,000 km - 137, 500 miles. This means that there are statistically 4 meters of sewer line for each West German citizen.

Table 1 shows the distribution of population and drainage systems in Germany. It can be clearly seen that the combined system is preferred in hilly southern West Germany. The separate sewage system predominates in the areas near the coast. The table shows the state of Missouri for the purpose of comparison. Notice that the concentration of population in West Germany is about 10 times greater.

## 2. The Present State of Stormwater Treatment in Separated Systems

As a rule, stormwater is not specially treated. A settling tank and/or oil separator are necessary only where surfaces are especially polluted, for example, in industrial areas, or in water protection zones. Stricter regulations may be expected over the next few years. Drainage from new highways, airports, etc., will be treated via settling and detention tanks.

One problem with separated systems that has not yet been solved is inproper connection, that is, the accidental or deliberate misconnection of sanitary and storm sewers.

## 3. Conventional Combined-Sewer Systems

For a long time it was customary to allow the storm runoff to discharge without treatment at combined sewer overflow structures ("storm overflows"). It was assumed that the overflow water would be diluted such that it would not harm open bodies of water. The dilution ratio of the overflow water was assumed to be between 1:5 and 1:7. However, the bodies of water remained severely polluted. The dilution theory was then discredited.

Since 1962, "storm overflows" have been dimensioned to cope with a critical storm runoff. Figure 1 shows a typical "storm overflow" /4/. The outflow to the WWTP is limited by a throttle pipe. The side weir is set much higher than it was with the prior practice. The height of the weir crest is selected so that overflow does not begin until the diverted rainfall per second per unit of area has reached 7 to 15 l/s per ha (0.1 to 0.2 cfs/ac) of the drained surface. The relatively long sill permits overflow from approx. 80 - 120 l/s per ha with very little backwater. Scum boards for trapping floating matter or even screens are practically non-existent.

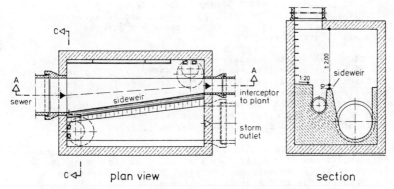

Fig. 1: Storm Overflow Design in Germany since 1962

The newer "storm overflows" are activated less often and, in rainy weather, provide additional channel storage. However, they have become the main source of pollution from combined-sewerage systems since highly efficient secondary WWTPs have become so widespread. The "storm overflows" are very common. I estimate that there are still 50,000 to 70,000 in West Germany.

## 4. Modern Storm Water Treatment in the Combined-Sewer System

In West Germany there is a WWTP that is also part of a university. This "academic sewage plant" is at the University of Stuttgart and handles the drainage for the suburb of "Buesnau". The drainage area is 31 ha (sealed area) with a population of 4000 people. This WWTP is equipped with the latest in measuring facilities.

Investigations carried out in Northampton from 1960 to 1962 /5/ showed that, when it began to rain, large quantities of polluted lines and that most of these "first flushes" escaped at the "storm overflows". This could be a reason for the bad reputation of the dilution theory. It was only natural that we use the university WWTP to trace this effect.

At the end of the 60's, KRAUTH made extensive measurements of the WWTPs influent charactersitics and actually succeeded in demonstrating the "first-flush" effect. In 1971, he published the results /6/ and, at the same time, suggested how the degree of efficiency of the overall WWTP/collection system could be improved. The inflow to the WWTP during rainfall should be limited to twice the dry weather flow and additional detention tanks should be installed upstream in the collection system. The first suggestion was intented to protect the WWTP from hydraulic overload. The additional storage capacity would compensate for the excess flow generated in the collecting system as a result of the flow throttling. At the same time, the "first flush" would be captured in the detention tanks and kept separate from the overflow. This was the beginning of the "storm-overflow tank".

Soon afterwards, in 1972, KRAUTH's idea was adopted by the state of Baden-Wuerttemberg and incorporated in the state ordinances. In 1977, the German Water-Pollution Control Federation (ATV) issued the "Guideline for the dimensioning and design of storm-water discharges in the combined-sewer system" /7/, called "work sheet A128" for short. Since that time, this guideline is considered to be state-of- the-art in all of West Germany.

## 4.1 First-Flush Tanks

The Guideline /7/ defines four basic types of storm-overflow tanks (see Fig. 2). First let's take a look at the left two types, the "first-flush tanks".

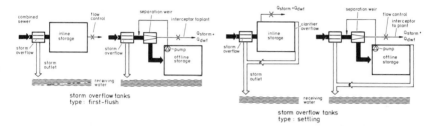

Fig. 2: Arrangements of "storm-overflow-tanks" in combined systems

As the picture on the very left shows, when there is a good system gradient it is economical to place the storage capacity in-line. Even when the weather is dry, sewage continues to flow through this storm-overflow tank and the tank is empty. When it rains, the outlet-flow controller restricts the flow to the WWTP at a level equal to $Q_{storm}+Q_{dwf} \approx 2*Q_{dwf}$. Even a light rain will fill the storage tank to a greater or lesser extent. The WWTP receives an even flow of sewage. If it rains continuously or heavily, the tank overflows. The overflow is upstream from the storage tank. If a "first flush" does occur, the worst of the polluting matter is immediately caught in the storage tank. The overflow water which follows is relatively clean. During his investigation in Buesnau, KRAUTH found 8 m³ storage capacity per 1 ha (114 c.f./ac) drainage surface area was sufficienct to keep 90% of the suspended solids from the receiving water in rainy weather. Once the rain has stopped, the tank runs empty again.

If the downgrade slope conditions are unsuitable or when there is need for a large tank volume, the storage capacity must often be placed off-line to the collector, see Fig. 2. This requires an additional separation weir. The tank is filled by the flow over this weir as soon as the inflow exceeds the outflow. The tank capacity is the same as that of the in-line. This means that the level of water in the tank reaches tank overflow at the same time as the level in the in-line tank. The "first flush" is caught just as effectively. The tank can sometimes be partially emptied by gravity otherwise the balance must be removed with pumps. Energy costs are minimized since these pumps operate only after the rain has ended.

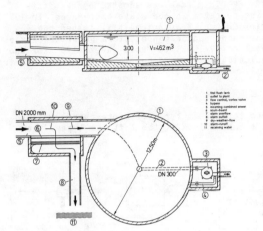

Fig. 3:
In-line storm-overflow-tank type "First-Flush"

Fig. 3 shows a common, typical in-line "first-flush" tank. The storage tank is circular in plan with a tangential feed. The outflow to WWTP is in the center. When the tank becomes filled with stormwater, a whirlpool effect is created that effectively transports the polluting matter and cleans out the storage space. The overflow is upstream the storage tank so that the overflow water cannot re-mix with the captured "first flush".

## 4.2 Settling Tanks

If there is no reason to expect a pronounced "first flush", it is better to choose a "storm-overflow tank" of the settling type. Although experiences have often been contradictory, they show that the "first flush" is no longer severe when the size of the drainage area is such that the time of concentration exceeds 15 minutes. If there are additional tanks or "storm overflows" above the storm tank, the likelihood of a "first flush" is slight. The same comment applies to drainage areas that are very flat where the stormwater flows sluggishly throughout the piping system.

A settling tank functions similarly to a primary settling tank in a WWTP. The settleable matter settles to the bottom and only pre-treated water discharges to the overflow. The hydraulic input to the settling tank is restricted to a critical rain discharge of between 10 and 15 l/s per ha. If more water flows in, the "storm overflow" above the tank is activated. As with the "first-flush" tanks, there are two variations of this tank type, in-line and parallel or off-line (see Fig. 2 right).

Fig. 4 shows a typical settling tank. The tank is rectangular in plan. The front wall is used as a tank overflow and the back wall as a clarified overflow. The tank can be partially emptied by gravity through the check flap and for the remaining volume is removed by a pump discharging directly above the outflow controller. The clarified overflow is controlled by a throttle orifice in order to prevent scour of the sludge settled pollutant.

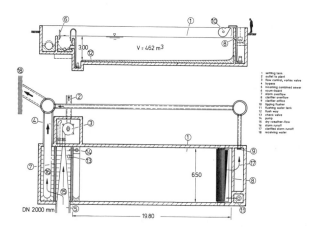

Fig. 4: Off-line storm overflow tank type "settling"

## 5. Storm-Overflow Tank Dimensioning

KRAUTH's measurements /6/ were adopted as design rule in the Guideline /7/. The dimensioning of the reservoir is quite simple. The requisite tank volume V is:

$V = V_{sr} * a * A_{red}$ in $m^3$, where
$V_{sr}$ is the specific storage volume in $m^3$ per ha,
$A_{red}$ is the reduced impervious area in ha tributary to the tank, and
a an empirical flow-time factor, dimensionless

The specific storage and the flow-time factor are taken from fig. 5. The storm-outflow intensity $r_{storm}$ is the specific extra flow charging the WWTP during a storm calculated from:

$Q_{storm} = Q_{out} - Q_{dwf}$ in l/s, where
$Q_{out}$ is the discharge from the tank outlet to the WWTP under full load in l/s, and
$Q_{dwf}$ is the dry-weather flow through the tank in l/s, and
$r_{storm} = Q_{storm}/A_{red}$ is rainfall intensity in l/s per ha.

Practical experience has shown, that the majority of german "storm-overflow tanks" are designed for a small range of rainfall intensities, $r_{storm}$ of between 0.2 and 2 l/s per ha. As can be seen in Fig. 5, this corresponds to a specific volume of 8 to 18 $m^3$ per hectar. The average population density in West Germany in relation to the paved areas is approx. 50 people per hectar. Based on a mean flow-time

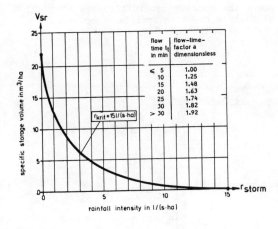

Fig. 5: Dimensioning curve to size the volume of storm-overflow-tanks

factor of a=1.5, from 240 to 600 liters or an average of 420 liters of additional storage capacity per capita must be installed in the sewer system.

For the storm tanks to actually collect "first flush", the catchment area must not be too great. A statistical study /8/ shows the frequency distribution of the storage capacities of 3310 "storm overflow tanks" (see Fig. 6). Most storage facilities have a capacity of about 100 m$^3$ and the arithmetical mean is 462 m$^3$. If we convert this volume to 420 liters per person connected to the system, such a medium-sized tank would serve 1000 persons on approx. 20 hectars of populated area.

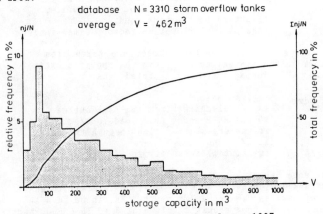

Fig. 6: Distribution of tank volumes 1987

This dimensioning procedure is no longer valid if there are many "storm-overflow tanks" in the upstream system. More elaborate empirical procedures or mathematic simulation models must instead be used. However, these models are subject of controversy.

## 6. Number of Storm-Overflow Tanks

Ten years after the introduction of the new Guidelines /7/ for overflow reduction at storm outfalls, there are more than 8000 "storm-overflow tanks" in West Germany with an effective capacity of almost 5 million cubic meters (see Table 2). On the average, one cubic meter of storage space costs approx. 1500 German marks ($3.40/gallon). The overall investment has been about 7 billion DM.

The right-hand column of Table 2 shows that the different German states have followed the Guideline with varying degrees of commitment. In Baden-Wuerttemberg, the figure has risen to within 40 % of the mean target volume of 420 liters per capita. Other Federal states need to do much more to attain this goal. In Germany as a whole, about 20% of the storage capacity required by the Guideline has been provided.

| No. | Federal State name | Number of existing storm-overflow tanks - | Existing storm-overflow-storage $m^3$ | Existing average storage per citizen liters/people |
|---|---|---|---|---|
| 1 | Schleswig-Holstein | 4 * | 1,900 * | 4 |
| 2 | Hamburg | 6 * | 2,800 * | 4 |
| 3 | Niedersachsen | 60 * | 28,000 * | 27 |
| 4 | Bremen | 6 | 66,500 | 179 |
| 5 | Nordrhein-Westfalen | 950 | 825,000 | 67 |
| 6 | Hessen | 220 | 310,000 * | 64 |
| 7 | Rheinland-Pfalz | 400 | 200,000 | 60 |
| 8 | Baden-Württemberg | 4,140 | 1,909,000 | 238 |
| 9 | Bayern | 2,400 | 1,200,000 | 133 |
| 10 | Saarland | 119 | 55,000 * | 52 |
| 11 | West-Berlin | 6 | 20,000 | 26 |
| | Federal Republic of Germany | 8,311 | 4,618,000 | 75 |

*) personal estimation, no official data available.

Table 2: Existing Storm-Overflow-Tanks in combined sewers 1987 Federal Republic of Germany

It is difficult to quantify the success of this enormous outlay. Sometimes it has also evoked critisism, especially from the communities that have had to pay for these measures. The water quality of surface waters in the Federal Republic of Germany is documented yearly /9/. In the past few years, the water quality of rivers, streams and lakes has improved considerably. An increase in the amount of sludge produced at the WWTPs has been noted.

## 7. Outflow Control

The outflow from conventional storm overflows was relatively high. For this reason, throttle pipes were usually adequate. The new storm-tank engineering methods place much greater demands on outflow control. The flow is greatly reduced due to the imposed hydraulic limit to the WWTP equal to twice the dry weather flow.

Here is an example using the figures from Section 5. A population of 1000 generates a sewage water flow of about 5 l/s. If we assume that an additional 5 l/s of infiltration, the dry weather flow at the storm tank is only 10 l/s. This quantity of water must flow through the tank without backing up. When it rains, no more than 20 l/s may flow out of the tank, which can fill to a storage level of perhaps 2 m. At these flow rates and pressures, a throttle pipe will never function correctly.

The industry has met this challenge with an entire spectrum of innovative outflow controllers. Two different throttle types are described here as examples.

### 7.1 Vortex Valves

Vortex valves are derived from a new technology called FLUIDICS (Fluid Flow Logic). Fluidics was created in the 70's in the US when extremely reliable switching elements were sought for controlling delicate processes in the nuclear and weapons industries. Fluidics components function without any moving parts and are therefore extremely reliable. The idea of Fluidics was adopted by the author about 15 years ago and transferred to the field of sewer flow control.

The vortex valve operates by flow-effect alone (see Fig. 7). A rigid, funnel-shaped swirl chamber set at an angle is fed from below via a feed pipe. The outflow is at the tip of the funnel. As long as the water pressure in the feed pipe is low, e.g. during dry weather, the water is easily channeled in a gentle arc and flows unobstructed through the outflow opening. If the inflow pressure increases due to runoff filling the storm tank, vortex action is created. This swirl flow forms an air-filled vortex core which blocks most of the drain. Despite an unrestricted cross-section, the flow resistance in this state is enormous. With a free passage of 200 mm, the unit can cope with outflows of 20 l/s at a pressure drop of 2 m. When the pressure reduces as the storm tank empties, the vortex deteriorates and the flow resistance disappears. A detailed description of vortex-valve flow mechanics can be found in /10/.

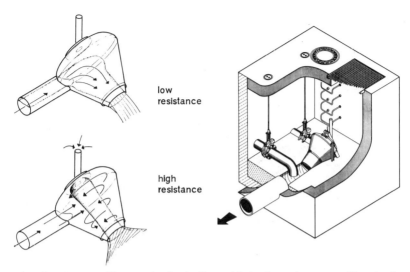

Fig. 7: Vortex flow control at the outlet of a storm-overflow-tank

Fig. 7, right, shows just such a vortex valve downstream from a storm tank. Normally, the valve is contained in a separate manhole shaft that is accessible at all times. The valve is manufactured from stainless steel. At the output end of the vortex chamber, there is a replaceable orifice that allows subsequent adjustment of outflow without tools. This is important to adopt later changes in the catchment which will affect the amount of dry weather flow.

Vortex valves have been used for more than a decade and more than 1000 storm tanks operate with these devices in Germany. Operating experience confirms the high degree of reliability and durability of this technique.

### 7.2 Inductive Flowmeter

The throttle system in Fig. 8 is much more complex and requires a power supply. The storage space in the example is a tubular tank two meters in diameter. The overflow is at the upstream end of the tank and is not shown.

The outflow from the storage space is measured with an inverted-syphon inductive flowmeter. If hydraulic design is correct, such flowmeters have proved to be extremely efficient. The measuring signal is directed via the manhole shaft to the control system housed in an outdoor control cabinet, where variables such as throughflow are recorded and if the set nominal flow is exceeded due to rain, the slide valve downstream from the flowmeter closes until the set value is again reached. This is a closed control loop. A control loop can also recognize and eliminate

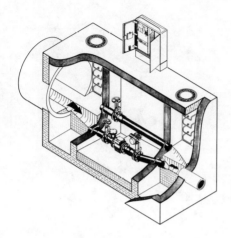

Fig. 8:
Flow-control by means of an inductive flow-meter plus electronic control and knife valve

faults. If, for example, the slide valve becomes obstructed, flow is slowed down and the slide valve automatically opens until the obstruction has been flushed out. Then the flow is, of course, too high and the slide valve recloses. The flow can be changed within very wide margins by adjusting the set value. This can also be done from a main control at the sewage plant.

In addition, the flow and fill level of the storm tank at any given time can be displayed at the sewage plant. The costs of sewage treatment can also be invoiced on the basis of the quantity of water discharged.

Due to the many functions of this throttle system, a small process computer is used, called "stored program control". The program is stored on an EPROM and is not lost, even if there is a power failure. It is usually connected to the WWTP by telephone and automatic modem.

In West Germany there are an estimated one hundred such systems. However, successful operation requires installation and maintenance by qualified engineers and, unfortunately, this poses a problem for public works authorities who concentrate on construction facilities rather than on operation. Problems are also created by thunderstorms and moisture. The electronics industry has underestimated the difficulties associated with the underground sewer environment. However, the working capacity of this system has never been surpassed and justifies the high expense in places where demands are high, e.g. in overloaded sewer networks.

## 8. Cleaning the Storm Tanks

A "storm-overflow tank" must avoid discharging pollutants to receiving waters. When this is accomplished, a large portion of the stormwater pollutants remain in the storm tank. It is even said that the amount of dirt in a storm tank is a measure of its efficiency. But with this success, there have been significant difficulties. Some tanks contained a pollutant deposit up to 10 cm thick after each rainfall. If the sludge and debris is not removed, the deposit could build up to half a meter. Such quantities of polluted matter endanger the sewer system downstream.

In the beginning, the deposits were manually removed with shovels and flushing hoses from the storm tank. However, this is very unpleasant work and must be repeated after each rainfall. Attempts to use intergral stationary spray nozzles failed because a great deal of flow and pressure are needed to dislodge the deposits. Mechanical rakers such as those used in clarification tanks were also tried but were found to be too expensive and susceptible to failure.

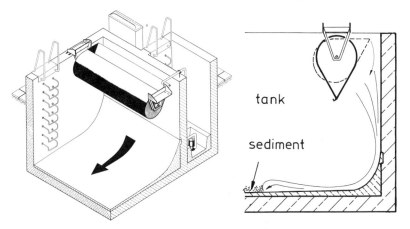

Fig. 9: Tipping flushers to sweep storm-tanks

The idea of the "Tipping Flusher" came from Switzerland (see Fig. 9). This is a container slung on an off-center swivel mounting with a side ejection opening along its entire length. This tipping flusher is suspended at the front wall of the storm tank. When the storm tank empties after rainfall, the flusher is automatically filled with ground water or tap water. This takes about 10 minutes. Because of the off-center bearing point, the center of gravity of the flusher and its contents moves through the swivel bearing just before the container is filled. The flusher suddenly turns and allows a massive surge of water to fall.

Standard flusher sizes are between 300 and 1500 liters per meter of flusher length. When the water falls from a height of two to four meters, a 50 m storm tank is easily flushed clean. Wider tanks are divided into a number of parallel flush ways and flushed one after the other.

It is simple to choose the correct flusher size. There is a hydraulic measuring procedure that was calibrated by measurement /11/. The tipping flushers are made of stainless steel. They have proved to be excellent and are now being retrofitted on many older storm tanks. An estimated some hundred storm tanks are presently equipped with tipping flushers.

## 9. Monitoring the Storm Tanks

With a conventional "storm overflow", it was of little importance when it overflowed, or how often or how long. All that mattered was that it didn't become blocked, although that often happened without detection. A modern storm tank is a completely different matter. In this case, if the tank outflow were to be obstructed, the tank would soon overflow and sewage would be released even in dry weather. Since the outflow from storm tanks is closely regulated by a throttle valve, the danger of an obstruction is much greater than with a simple storm overflow.

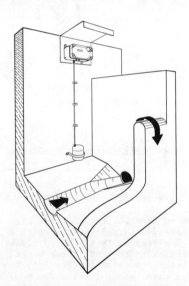

Fig. 10: Battery powered microprocessor to record overflow frequency and duration

There is much more technology involved in a storm tank than in a conventional overflow. There are more overflows, throttle valves, flushing devices, remote control systems, etc. These must occasionally be checked to make sure they are functioning properly. Storm tanks are also much more expensive than simple storm overflows. It is therefore justified to question the effectiveness of the investment. The problem arose that was expected by experts but long ignored by operators, namely, how would these new storm tanks be monitored and be maintained, and who would do it.

The maintenance of such tanks can be organized without too much trouble, but the number of personnel must be increased. Every large storm tank should be checked once a week. It was difficult to prove the efficiency of the tank as overflow frequency and duration have to be known. However, rainfall is not limited to the maintenance personnel's working hours. The maintenance personnel can tell that the storm tank was activated only by the traces left by the polluting matter. One survey involving 25 storm tanks over an entire year, revealed that the maintenance personnel had greatly underestimated the frequency of overflow.

This informational gap is filled by the device in Fig. 10. A sensor is attached to the wall near the overflow sill. When the water level in the tank rises, the pressure is transferred to a instrument box. The box contains a battery-powered system that measures the pressure every 60 seconds. A microprocessor immediately calculates the number and duration of levels exceeded for 256 water stages. The accumulated data can be read manually on the display or by means of an electronic memory card. If the hydraulic characteristic values for the overflow sill are known, the amount of overflow water can be calculated with the greatest of ease and accuracy.

Such devices were not developed until recently. However, it is expected that they will supply valuable data for planners, operators and the supervising authorities. For this reason, some West German states require that new tanks be equipped with this device.

## 10. Future Developments

Recently the Guideline /7/ has come under criticism. It is said that it did not give enough attention to varying local conditions such as gradients in the collecting system and the degree of contamination. Furthermore, the Guideline was of little use in calculating multiple storm tanks in series. This led to the introduction of the mathematical simulation models. Unfortunately, these were often misused in order to "bring down" the costs of the expensive storage capacities. For this reason, the Guideline is due for revision in order to correct the obvious deficiencies.

It will never be possible to design an optimum storm-water handling system for large and complex sewer networks using a simple guideline. Simulation models are needed. In West Germany, there are about a dozen of these models which are sometimes very close to SWMM and STORM from the US. Quality prediction of such models is a subject of controversy. One major complaint is the lack of basic knowledge in creating models of pollutant accumulation, erosion and transport in the collector network. Research programs are on their way to correcting these deficiencies. Alternative techniques for storm-water handling, such as inlet control, swirl separation and infiltration, are being seriously discussed. However, these alternatives are not yet considfered feasible.

## 11. Acknowledgement

The support of the ministries for Environmental Control of the West German States in collecting actual data on the number and size of storm tanks is gratefully acknowledged.

## References:

/1/ NN: Lehr- und Handbuch der Abwassertechnik, Band 1. Dritte Auflage, Verlag W. Ernst & Sohn, Berlin, München, 1982.

/2/ Annen, G.: Karl Imhoff. Schriftenreihe der VDI-Gesellschaft Bautechnik Heft 4, VDI-Verlag, Düsseldorf, 1986.

/3/ Gilles, J.: Öffentliche Abwasserbeseitigung im Spiegel der Statistik. Korrespondenz Abwasser, Heft 5, S. 414-437, 1987.

/4/ NN: Bauwerke der Ortsentwässerung. ATV Arbeitsblatt A 241, 1978.

/5/ Gameson, A.L.H. et al: Storm-Water Investigations at Northampton. The Institute of Sewage Purification, Llandudno, UK, 1962.

/6/ Krauth, K.: Der Abfluß und die Verschmutzung des Abflusses in Mischwasserkanalisationen bei Regen. Stuttgarter Berichte zur Siedlungswasserwirtschaft, Heft 45, Oldenbourg Verlag, München, 1971.

/7/ NN: Richtlinien für die Bemessung und Gestaltung von Regenentlastungen in Mischwasserkanälen. Arbeitsblatt A128, Abwassertechnische Vereinigung (ATV), 1977.

/8/ NN: Regenwasserbehandlung in Baden-Württemberg. Ministerium für Umwelt, Heft 20, 1987.

/9/ NN: Daten zur Umwelt 1986/87. Umweltbundesamt. Erich Schmidt Verlag, Berlin, 1986.

/10/ Brombach, H.: Vortex Flow Controllers in Sanitary Engineering. Journal of Dynamic Systems, Measurement and Control. Trans. ASME, Vol.106, p. 129-133, New York, 1984.

/11/ Brombach, H.: Funktion und Bemessung von Spülkippen. Broschüre der Firma Umwelt- und Fluid-Technik GmbH, Bad Mergentheim, West-Germany, 1987.

13.03.1988,
13.05.1988,
ASCE88.ENG

Question:

Do yo have any statistics on how the use of these tanks has changed the upgrading or expanding schedules for existing sludge-handling facilities?

Answer:

Not really.

Question;

Have the existing facilities being able to handle such increased sludge volumes as may have been produced or captured through use of the tank systems?

Answer:

Yes.

Question:

Doesn't your artificial flushing of the (empty) storage tanks cause a new type of first flush?

Answer:

Yes, flushers lead to a new first flush, but it is a contained, captured flush.

Question:

After the artificial flush, is there a sedimentation problem in the dry-weather flow sewers?

Answer:

Yes, throughout the country, even without the tank-and-flusher systems, there is a sewer sedimentation problem. Because the flushers are used on off-line tanks only, however, what they flush back as far as the dry-weather sewer are only the lighter particles.

Question:

Would these tank systems be effective for storm water alone, as opposed to combined sewage?

Answer:

Should not be much different, but there must be a downstream treatment device at some point to capture the temporarily held solids in the tank(s).

Question:

What is the effective length limit for a flusher?

Answer:

That is controlled by two things: (1) the head or height above the tank floor, and (2) the gradient down which the flush of water must flow, which is a curved surface we built into the end of the tank. The practical limit, though, appears to be 50 meters.

Question:

Have you performed any benefit-cost analyses for these devices?

Answer:

There is a monitoring program underway, but we're not really sure. Politicians have said we must being doing something, even if it's not optional. Perhaps the benefits will be countable in the next decade.

Question:

How much does this tank system improve the number of overflows?

Answer:

Where we added storage tanks sized for 20 cubic meters per hectare of drainage area, we have reduced overflows from 50-80 per year down to the order of one per month, or 12-15 per year.

## Discussion of Dr. Brombach's Paper

Question:

How many points of overflow are there in the sewer systems you have addressed? Many? A few?

Answer:

In Stuttgart, 40% of the sewers overflow. There are 200 to 300 overflows, and we expect to put in 150 tanks.

Question:

Do German cities chlorinate overflows?

Answer:

No. Nobody in Europe does that; only in America.

Question:

You discussed four basic tank configurations, two first-flush types, and two settling types. What is the real benefit of the first type over the second?

Answer:

If nothing overflows, there is no difference. The benefit of the first type exists in steeper areas where first-flush truly occurs. The benefits of each type must be derived from case-specific conditions. We have other types as well, which are more complex and specialized than the four basic types I showed.

Question:

Are there odor complaints from areas under the (open) tanks?

Answer:

Yes, this has been a problem. In future applications, the tanks will be underground.

Question:

What do you do with accumulted sludge in these tanks?

Answer:

It is flushed into the sewer to the treatment plant.

NONPOINT POLLUTION FIRST STEP IN CONTROL

Introduction by James Murray*

The following is a report on the Washtenaw County, Michigan Huron River Pollution Abatement Program. This report was prepared under the direction of James Murray for the Washtenaw County Statutory Drainage Board and the Michigan Water Resources Commission. Mr. Murray chaired the Board and the Commission at the time of this report and is currently the Director of Public Works for Wayne County, Michigan.

## The Magnitude of Improper Waste Discharges
## Urban Stormwater System

by Stacy D. Schmidt, Douglas R. Spencer
Washtenaw County Health Department
Ann Arbor, Michigan

The three most significant sources of water pollution in the U.S. are industrial point-source discharges, municipal waste treatment plant effluents, and nonpoint source discharges. Wasteload allocations and receiving water impacts are also a major concern. However, urban stormwater runoff has not always been seriously considered as a major contributor of pollutants, even though under certain conditions it can dictate the quality of the receiving water. Therefore, urban stormwater runoff as well as point sources of pollution must be considered and controlled in order to meet national water quality objectives.

Historically, urban nonpoint pollution sources have been overlooked by local, state, and federal agencies charged with surface water regulation. It was speculated that curbing urban nonpoint pollution would prove to be expensive and unfeasible in many cases. Therefore, efforts to abate surface water quality degradation have targeted point sources which were simpler to regulate. Urban nonpoint efforts have focused on public education, recycling waste oil, and street sweeping, among other activities. These programs have produced some positive results but have been limited in their scope. This study demonstrated that direct connection of individual, commercial and industrial pollutant discharges to the storm drain system is a major

------------------------------
*Director, Wayne County Public Works, 415 Clifford St., Detroit, Michigan 48226.

contributor of urban nonpoint pollution. The magnitude of these discharges on urban nonpoint pollution had not been adequately addressed prior to this study.

## PROBLEM OVERVIEW

One of the first efforts to determine and compare the effects of combined and separate sewer discharges was a 1963-65 study by the Federal Water Pollution Control Administration (the forerunner of the U.S. Environmental Protection Agency), which demonstrated that separate storm sewer discharges have a definite pollutional effect on receiving waters. Coincidentally, those studies were performed on the same drainage basin evaluated in this study. Much of the research done since the publication of those studies involved the characterization of urban runoff in various watersheds. However, relatively little work has been done on the specific sources and, more importantly, control of toxic pollutants carried in urban runoff.

Several researchers have emphasized the importance of petroleum hydrocarbons and heavy metals as major components of urban stormwater. Hunter et al. demonstrated stormwater runoff from metropolitan areas is an extremely important source of oil pollution to receiving waters and suggested that the hydrocarbons detected in urban runoff could come from such sources as accidental spills, deliberate dumping, and crankcase drippings onto road surfaces. They further suggested that the most likely suspect for the majority of the petroleum inputs, both deliberate and accidential, would be crankcase oil. In fact, they found that the carbon ranges detected in the stormwater samples most closely resembled those of heavy lubricating oils. Likewise, other researchers found close agreement between the molecular composition of hydrocarbons in used automobile crankcase oil and those present in urban stormwater runoff. Street sweepings from industrial areas had much higher heavy metal concentrations than those from residential areas. It was suggested that this was a reflection of the relative number of automobile-related facilities located in the industrial portion of the drainage basin. These studies assumed that urban stormwater is composed almost solely of road runoff. The possibility that illegal and improper storm drain connections as contributed to the pollution problem was not directly addressed.

Very few studies have considered the contribution of both permitted and illegal discharges to urban stormwater collection systems. Whipple and Hunter suggested that because oily wastes are disposed by individuals, automotive garages, car washes and other industries, water pollution control programs that take these sources into account may be warranted. Furthermore, Lashkari et al., pointed out

that the waste assimilation and transport capacities of
many receiving waters have been overloaded because of unre-
gulated discharges of urban and industrial wastes. Other
reports also mention the contributions of pollutant inputs
from commercial and industrial facilities. However, none
of these studies attempted to quantify the extent of the
problem.

It has only recently been realized that both non-
permitted and National Pollutant Discharge Elimination
System (NPDES)-permitted waste discharges from industrial
and commercial facilities have a definite impact on urban
water quality. This study was conducted to isolate individual pollutant or waste dischargers and initiate appropriate source control methods to eliminate further input of
contaminants to urban stormwater systems and receiving
waters.

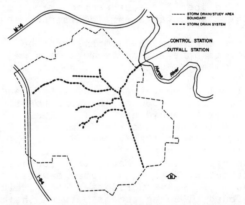

Figure 1--Allen Creek drainage basin
City of Ann Arbor, Michigan

## STUDY AREA

The Allen Creek storm drain system, one of six major
storm drains serving Ann Arbor, Michigan (Figure 1) was
studied. Ann Arbor is a growing community of 110,000 persons and is located approximately 64.4 km west of Detroit.

The Allen Creek storm drain receives runoff from a
relatively large urban area (15,500,000 $m^2$) in western Ann
Arbor and discharges into the Huron River. Land-use studies of the drainage basin indicate a typical urban area
with approximately 74% residential, 17% commercial and
light industrial, and 9% undeveloped lands. The drainage
basin covers downtown Ann Arbor and the majority of the
central business district. It is interesting to note that
much of the earliest commercial and residential development
in the City of Ann Arbor occurred within the Allen Creek

drainage basin, and, in fact, along the Allen Creek.

Except for a few hundred feet along one of the lateral branches, the Allen Creek storm drain is completely enclosed and below grade. The storm drain has a constant discharge (.085-5.4 $m^3/s$) into the Huron River from a 3.7-m diameter outfall pipe. There are currently two NPDES-permitted facilities in the drainage basin that may discharge a combined total of 450 $m^3/d$ of process and non-contact cooling water to the Allen Creek storm drain system.

## PROGRAM METHODOLOGY

The four project phases of the Allen Creek Drain water quality survey are in various stages of implementation.

**Comprehensive Drain Survey (Phase I).** The main stem and accessible lateral branches were examined to identify the origin of suspicious connections to the storm drain first detected by consulting engineers during the 1970s. Beginning at the outfall and working upstream, county personnel both inside the storm drain and on street level attempted to identify the origin of all connections to the storm drain. Individuals inside the storm drain were supplied with self-contained breathing apparatus, flashlights, and short-wave radios. All suspicious connections (those not associated with catch basins or those that demonstrated a continuous discharge during dry weather conditions) were recorded, and, if possible, dye-tested immediately. The list of suspicious connections that resulted served as a target for future pollution control investigations.

**Intensive Bacteriological Study (Phase II).** The main stem, including the outfall, and all major lateral branches of the storm drain were sampled for fecal coliform and fecal streptococci at twenty-eight sampling stations on a weekly basis over a nine month period. The large number of sampling stations was used to better isolate individual sources of fecal pollution to the storm drain and minimize the effects of dilution on bacterial concentrations in the Allen Creek Drain.

**Business Survey and Dye-Testing Program (Phase III).** Non-residential premises in the Allen Creek drainage basin were prioritized based on the extent of petroleum or other potentially hazardous substances stored or used on the site. Top priority was given to businesses directly involved with the use of petroleum products and other potentially hazardous chemicals. The "target list" was generated from a listing of Ann Arbor businesses by type provided by the city planning department.

Individual businesses were visited to determine the type, use, storage, and disposal of chemicals that may be discharged to the storm drain, and whether plumbing fixtures and floor drains showed any signs of chemicals being discharged to the storm drain. Survey results formed the basis for dye-testing priorities. Fluorescent tracing dyes were used so that discharged wastewaters would be brightly colored and easily distinguished in manhole structures. Facilities discharging to the storm drain were notified and referred to the City Building Department for enforcement action aimed at eliminating the improper connections.

Chemical Sampling (Phase IV). Thirty-eight physical and chemical parameters were used to evaluate the chemical contamination of the storm drain. Quarterly samples were taken beginning in the third quarter of 1984. The main stem and major lateral branches were sampled during both dry and wet weather at eight stations.

## WATER QUALITY SURVEY

The comprehensive drain survey performed in Phase I demonstrated that human fecal inputs to the drain are not a common problem. Instead, the survey showed that the majority of observed storm drain connections originated in automobile-related facilities. Thus, chemical pollutants such as detergents, oil, grease, radiator wastes, and solvents were much more problematic than bacteriological contaminants.

Phase II of the project, an intensive bacteriological study of nearly 30 sampling locations on the Allen Creek Drain, was important for two reasons. First, previous sampling of the drain indicated it to be highly contaiminated and fecal-associated microorganisms and a significant source of bacterial pollution for the Huron River. Second, Phase I of this project indicated that chemical contamination from automobile-related facilities with storm drain connections should be of greater concern than sanitary connections that contribute human fecal wastes to the stormwater system.

While the results of the intensive bacteriological study do not indicate that the Allen Creek Drain is completely free of human fecal contamination, they do show that fecal contamination of the Allen Creek Drain is not a significant pollution problem. The overall trend in the data is one of relatively constant bacterial concentrations, generally less than 5000 fecal coliforms/100 mL, while the ratio of fecal coliform to fecal streptococci has been less than 4.0 for 90% of the samples taken.

As the initial project phases had failed to demonstrate a significant fecal contamination problem, emphasis was shifted toward evaluating the extent of chemical contamination in the Allen Creek Drain. The chemical pollution problem has been recognized for several years, but has not received much attention until now. The outfall is visibly affected and smells of petroleum. Preliminary chemical sampling results elevated concentrations for a number of constituents, particularly at the outfall sampling station. Table 1 gives the range and mean of the results for the outfall station and the upstream control (Argo Park) station. A cursory analysis of the data shows relatively good water quality at the upstream control station. At the outfall station concentrations exceed recommended surface water quality standards for many of the parameters monitored. A comparison of chemical results obtained at the outfall and the upstream control demonstrates that the outfall station means exceed the control station means 87% of the time. In fact, in several cases the outfall mean is an order of magnitude greater than the upstream control mean.

## DYE-TESTING RESULTS

Seventy-two Priority I businesses were targeted for investigation in Phase III. Priority I facilities were those known to use or store petroleum products or other hazardous materials. Included in this top priority group were automobile dealerships, service stations, car washes, body shops, and light industrial facilities. Of these, 43 (60%) discharged wastes to the storm drain system. The 29 remaining top priority businesses were properly connected to the sanitary sewer.

Two hundred sixty Priority II businesses were identified. These facilities use or store lesser quantities of petroleum products or other hazardous materials. Included in this second group were photographic processing labs, dry cleaners, printers, utility companies, and chemical labs and warehouses. This phase of the project is not yet complete; however, 25% of those businesses already investigated have confirmed storm drain connections. In addition, 17 nonprioritized businesses have been evaluated; 11 (65%) of them have had storm drain connections.

Results from the dye-testing phase of the project showed that 75% of the observed storm drain connections identified to date originate in automobile-related facilities. See Table 2 for a summary of business types and connection rates. Among the various chemical categories exist: oils and greases, radiator fluids, detergents, and solvents.

Table 1--Range and mean concentrations[a] of selected constitutents for Allen Creek storm drain outfall and control stations.

| Constituent | Outfall Range | Outfall Mean | Control Range | Control Mean |
|---|---|---|---|---|
| $NH_3$ as N | 0.07-0.86 | 0.488 | 0.08-0.21 | 0.150 |
| $NO_3$ as N | 0.01-1.20 | 0.610 | 0.10-0.72 | 0.305 |
| Ortho $PO_4$ as P | 0.01-0.51 | 0.121 | 0.01-0.06 | 0.023 |
| Sulfates | 70-91 | 83.3 | - | 44.0 |
| TKN | 0.69-2.00 | 1.430 | - | 0.980 |
| Total P | 0.01-0.62 | 0.231 | 0.02-0.10 | 0.068 |
| Alkalinity | 72-250 | 171.8 | 110-200 | 162.5 |
| Conductivity, umho/cm | 420-1200 | 830.0 | 540-590 | 555.0 |
| pH | 6.9-8.5 | 7.7 | 7.4-8.6 | 8.0 |
| TDS | 150-790 | 545.0 | 210-420 | 342.5 |
| TSS | 8-340 | 77.4 | 3-8 | 5.0 |
| TS | 470-1200 | 661.1 | 340-430 | 387.5 |
| Turbidity, NTU | 2.7-90 | 23.5 | 2.1-3.7 | 2.9 |
| MBAS | 0.01-0.63 | .178 | 0.01-0.03 | 0.013 |
| Oil & Grease | 1-19 | 6.4 | 1-3 | 1.5 |
| Phenols | 0.00-0.14 | 0.028 | - | 0.004 |
| TOC | 2-64 | 18.4 | 2-17 | 9.3 |
| Al | 0.10-6.50 | 1.300 | 0.05-0.12 | 0.093 |
| As | 0.00-0.01 | 0.004 | - | 0.002 |
| Ba | 0.02-0.14 | 0.070 | - | 0.060 |
| Cd | 0.00-0.04 | 0.011 | - | 0.010 |
| Cr | 0.00-0.04 | 0.020 | - | 0.005 |
| Cu | 0.01-0.08 | 0.041 | - | 0.005 |
| Fe | 0.71-26.0 | 5.4 | 0.12-0.25 | 0.183 |
| Pb | 0.01-0.32 | 0.079 | - | 0.050 |
| Mn | 0.08-0.72 | 0.208 | 0.02-0.08 | 0.044 |
| Hg | - | ND[b] | - | ND |
| Ni | 0.01-0.04 | 0.023 | - | 0.020 |
| Se | - | ND | - | ND |
| Ag | 0.00-0.04 | 0.019 | - | 0.020 |
| Zn | 0.09-1.20 | 0.561 | 0.03-0.10 | 0.068 |
| Chlorides | 87-220 | 139.2 | 42-61 | 50.5 |
| Cyanides | 0.01-0.03 | 0.015 | - | 0.010 |
| Fluorides | 0.35-0.60 | 0.483 | - | 0.200 |
| Na | 51-60 | 56.7 | - | 27.0 |
| TOX-Brominated | 0.00-0.02 | 0.009 | 0.01-0.02 | 0.011 |
| TOX-Chlorinated | 0.02-1.20 | 0.357 | 0.01-0.15 | 0.080 |
| TOX-Iodated | 0.00-0.01 | 0.008 | 0.00-0.01 | 0.008 |

[a] mg/L except where noted
[b] not detected

Table 2--Summary of storm drain connections by business type for Allen Creek Drainage Basin, Ann Arbor, Michigan

| Business Type | Priority | Businesses | Connected to |
|---|---|---|---|
| Auto repair shops/tire stores | I | 26 | 65 |
| Service Stations | I | 16 | 63 |
| Printers/copiers | II | 11 | 9 |
| Manufacturers[a] | I,II | 9 | 56 |
| Dry cleaners/laundries | II | 9 | 0 |
| Government Facilities[a] | I,II | 5 | 80 |
| Auto Parts Stores | I | 5 | 40 |
| Auto Body Shops | I | 4 | 75 |
| University Facilities[a] | | 4 | 75 |
| Muffler/Transmission Shops | I | 4 | 50 |
| Car Washes | I | 4 | 50 |
| Auto Dealerships | I | 3 | 100 |
| Auto/Truck Rental Agencies | I | 3 | 33 |
| Photographic Processors | II | 3 | 33 |
| Utilities | II | 3 | 33 |
| Private Service Agencies[a] | | 2 | 50 |
| Train/Bus Stations | I | 2 | 0 |
| Paint Stores[a] | II | 2 | 0 |
| Plating Shops | II | 1 | 100 |
| Water Conditioning Companies | II | 1 | 100 |
| Party Stores[a] | | 1 | 100 |
| Private Homes[a] | | 1 | 100 |
| Chemical Laboratories | II | 1 | 0 |
| Construction Companies | II | 1 | 0 |

[a] Non-prioritized

## MUNICIPAL IMPLICATIONS

The project has already demonstrated a surprising number of storm drain connections. Local government officials had allowed, and in some cases even recommended, storm drain connections up until 10 years ago. The rationale for this policy was an aging wastewater treatment plant that was significantly overburdened. While some of the problems discovered in this study were the result of improper plumbing or illegal connections, a majority were approved connections at the time they were built. Because Ann Arbor is not expected to be that much different from other communities of its size, it is likely that similar problems exist elsewhere in the U.S. Therefore, because urban storm runoff is polluted in and of itself, it is important that illegal and improper discharges to storm drains like the ones identified in this study be eliminated.

The solution to the Allen Creek storm drain pollution problem is source control. Oliver and Grigoropoulos pointed out that the most effective control effort is to eliminate pollutant sources before the runoff enters the drainage basin or receiving water. Thus, the importance of removing illicit taps from storm drains cannot be overemphasized. Since the initiation of the project, 65 Allen Creek drainage basin businesses have been identified to have at least one storm drain connection. In the 35 months since the dye-testing began, 35 (54%) of the businesses have corrected their storm drain connections. In addition, the chemical sampling of the storm drain outfall has shown a decrease for many constituents since the initiation of the project. A continued decrease in pollutant concentrations can be expected as additional contributors to the pollution problem are eliminated.

It is important to note that both of the NPDES-permitted dischargers to the Allen Creek storm drain had problems with their operations. One permitted facility had 14 storm drain connections in addition to their permitted discharge, including three restrooms that were connected directly to the storm drain. The other NPDES permitted facility has had problems with a sanitary waste discharge that overflows into the storm drain when a pump unit fails. Such problems point to the need to more closely monitor permittees and to evaluate the impact of permitted discharges to the storm drain system.

## CONCLUSIONS

The Federal Water Pollution Control Act (PL 92-500) mandates that all dischargers to the navigable waters of the U.S. be permitted through NPDES. Furthermore, the Act

(Section 305) requires that nonpoint source pollution be described and control methods recommended. This study shows that the following should be considered when identifying nonpoint pollution sources:

• Urban nonpoint pollution is not strictly limited to road runoff. It contains wastes from permitted and illegal discharges, many that were previously unsuspected.

• Urban stormwater runoff is a major contributor of a variety of pollutants and should be considered when assessing wasteload allocations to receiving streams. In fact, the inclusion of urban nonpoint impacts may allow a reduction in the level of treatment which municipalities and industry must provide. This may reduce their financial burden currently required to meet existing water quality standards.

• The present NPDES permitting system may have to be modified to include a more extensive review of applicants for discharge permits and renewals. The agency that oversees this permitting system in each state should not only investigate businesses that apply for discharge permits, but also other potential polluters targeted by the nature of their business.

• Most plans to eliminate the pollution of surface waters by stormwater runoff have centered on treatment of the contaminated stormwater. However, treatment is very expensive and beyond the means of most communities. This project has demonstrated the success that can be attained through source control of improper and illegal waste discharges. With this approach, the cost for water quality improvement has not been put on the taxpayers, but rather on the individual dischargers.

• The need for a more detailed examination of the effect of illegal and permitted discharges as a significant contributor to urban nonpoint pollution has been demonstrated. It is hoped that the information gathered in Ann Arbor will have an impact on local, state, and national strategies developed for the control of urban nonpoint pollution.

## Discussion of Mr. Murray's Paper

Question:

When you hve found hundreds of "bootleggers" who have tapped illicit, nonstormwater sewers into the storm drains, why not just seal them up?

Answer:

That is the instant response, and in our frustration, we would probably all love to do that. But everybody discharging to pipes that end up in storm drains isn't responsible. The only responsible way to proceed is the more tedious method of tracing them, one by one, back to the source.

Question:

Have you tried smoke testing?

Answer:

We talked about it, and we finally concluded that smoke testing won't work for these storm drains.

Question:

Did you consider redefining the storm sewers with all the illicit connections as combined sewers and routing the DWFs to a treatment plant?

Answer:

That was not really feasible in this case. I will say that combined sewers in our area appear to collect legal and illegal discharges plus the first flush of low-flow, small storm-event flows.

Question:

Does law or regulation allow you to have the owner of an illegal connection pay for the fix?

Answer:

Yes, or he can apply for permit, but he won't get one.

Question:

Are there political problems associated with forcing the disconnection of somebody who has been on the wrong sewer for 40 years?

Answer:

Yes, sometimes. But we work with those people and get the problem solved.

Question:

Did you find deliberate cross-connections from the sanitary sewer system to the storm sewer?

Answer:

Yes. Seven.

Question:

Has there been a **hazardous** waste control problem associated with these storm drains or illicit connections? If so, what has been industry's response?

Answer:

We have had lots of hazardous **materials** (not wastes, per say) stored on industrial sites whose runoff runs into these storm drains. One cleaners who had some barrels spill in the backyard had to remove an 80 x 30 x 20 foot hole's worth of soil. We think there is a 1/100,000 chance of getting cancer from eating fish caught in receiving waters below such places. It's very difficult to set standards to control situations like these.

Question:

Has all this disconnect program been through a public hearing process?

Answer:

We did mailers, questionnaires. We sent out 30,000 and got 170 back. There were seven opposing. We had one volunteer to get disconnected.

# SWIRL CONCENTRATORS REVISITED.
## The American Experience and New German Technology

William C. Pisano*

## Abstract

The Swirl Concentrator is a small, compact, "no moving parts" solids separation device. Its function is to act as as a static combined sewer overflow (CSO) regulation chamber. Only dry weather flow is intercepted for discharge to downstream wastewater treatment plant (WWTP). During wet weather the unit's outflow is throttled, causing the unit to fill and to self-induce a "swirling" vortex-like operation. Secondary flow currents rapidly separate "first flush" settleable grit and floatable matter. Concentrated foul matter are intercepted for treatment while the cleaner, treated flow discharges to receiving waters. The device is intended to operate under extremely high flow regimes which are an order of magnitude greater than conventional sedimentation rates, and is therefore extremely cost-effective. The US swirl is actually an adoption of an English design developed in the early 60's. In the 70's the Federal Water Program conducted a long term research program evaluating this technology and generated many conflicting claims and unanswered questions which are beginning to be clarified from recently implemented facilities. This paper overviews the American experience and recomends corrective measures. In the mid 80's, English and German investigators developed insights into the American swirl performance problems, and have developed new improved versions similar to the earlier English design. One prototype installation in West Germany was implemented in August,1987 and is currently being tested. Highlights of two US projects using this new German technology (design phase) are also presented.

## 1. INTRODUCTION AND HISTORICAL PERSPECTIVE

An increasing number of urbanized areas within the US are being pressed to control CSO. At the same time federal financial aid is rapidly diminishing. Cost-effective technolgies are needed to solve these problems. The urgency for such solutions is the same today as it was in the 60's when the Federal Water Program recognized that conventional storage with drainage to WWTP would result in excessive national cleanup costs. Federal research focused its efforts on developing new innovative cost saving technologies for abating "first flush" urban runoff pollution. The concept of the Swirl Concentrator emerged during that era.

In many other areas of the country, pollutant emissions from separated storm sewered systems impact fragile water bodies. Although the concept of wet detention/retention ponds are ideal, they may not be a practical solution for highly congested land-constrained areas. Small-scale compact remotely-operated treatment controls for removing "first flush" filth in stormwater are still needed for certain types of problems.

### A. First Vortex Solids Separator, Bristol, England

The pioneer of this field, Bernard Smisson of Bristol, England conceived the notion of circulating stormwater in a vortex chamber to remove settleable solids (vortex chambers in the UK were prior used for flow control at overflows - similar use in Portland,Oregon). The resulting vortex flow condition produces a solids separation effect in which settleable solids are swept by secondary flow currents down to a gutter draining to an off-centered foul sewer outlet flowing to a

---

* President, Environmental Design & Planning, Inc., 253 Washington St. Belmont, MA 02178
(617) 484-8087.

downstream WWTP. The clearer flow spills over a central concentric weir to a conduit below the chamber floor leading to receiving stream.

In 1963, two such vortex chambers, 5 m (18 feet) in diameter, were built of brick in Bristol and still successfully operate today /1/. Smisson noted by observation and by measurement that settleable solids separation efficiencies during low storm flow conditions (less than 30% design flow) were nearly twice the desired separation level (35% settleable soldis removal) at peak design flow conditions.

### B. US Swirl Program Development

The favorable U.K. experience in Bristol motivated U.S.EPA and American Pubic Works Association in the late 60's to begin a 12 year R&D program investigating the viability of this technology for application in the U.S. The resultant product of this research program was the development of a design manual proscribing vessel dimensions as a function of design flow rate for varying settleable solids separation levels /2/. The modified US design incorporated internal baffles and spoilers meant to impede free vortex conditions which were believed to limit solids separation efficiency. The device was called the "Swirl Concentrator", see Fig. 1 /3/. The technology seemed promising since model tests showed that substantial settleable solids/floatables removal could be accomplished within a small physical space.

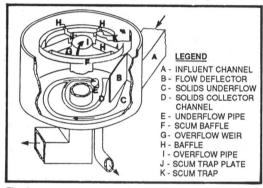

LEGEND
A - INFLUENT CHANNEL
B - FLOW DEFLECTOR
C - SOLIDS UNDERFLOW
D - SOLIDS COLLECTOR CHANNEL
E - UNDERFLOW PIPE
F - SCUM BAFFLE
G - OVERFLOW WEIR
H - BAFFLE
I - OVERFLOW PIPE
J - SCUM TRAP PLATE
K - SCUM TRAP

Fig. 1
**Schematic View of Swirl Concentrator**

The device seemed perfect for remote and inexpensive control of "first flush" filth and what only remained was to test the device within the American setting.

### C. Syracuse, New York - Demonstration Swirl For CSO

The first US swirl concentrator (12 ft diameter - 3.8 m) demonstration project was conducted in Syracuse, New York in the mid 70's. The project motivated further investigations as the reported results were extremely impressive. Total Suspended Solids (TSS) and Biochemical Oxygen Demand (BOD5) concentration efficiencies in excess of 60% were reported /4/. The results created a high degree of expectation for the ensuing swirl demonstration projects that unfortunately never materialized. The average storm flows during the test period were less than 30% of the design flow used to size and dimension the unit. In retrospect, the results reported could be expected, but the unit's operational performance was not tested in the critical design flow range (much higher), used in the earlier bench model studies to develop swirl dimensions for varying settleable solids performance standards. The results however, are valid for the flow range or conditions actually encountered.

### D. Lancaster, PA. Demonstration Swirl For CSO

In the mid 70's, a full scale 24 foot diameter (7.3 m) swirl concentrator was constructed in Lancaster, PA. for handling CSO from a 250 acre watershed having steep topography with expected high degrees of "first flush". The unit was deliberately sized such that it would be frequently stressed at or above design flow conditions (40 cfs - 1130 l/s) to permit repetitive evaluative testing .

Solids separation efficiency for all forms (TSS, settleable and volatile solids) performance were generally, but not always, favorable for most measured events during the one year evaluation at Lancaster. Solids concentration separation generally exceeded 60% during obvious "first flush" conditions /6/. Most of the settleable matter noted during the extremely peaked "first flush" periods was inorganic. BOD5 testing was not performed and solids removals during runoff periods subsequent to "first flush" were negligble. Settling tests could not confirm whether "first flush" materials encountered were similar to assumed particle fractions and settling velocities used in the model experiments (sample collection problems).

On several occasions during intense summer thundershowers, turbulence within the chamber during presumed peak flow conditions was visibly dramatic and appeared excessive. Flow metering problems precluded documentation of mass rate separation efficiency. It was probable (but not certain) that the unit was evaluated at design flow conditions during several storm events. The evaluation was encouraging but left many unanswered questions.

### E. West Roxbury, Boston Swirl For Stormwater

During the late 70's the Federal government commenced the 208 Areawide Wastewater Management program aimed at defining potential control strategies for nonpoint source pollution problems. At the same time there was interest in developing a permit system for storm sewered discharges. Performance standards and technolgy controls were to be part of this grand scheme.

The government became interested in the swirl concept as a potential "stand-alone", self-activated, inexpensive stormwater pollution control device meant to treat direct discharge from storm sewer outfalls. The US EPA, the State of Massachusetts and the Boston Water & Sewer Commission collectively funded and directed a R&D program in Boston aimed at testing the effectiveness of a reasonably full-scale swirl along with another secondary current flow treatment device called the "Helical Bend Regulator" for removing settleable solids and floatable matter from stormwater runoff.

At the West Roxbury facility (westerly suburb community of Boston), stormwater runoff from 160 separated acres (65ha) entered the test facility. A portion of the inflow was diverted by gravity and was spilt evenly to the two units. The topography of the tributary area is rolly to mild. The land use is mostly low density single family homes with some open area, and the soils are a mix of sand/gravel and ledge which are densely covered with high percentage of decidous trees. Design flow into each unit was set at 6 cfs (170 l/s) each with surcharge limit of 12 cfs each. The swirl's diameter was 12 feet (3.7 m ) and was of steel construction with epoxy coating. The device was sized as per the EPA manual /2/ to remove 80% theoretical settleable solids. As with Lancaster, the aim was to frequently stress the unit at design flow conditions during the evaluation period.

The evaluation program began in late 1979 and concluded in the fall of 1981. Table 1 summarizes the results of the test program which shows that the TSS separation efficiency of the swirl ranged from 5% to 30% /5/ ("removals" reflecting underflow effects are higher). TSS efficiencies for the swirl attained levels as high as 50% for short intervals during several of the

TABLE I  Summary: West Roxbury Evaluation; Suspended Solids Removal and Efficiency Rates

| Date | Flow cfs | Removal % | Efficiency % |
|---|---|---|---|
| Swirl Concentrator | | | |
| 6/29/80 | 2.0 | 21.0 | 8.2 |
| 7/29/80 | 3.0 | 35.3 | 27.5 |
| 10/3/80 | 6.0 | 36.0 | 32.0 |
| 10/25/80 | 3.0 | 29.7 | 21.4 |
| 6/9/81 | 0.8 | 34.0 | 2.75 |
|  | 3.0 | 27.0 | 19.0 |
| 6/22/81 | 2.2 | 27.0 | 16.0 |
|  | 2.0 | 33.0 | 21.0 |
| 8/4/81 | 6.0 | 9.5 | 5.5 |
|  | 12.0 | 5.8 | 5.7 |

storm events. Settleable solids removals were within the same range. Floatables removal visually appeared good, except during the heavy fall leaf generation period.

Unexpected microscale watershed effects impacted the test program. First, the relative preponderance of grit and runoff peaked ness varied during the program and depended to a significant extent when the large diameter catchbasins with deep sumps in the catchment were cleaned.

The second complication was an intermittent overflow to the storm sewer system in the catchment from an undersized holding lagoon for re-cycled washwater from a sand, gravel and stone crushing operation. The operation discharged with no regularity for several days at a time, about 1-2 cfs containing high concentrations (300-400 mg/l TSS) of very fine "rock flour" particles. When present, the high concentration of fine quarry rock were not removable in the swirl and tended to "mask" and obscure performance results.

Simple visual dye experiments were conducted for a number of storm events covering a wide variety of flow conditions. In all cases, the influent plume immediately rose to the surface just beyond the inlet deflector wall and as expected, the time for complete circulation varied with flow rate. Visual turbulence levels became dramatic near and above design flow conditions and were similar to those noted in Lancaster.

## 2. MAJOR MUNICIPAL SWIRL FACILITIES

During the early 80's, a number of swirl concentrator complexes nevertheless proceeded to design and eventually were constructed. Facilities were constructed at: Toledo, Ohio; Presque Isle, Maine; Decatur, Ill; and, Auburn, Indiana. A major facility is presently under construction and nearing completion in Washington D.C. for removing particulate BOD5 from a large catchment of the District to lessen adverse sediment oxygen demand impacting the Anacostia River. A full evaluation program with elaborate sampling procedures is proposed.

With the exception of the McKinley Ave. facility in Decatur, performance evaluations have been meager and reported results have been mixed. The Presque Isle facility performance appears to be promising as average TSS concentration efficiency for 9 storms (1986-1987) exceeds 45% (flow levels are not known: instrumentation problems) /6/, while the Toledo swirl performance has been much lower, TSS and BOD5 removal ranges of 16-39% and 0-17%, respectively /7/. The Decatur, Ill. evaluation program monitored for the very first time ever, TSS and BOD5 performance chacteristics covering the full flow range (well below, at, and above design conditions) where discharge was carefully measured.

### A. Presque Isle, Maine Swirl Facility,

The Presque Isle swirl concentrator treats bypassed separate sanitary flow impacted by excessive wet-weather infiltration and sand inflow. The swirl has a diameter of 18.5 feet (5.5 m). The inner assembly is constructed of stainless steel plate within a concrete shell. The device handles flows in excess of 5 MGD (219 l/s) - the maximum WWTP rate. Swirl design flow is 6 MGD (262 l/s) for 6 month storm and the maximum flow projected is 7.4 MGD (324 l/s) expected for a 50 year event.

The facility's design is of interest since it is an exception to American practice to date. The sizing does not conform to EPA design handbook dimensioning /8/ for settleable solids removal as the unit was deliberately over-sized (the EPA handbook would recommend a 13.5 foot (4.1m) diameter unit for the 50 year storm flow for 90% settleable solids removal). Motivation for this design change stems from the successful. Syracuse swirl results at reduced flow levels, the visual observations of excessive turbulence during the Lancaster evaluations, and solids settleability tests. Problems with again measuring discharge preclude definitive performance conclusions.

## B. McKinley Ave, Decatur, Illinois CSO Facility

The most definitive results regarding the solids separation effectiveness of the US swirl occurred last year during the performance evaluation of the 40 MGD (1750 l/s) swirl in Decatur. The facility treats CSO from a 661 acre (267 ha) catchment on the north side of the city. The land use in the area is mostly residential with some commercial. The topography is generally rolling.

The layout is shown in Fig. 2 /9/. Diverted CSO is first screened by passage through a 1-inch spacing mechanically cleaned, automatically controlled, catenary screen. Bypass flows through a manually cleaned bar screen. Downstream of the screen chamber are two liquid-level actuated, motorized sluice gates directing flow into the "first flush" retention tank, or alternatively to the the swirl. "First flush" volume is defined as CSO with TSS, BOD5, and volative SS levels greater than longterm averages at downstream WWTP. Illinois standards require secondary treatment of this quantity for one year-one hour storm event. CSO with lesser levels require preliminary treatment before discharge.

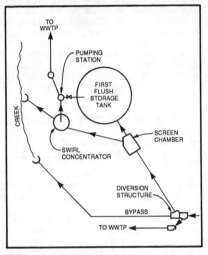

Fig. 2
McKinley Ave. Facility

The first 0.5 MG (1895 cu.m) of CSO or "first flush" is directed to the 0.632 MG (2395 cu.m) open storage tank. This tank is 85 feet (25.9 m) in diameter, 24.8 feet (7.6 m) deep, equipped with mixer and aerators, and the tank's floor slopes to a circular gutter 46 feet (14 m) in diameter draining to a pumping station. Excess flows are then diverted to the swirl concentrator. The swirl is 25 feet (7.6 m) in diameter and 19 feet (5.8 m) deep with an operating water depth of about 6.8 feet (2.07 m). The swirl underflow drains to the foul sewer pumping station where two 350 GPM (22 l/s) submersible pumps discharge underflow either to the interceptor (if capacity available) or to the "first flush" tank. Treated flows from the swirl discharge by gravity to a nearby creek. All flows beyond the swirl design flow of 40 MGD (1750 l/s) overtop a high level weir in the diversion chamber and flow directly to nearby creek.

The design assumption for the McKinley Ave. swirl is that the device is to remove at least 15% of all BOD5 and TSS loadings, bypassing the "first flush" tank for the 1 year-1 hour design event (1.2 in./30.5 mm rainfall). Flow in excess of 40 MGD is directly bypassed to creek. It was further assumed that this level of control could be achieved by a swirl sized (from the US EPA handbook) to remove 90 % settleable solids at the design flow condition of 40 MGD. Note that the swirl is to handle CSO not captured by the "first flush" tank.

The evaluation program was conducted over 1987 covering 30 events into the facility. The initial mode of operation was as designed, that is, flow was first directed to the "first flush" tank, and then when storage allocated to "first flush" was filled, flow was then diverted to the swirl. This method continued until July handling 18 events. Both the "first flush" tank and the swirl were used for only seven of the events. Evaluation of the swirl was critical as the City is under a court order performance-based program, and is ready to implement four more (designed) CSO configurations based on similar design concepts. The "first flush" tank was bypassed for the remainder of the evaluation so that the swirl could be repetitively tested.

A total of four storms were monitored for TSS, VSS, BOD5, rainfall, and discharge during the remainder of the evaluation program. Performance efficiencies were based on samples taken upstream of the mechanical screen near invert and from a location just below the swirl clearwater weir /10/. Samples were taken with automatic equipment at frequent time intervals (typically one-two minutes). Discharge levels were noted at concurrent time intervals. TSS, non-settleable solids (after one hr settling), total settleable solids, volative settleable solids, and unfiltered BOD determinations were made.

Summarized pollutant mass removal results attributable to the swirl operation for the four storms events are given in Table 2. Average pollutant mass removals per event are presented for three data categories: A)"entire storm", B) "the duration of storm less obvious "first flush" periods", and C) "first flush" only. Average flow conditions, number of samples and duration are also noted. Generally the amount of pollutants removed during the "first flush" periods (Category C) was about a third of overall total event removals (not shown in Table).

From inspection of Category A in Table 2, it appears that average event performance generally decreases with increasing flow magnitude. Exclusion of the "first flush" data (Category B) indicates that little treatment for all pollutant forms is achieved when discharge exceeds about 25% of design flow. Category B results would be generally typical of the intended facility concept. It is unlikely that the US swirl design would satisfy the project's performance standard given normal method of operation.

Results depicted by Category C generally show that the US swirl is capable of substantial removals of "first flush" related contaminants, but not at the levels assumed in the design manual. A careful examination of mass removals for the event of 7/26/87 indicated that the swirl was twice stressed at the design flow of 40 MGD. The peak rate of "first flush" mass inflow coincidentally occurred at design rate of 40 MGD, about 10 minutes into the storm event (rising limb). Settleable solids and BOD removals were about 40% at this rate. At 15 minutes the storm peaked at about 70 MGD (no removal), and then uniformly declined to zero 30 minutes later. Settleable solids and BOD removals were negligible (SS = 12%, BOD = 0) at the 40 MGD rate (falling limb), but again rose to substantial levels (30-50%) when flow was less than 10 MGD. It appears that the unit's performance is extremely sensitive to solids composition and deteriorates once "first flush" conditions are passed. The unit does not however remove desired level of settleables at design conditions.

Careful visual inspection of the swirl's water surface around design flow of 40 MGD indicated significant turbulence in the tank. The turbulence is evident in an area from approximately where the flow deflector wall ends to about 120 degress downstream of the wall and appears as "boils" of turbulence at the tank's surface. These "boils" indicate vertical flow patterns in the tank. This vertical flow is sufficient to carry and sweep lighter material under the scum baffle and over the clearwater weir. It is felt that this turbulence is impairing the removal efficiency /10/.

## 3. SUMMARY - AMERICAN SWIRL RESULTS

The US swirl urban runoff evaluation experience is summarized as follows. The Syracuse evaluation showed that the swirl was capable of high solids separation performance at flow conditions much lower than design flow. The results are valid for the conditions encountered, but not a true test of the handbook procedure used to dimension the Syracuse swirl.

The Lancaster evaluations were overall favorable, but not consistent and the actual flow levels were generally unknown (although "perceived "high flows during evaluation did occur). The unit was capable of significantly reducing high concentrations of settleable inorganic solids occurring during the onset of flash, highly peaked storm runoff conditions.Removal rates rapidly fell as "first flush" passed. Absence of reliable flow measurements precludes removal performance conclusions. Substantial surface turbulence was visually observed during presumably high flow conditions.

## TABLE 2

MCKINLEY AVENUE FIRST FLUSH TREATMENT FACILITY: DECATUR, ILL.
SUMMARY OF SWIRL CONCENTRATOR EVALUATION (7/26/87- 9/29/87)

| EVENT DATE | AVG. FLOW[1] (MGD) | TSS[2] (%) | SS[3] (%) | VSS[4] (%) | BOD[5] (%) | NO. OBSER/TIME (MIN) |
|---|---|---|---|---|---|---|
| **A. MASS POLLUTANT REMOVAL EFFECTIVENESS - ENTIRE STORM** | | | | | | |
| 8/08/87 | 10.1 | 33.1 | 36.0 | 23.3 | 44.4 | 17/54 |
| 9/16/87 | 13.1 | 16.4 | 20.2 | 20.8 | 42.7 | 24/53 |
| 9/29/87 | 24.0 | 23.2 | N.A. | N.A. | N.A. | 29/65 |
| 7/26/87 | 31.4 | 22.9 | 27.2 | 23.3 | 13.1 | 28/63 |
| **B. MASS POLLUTANT REMOVAL EFFECTIVENESS - EXCLUDE FIRST FLUSH** | | | | | | |
| 8/08/87 | 9.3 | 36.7 | 38.8 | 45.4 | 44.8 | 9/48 |
| 9/16/87 | 15.0 | 33.9 | 7.8 | 8.7 | 41.1 | 4/32 |
| 9/29/87 | 26.4 | 5.7 | N.A. | N.A. | 9.3 | 21/51 |
| 7/26/87 | 32.4 | 5.2 | 7.9 | 0.1 | 0.1 | 22/53 |
| **C. MASS POLLUTANT REMOVAL EFFECTIVENESS - ONLY FIRST FLUSH** | | | | | | |
| 8/08/87 | 9.0 | 43.6 | 43.7 | 56.6 | 43.8 | 4/6 |
| 9/29/87 | 9.8 | 80.6 | N.A. | N.A. | 45.0 | 8/14 |
| 9/16/87 | 10.4 | 32.6 | 36.0 | 36.7 | 43.9 | 13/21 |
| 7/26/87 | 26.5 | 56.4 | 62.1 | 55.0 | 34.2 | 6/10 |

NOTES:

(1) AVG. FLOW = TIME AVERAGE (ONSET TO DECLINE OF ABOUT 1 MGD)

(2) TSS = TOTAL SUSPENDED SOLIDS

(3) SS = TOTAL SETTLEABLE SOLIDS DETERMINED BY DIFFERENCE OF INITIAL SUSPENDED SOLIDS - DECANTED SUSPENDED SOLIDS AFTER ONE HOUR SETTLING

(4) VSS = VOLATIVE SETTLEABLE SOLIDS

(5) BOD = BIOCHEMICAL OXYGEN DEMAND (5-DAY) (UNFILTERED)

* EFFICIENCY = CUMULATIVE MASS INFLOW/CLEAR WATER MASS OUTFLOW (WITH FOUL SEWER ADJUSTMENT )
* POLLUTANT REMOVALS ARE MASS AVERAGE REMOVALS ADJUSTED FOR UNDER-FLOW OF 1 MGD
* INTERVAL OF FIRST FLUSH WAS DECIDED BY RAPID RISE OF INFLUENT TSS CONCENTRATION FROM THE STORM AVERAGE AND SUBSEQUENT DECLINE TO SIMILAR LEVEL . NOTE THAT "FIRST FLUSH" USED IN ANALYSIS IS NOT RELATED TO CONCENTRATIONS IN EXCESS OF THE LONG-TERM DOWNSTREAM WWTP AVERAGE, BUT TO STORM AVERAGE.
* LAST ROW SHOWS NUMBER OF OBSERVATIONS AND PERIOD OF SAMPLING OR DURATION IN MINUTES.

The West Roxbury swirl evaluation showed that moderate degree of "preliminary" TSS removals could be achieved in a separated sewer setting. Erratic watershed effects precluded definitive settleable solids evaluation at design flow conditions. The unit always exhibited visual surface turbulence (dye tests showed this to occur at all flow conditions). The swirl concept is probably reasonable for expected urban stormwater solids/floatables reductions in difficult, space constrained and congested areas where no other controls are feasible.

The Presque Isle swirl was deliberately oversized and seems to consistently work. It is highly unlikely that flow levels could ever achieve the "settleable solids design flow conditions" established in the manual.

The McKinley Ave. swirl verifies the earlier Syracuse findings that high BOD5 and TSS removals are achieveable, but only at flow conditions a fraction of design flow . At flows around design conditions, the results are mixed: the separation efficiencies are about half of projected design levels during obvious "first flush" conditions and very low in the absence of "first flush". Solids separation performance rapidly falls to zero as soon as design flow is exceeded. Visual surface turbulence at design flow conditions seems to coincide with the monitored low separation levels.

In summary, it appears subject to further verification, that the design handbook proscribes vessel dimensions that simply permit too much flow (energy) at design flow conditions to be concentrated into a small vessel whose rotational section is disturbed by eccentric gutters, spoilers, and a deflector wall.

## 4. EUROPEAN SWIRL R&D EXPERIENCE

The American experience with swirl concentrators has been closely observed by European designers over the last decade. The inclusion of additional deflectors, spoilers, gutters and scum baffles into the U.S. design during the 70's R&D program to improve perceived performance is questionable. Recent experience indicates that flow in a vortex chamber with such jagged protruding intervals is relatively turbulent so that separation efficiency is reduced /11/.

In the early 80's Balmford /12/ performed model experiments using simplified vortex overflows which again had a smooth rotationally symmetric vortex chamber with central outlet for concentrated flows. Clear water overflow from the chamber occurred through a section of the outside wall. One practical application is currently being implemented.

In the mid 80's, Brombach concluded that the separation efficiency of current vortex separators could be improved and process performance uncertainty reduced. This conclusion was strongly motivated by earlier research work measuring actual turbulence levels within vortex

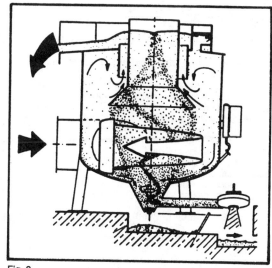

Fig. 3
Idealized View of Vortex Separator During High Flow Operation

chambers /13/ using laser-doppler-anemometry methods. Low turbulence levels were noted in smoothly shaped vortex chambers and is believed to be a requiste condition for solids separation within vortex chambers. Brombach sought to improve Smisson's original vortex design by improving the secondary current solids separation effects while still maintaining free vortex conditions. Details of his experimental work are provided elsewhere /14/, summarized /15/ and will be only capsulated here. Brombach's aim was to develop a vortex chamber for operating at high hydraulic loadings and to substantially remove settleable and floatable solids. Since storage is primary consideration in German urban runoff control philosopy, the vessel volume was to be maximized so that the chamber would first provide detentive storage (for small storms) and then act a solids separator for higher flow rates.

A schematic representation of the new German vortex separator is shown in Figure 3. The rotationally symmetric vortex chamber with a well-rounded bottom edge has a conical bottom sloping to the center. The diameter is greater than the height to maximize volume. The foul sewer outlet is located on the rotational axis and the inlet is tangential to the vessel. A guiding screen and scum board are attached to the cover of the vortex chamber. The guiding screen.prevents particles which are unavoidably entrained upward by the secondary flow currents from entering the overflow section. The final geometry of the guiding screen was selected after many "trial and error" attempts. Overflow leaves the unit through an annulus between the scumboard and the guiding screen and flows on roof top to discharge. The under-flow is throttled to allow operation during wet weather. Differences from US swirl are the center-placed foul outlet, clear water roof discharge,conical floor (no gutters) and no inner assembly section/components transverse to rotational flow axis.

In August, 1987 a treatment facility with two parallel German vortex separators commenced operation treating CSO from the Village of Tengen, a small farming community near the German/Swiss border. Treated overflow discharges to a nearly brook. Design inflow to the facility requiring treatment is 34 cfs (950 l/s) with maximum installation discharge of 104 cfs ( 2929 l/s). Maximum allowable foul sewer flow is 1.3 cfs (35 l/s) which is processed by WWTP located along the Rhine River. Level sensing and recording equipment are in use to develop long- term volume/level statistics. Evaluation of the configuration will be conducted over the next two years.

## 5. NEW US INSTALLATIONS USING GERMAN VORTEX SEPARATOR

### A. Seventh Ward CSO Facility; Decatur, Illinois

Over the last four years construction documents for four other CSO treatment facilities have been prepared using concepts similar to McKinley Ave configuration. The 7th Ward CSO facility on the south side of the city near the WWTP has been next slated for construction. The "first flush" storage tank will be 1.2 MG (4550 cu.m) and the design swirl capacity is 113 MG (5000 l/s). The catchment is rolly to mild terrain and the land use is mostly residential. A US swirl with diameter of 44 feet (13.5 m) had been proposed to handle "first flush" storage excess. Foundation construction is a significant factor in this design due to groundwater. Two German vortex separators in parallel are proposed in lieu of the US swirl in view of process uncertainty, particularly during periods following "first flush". Another factor favoring the German design is the high foundation/clear water conduit cost associated with US swirl (clear water discharge for German device is near ground level - significant cost saving) .

### B. 14th St. Pumping Plant CSO Facility; Saginaw, Michigan

The second proposed CSO Facility employing the new German type vortex solids separator is in Saginaw, Michigan. The Saginaw River about evenly divides the 10,437 acre (4200 ha) City which is serviced by a combined sewer system. Overflows may occur at 34 regulator chambers. Five pump stations relieve the system during flooding conditions. The intercepting tunnel sewers run along both sides of the river with a river crossing and WWTP on the northeast side of the City.

The City's CSO effort has been described elsewhere /16/. The program started over 18 years ago with a CSO facility concept of pumped overflow consolidation to detention and sedimentation facilities located at the major pumping plants. The West Side 3.6 MG (13,640 cu.m) detention and sedimentation facility at the Hancock St. Pump Station was completed in 1976 and has been followed by rehabilitation of 34 regulation chambers in 1984-1986 using vortex valves and chamber weir modifications to enhance in-line storage. Another West Side 6 MG (22,740 cu.m) retention facility is presently in the design phase.

The design concept for the new 14th Street facility folds together the 1972 CSO facility plan technology (detention/sedimentation) and the new vortex solids separator technology. Packaging vortex separators with conventional facilities would provide greatest adaptive operational flexibility; reduce the cost of maintenance and problems in flushing settleable solids and debris from detention tanks; and would reduce overall disinfection costs as the regulatory agency is presently requiring set-aside space at new CSO treatment installations for future de-chlorination facilities.

The 14th St. Station has three pumps each with 100 cfs (2830 l/s) capacity. Four low-level trunk sewers drain the 14th St. District discharge. The fifth main sewer drains the 16th St. District. Total flow capacity tributary to the facility is about 600 cfs (17,000 l/s) with an area of 1103 acres (460 ha). The Station pumps stormwater overflow to the river from the four low-level 14th St. sewers which otherwise flow to the East Side Interceptor by gravity during dry weather. Pumped station discharge and 16th St. trunk sewer overflow (spill over backwater gates) discharge into two concrete box culverts and then into an earthen ditch before discharging to the river. See Figure 4.

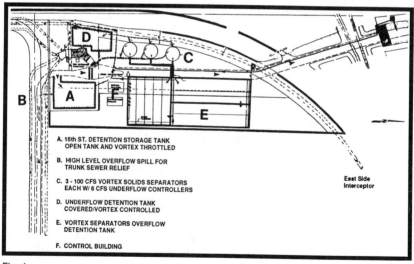

Fig. 4
**Proposed 14th St. Pumping Station CSO Facility**

Treatment of 14th St. pumped discharge occurs with initial processing by three new reinforced concrete German-type vortex solids separators about 30 feet (9.2 m) in diameter and each with maximum capacity of 100 cfs (2830 l/s).

Treated overflow from the vortex separators discharge in a channel over the top of existing box culverts and into a new 2 MG (7580 cu. m) detention and storage facility. This tank will retain 15 min. of "one pump-on" (common event) and will provide further treatment of clear water overflow from the vortex solids separators. The overflow from this treatment facility will then discharge back into the box culverts draining to ditch storage and ultimately the river. All retained flows will then be drained back to the WWTP.

The underflow from each vortex solids separator will be throttled and discharge into a new covered foul sewer detention tank. The tank's discharge will be controlled by an adjustable flow throttle to an existing tunnel drop shaft discharging to the East Side Interceptor.

Detention storage for 16th St. sewer is provided in a 0.65 MG (2470 cu.m) shallow open storage tank. This tank is to detain small volume frequently occurring overflows from the 16th St. sewer for bledback to the East Side Interceptor.

## 6. CONCLUSIONS

The spoilers, gutters, and deflector inlet baffle assembly within the US based swirl causes excessive vessel turbulence, which in turn reduces solids separation efficiency. Unless it can be shown by evaluation data, swirls build in accordance with established EPA methods should be operated at lesser hydraulic rates than designed for. New designs should be over-sized relative to handbook recommendations. Settleability testing of potential "first flush" characteristics is an essential requiste for rational basis of design.

A new improved German "swirl" design has emerged and is designed to operate at high flow rates for settleable solids separation. The design concept minimizes vessel turbulence and appears to be less costly to construct.

## 7. REFERENCES

/1/ Smisson, B.　　Design, Construction and Performance of Vortex Overflows. Proc. Symposium on Storm Sewage Overflows, Institution of Civil Engineering. pp. 81-92, London, 1967.

/2/ Sullivan, R. et al.　　Design Manual: Swirl and Helical Bend Pollution Control Devices. U.S. EPA-600/8-82-013, July, 1982.

/3/ Field, R.　　Design of Combined Sewer Overflow Regulator/Concentrator, J. WPCF, Vol 46 No 7 , July, 1984.

/4/ Drehwing, F. J. et al.　　Disinfection /Treatment of Combined Sewer .Overflows - Syracuse, New York. O'Brien & Gere Inc, Report No. EPA-600/2-79-134, Aug.1979

/5/ Pisano, W. et al.　　Swirl and Helical Bend Regulator/Concentrator for Storm and Combined Sewer Overflow Control. EPA Grant No. S805975-report, January, l982.

/6/ Pisano, W.　　Personal Communication, Wright-Pierce Engineers, William Brown, July, l987.

/7/ Wordelman　　Swirl Concentrators That Treat Excess Flow At S. Toledo, Ohio, Jones & Henry Engineers, Toledo, Ohio, June 1984.

| | |
|---|---|
| /8/ Pisano, W. et al. | Application of Swirl Technology, Presque Isle, Maine. Report prepared for Wright-Pierce Engineers, Environmental Design & Planning, July, 1980. |
| /9/ Heinking & Wiloxon | Use of a Swirl Concentrator for Combined Sewer Overflow Management. J WPCF, No. 5, 1985. |
| /10/ Hunsinger et al. | Combined Sewer Overflow, Operational Study, Bainbridge, Gee, Milanski Associates and Crawford, Murphy & Tully, Inc. ,Sanitary District of Decatur,Illinois, Dec.,1987 |
| /11/ Treutler | Absetzverfahren mit kontrollierter Hydraulik. Wasser und Boden, Heft 3, p. 95-96, 1985 |
| /12/ Balmforth et al. | Developement of a Vortex Storm Sewage Overflow withPeripheral Spill. Proc. 3rd International Conference on Urban Storm Drainage.pp. 107-16, Goteborg, Sweden, 1984. |
| /13/ Brombach, H. | Geschwindigkeitsmessungen mit dem Laser-Doppler Anemometer in Wirbelkammerverstaerkern.(ATM) Archiv fuer Technisches Messen,Blatt V 144-8,Heft 9, S. 263-300, 1976. |
| /14/ Brombach, H. | Liquid-Solid Separation at Vortex-Storm-Overflows Fourth International Conference On Urban Storm Drainage, Lausanne, Switzerland, 1987. |
| /15/ Pisano, W.&. Brombach, H. | Recent European Experience Provides New & Motivation for Utilization of Vortex Solids Separators for Combined Sewer Overflow Control", Oct. 1987, Water Pollution Control Federation,. Philadelphia, Pa. |
| /16/ Pisano ,W. | Combined Sewer Overflow Case Study: Reduction of Pollutant Loadings to Saginaw River."Conference on Water Quality, a Realistic Perspective", University of Michigan, Ann Arbor, Feb., 1987. |

## Discussion of Dr. Pisano's Paper

### Question:

Even if the swirl concentrator didn't work well in all the cases you have described, would a somewhat larger one, which your data appear to show would work, be superior to your next most expensive alternative?

### Answer:

Yes, I couldn't agree more. Just make these things a little bigger than the original design criteria curves indicated, and they will work very cost-effectively.

### Comment:

The original intent of this device was to act as a combination flow regulator/separator and a solids capture device. It was found that because there was little difference in the inertial properties of water and most sanitary sewage solids, that the device would perform well as a solids capturer only for solids of grit sizes and above and with a specific gravity of 2.5.

The key to designing these systems, then, was to get accurate setting velocity data for the solids from the municipal area of interest. And, naturally, the sample(s) must be taken such that the heavier particles are represented.

Now, simply making the device larger in an attempt to make it a more efficient solids removal device appears to oppose these basic ideas.

### Comment:

The swirl concentrator, based on the data shown, appears to work at its worst as a solids capture device when it is operating near its design flow rate. It appears to be an inappropriate device for treating storm runoff effectively, compared to other devices we have had introduced which appear to work reasonably well over a wide-range of flow rates.

SOURCE CONTROL OF OIL AND GREASE IN AN URBAN AREA
by
Gary S. Silverman[1] and Michael K. Stenstrom[2], M. ASCE

ABSTRACT: We explored a variety of techniques to remove oil and grease from urban stormwater runoff. Previous work has shown that relatively high concentrations and substantial mass loadings of oil and grease occur in stormwater runoff draining urban areas, with a substantial potential to adversely impact receiving waters. Furthermore, modeling has shown that waters off-shore from urban areas can be protected by implementing controls on a relatively small fraction of the watershed. However, traditional technologies designed for industrial and municipal wastewater do not have the capability to remove oil and grease from stormwater on a cost-effective basis. Four structural control technologies are described which have the potential for being practical tools for enhancing water quality through the removal of oil and grease from urban stormwater runoff.

INTRODUCTION

We have evaluated a variety of techniques for controlling oil and grease discharge from stormwater runoff draining urban areas (details of this work are published in Stenstrom et al. 1982). Devices designed for municipal and industrial wastewater treatment offer little potential for implementation, due to cost and performance inefficiencies resulting from the sporadic nature of stormwater flow. Thus, we have considered non-traditional methods for reducing the magnitude or impact of oil and grease discharge from urban stormwater runoff.

The significance of oil and grease in urban stormwater runoff has been defined in terms of relative contribution as an input source, but is less well understood in terms of environmental degradation. A variety of studies have shown that substantial concentrations and mass loadings of oil and grease typify urban systems. For example, Connel (1982), Whipple and Hunter (1979) and Stenstrom et al. (1984) reported that urban runoff is a substantial source

---

1. Director and Assistant Professor, Environmental Health Program, Bowling Green State University, Bowling Green, Ohio 43403. (419) 373-8242.
2. Professor, Civil Engineering, University of California, Los Angeles.

of oil and grease loading to the Hudson Estuary, Delaware Estuary, and San Francisco Bay, respectively. Future control of point sources is anticipated to increase the relative contribution of oil and grease from urban stormwater (Silverman et. al. 1985). However, while oil and grease in urban stormwater is known to contain trace quantities of many extremely toxic hydrocarbons (Eganhouse et al. 1981; Fam et al. 1986), very little is known about environmental degradation caused by discharge from this source.

Regulatory pressure to force control of urban runoff has been slow in developing, but appears inevitable. The Water Quality Act of 1987 postponed promulgation of U.S. Environmental Protection Agency (EPA) stormwater regulations, but established firm deadlines for development of a comprehensive program to control stormwater point source discharges. For example, U.S. EPA must promulgate regulations governing discharge from large municipalities by February 4, 1989. Implementation of such a program may result in a large demand for control measures to reduce oil and grease from urban stormwater. However, practical methods have not been demonstrated which can accomplish this reduction. In this paper, control measures are described which have the potential for use in urban areas to provide protection to receiving water from oil and grease in urban stormwater runoff.

## TRADITIONAL CONTROL TECHNOLOGY

A variety of traditional "end-of-pipe" treatment options were explored, with little potential seen for treating urban stormwater. Oil and water separators were not seen to be practical for a number of reasons. API-type oil/water separators are relatively expensive, with their use typically limited to waste streams with fairly high concentrations of oil and grease, such as found in refinery wastewater, that would foul other types of systems. Corrugated or parallel plate separators provide a more cost effective means of treating moderate concentrations of free oil and grease, but since 40-60% of oil and grease in urban stormwater typically is in a colloidal or dissolved state (Eganhouse and Kaplan 1981, Stenstrom et al. 1984), removal efficiencies would be low. Furthermore, frequent cleaning might be required due to a build up of silt and grit not found in most oily process water. Dissolved air flotation and high rate filtration offer the potential of higher efficiency and lower surface area requirements than separators, but at a higher cost. Furthermore, the use of coagulants would be required to obtain maximum efficiency, which would be impractical at most stormwater discharges.

Using conventional wastewater treatment plants is infeasible due to costs and inefficiencies of removal. Eganhouse and Kaplan (1982) and Eganhouse et al. (1981) found that most oil and grease in urban stormwater are hydrocarbons, and that hydrocarbons may be poorly removed in secondary treatment. A combined treatment/storage system would be more efficient (for separators as well as conventional treatment plants) than use of any of these treatment systems alone. Storage would decrease the size requirement for treatment facilities and provide additional retention time for oil and grease degradation prior to release. However, the storage facility would have to be large to account for the stochastic nature of stormwater flow, and would be expensive to build and maintain.

## OPPORTUNITIES FOR SOURCE CONTROL

Since conventional treatment technologies offer little potential for controlling oil and grease in stormwater, alternative technologies need to be developed. As a preliminary step to this development, it is useful to understand the origin of oil and grease found in urban stormwater to guide considerations of the potential for source versus end of pipe control.

We monitored oil and grease in runoff from 15 watersheds draining into San Francisco Bay, and correlated loading with land use. A model was used to estimate loading within the entire local drainage to San Francisco Bay, using census data to determine current and projected land use. Thus, we were able to predict loading as a function of growth and implementation of oil and grease control strategies (details of this work are published in Silverman et al. 1985). We also took an intensive look at one of these watersheds (in Richmond) and examined more closely the potential for reducing loading by implementing control measures for particular land uses (details of this work are published in Stenstrom et al. 1982).

Shown on Table 1 are results from simulating a reduction of oil and grease loading of 60% and 90% from specific land uses within the Richmond watershed. Reducing the contribution from commercial, parking and residential land uses shows the most potential for reducing loading from the watershed (columns 1 and 2). However, when taking into consideration the much greater land area devoted to residential uses in the watershed (74%) than to commercial and parking uses (12%), it appears that it would be much more practical to focus control strategies on commercial and parking areas. The ratio of percent reduction to

## Table 1. Mitigation Simulation for Various Land Uses

| Land Use (1) | Percent Reduction* in Oil & Grease From Watershed After Reducing Contribution From Land Use | | Ratio of Percent Reduction to Percent Area After Reducing Contribution From Land Use | |
|---|---|---|---|---|
| | 60% (2) | 90% (3) | 60% (4) | 90% (5) |
| Residential | 13.3 | 19.9 | 0.18 | 0.27 |
| Industrial | 5.5 | 8.3 | 1.28 | 1.93 |
| Commercial | 19.0 | 28.4 | 3.17 | 4.73 |
| Parking | 16.7 | 25.0 | 2.88 | 4.31 |
| Freeway & Railroad | 5.5 | 8.3 | 1.53 | 2.31 |
| Commercial & Parking | 35.7 | 53.4 | 3.04 | 4.53 |

\*= the Richmond watershed has annual average rainfall of 50 cm, with a watershed area of 6.58 km$^2$, and a projected annual average oil and grease load of 8200 kg if stormwater is not treated.

percent area for commercial and parking land uses is about 17 times greater than that for the residential area (columns 4 and 5). Thus, this model predicts that on an area basis, reducing the contribution of oil and grease from commercial and parking areas will be 17 times as effective as controlling residential land. In the Richmond watershed, a 90 percent reduction in oil and grease from only the 12 percent of land area classified as commercial and parking areas, would result in over 50 percent total reduction in oil and grease discharge from stormwater runoff.

## CONTROL TECHNOLOGY DESIGN

Potential control measures were evaluated with respect to efficiency, potential for implementation, and cost. In reviewing and developing suggested control measures, particular consideration was given to those areas characterized by higher inputs, since "end-of-pipe" treatment fails to take advantage of the heterogeneous distribution of oil and grease loading within a watershed.

Four control measures were identified as offering the best potential for source reduction of oil and grease from urban runoff: porous pavements, greenbelts, adsorbents in storm drain inlets, and surface cleaning using wet scrubbing. None of these methods have been adequately tested, and all may have limitations that would restrict their application. However, they all would allow installation focusing on areas where pollutant discharge is particularly high, limiting the need for basin wide controls.

Porous Pavements. In an urban area, precipitation will quickly wash most pollutants off relatively impervious surfaces such as asphalt and concrete. Runoff of oil and grease, and other pollutants, may be reduced by capturing these materials in relatively pervious paving materials.

Two different types of porous pavements are envisioned for possible application. One type of pavement, probably suited only for parking areas, has open areas within the paving materials. Examples of this type of material are open concrete blocks, paving stones, and turf-stones. These types of materials allow vegetative growth (probably grass) directly in the parking lot, which would require routine maintenance. Vegetative growth might also promote biological degradation of oil and grease deposits, and have aesthetic benefits. However, excessive deposition may prevent vegetative growth.

The second type of porous pavement would be asphalt pavements which do not include fine particles in their construction. Water and pollutants would infiltrate through a relatively shallow surface layer, where much of the pollutant load would be captured.

Diniz (1980) reported on the use of porous pavement at a demonstration site in Texas (Figure 1). A surface coat of approximately 6.5 cm of asphalt deficient in fine particles was used as a surface coat. Beneath the top coat, a stone top course of about 1.3 cm crushed rock with a thickness of about 5 cm was installed. A final base layer of larger crushed rock, about 6.4 cm, was used to provide substantial volume for temporary water storage, while retaining more strength when saturated than would smaller aggregate. The depth of the rock base for a particular site would need to be determined through consideration of the local water storage requirement, and the infiltration rate through the pavement.

Assessment of the utility of porous pavements are severely hampered by a lack of data on both their effectiveness in water treatment, and their durability and cost as a paving surface. Conceptually, the use of porous pavements has numerous advantages. Biological degradation of oil through soil cultivation is well documented as an effective treatment practice (Kincannon 1972); capturing oil and grease in pavement and subpavement, or in soils surrounding and underlying paving blocks, appears to present a analogous opportunity for degradation. Inorganic pollutants such as heavy metals may sorb to soil and

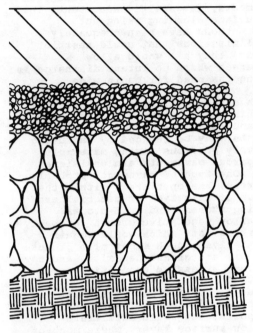

**Porous Asphalt Course**
1/2" to 3/4" Aggregate
Asphaltic Mix (1.27 - 1.91 cm.)

**Filter Course**
1/2" crushed stone (1.27 cm)
2" thick (5.08 cm)

**Reservoir Course**
(2.54 - 5.08 cm)
1" to 2" crushed stone voids
volume is designed for runoff
detention.

Thickness is based on storage
required and frost penetration.

**Existing Soil**
Minimal compaction to retain
porosity and permeability.

Figure 1. Porous Asphalt Paving Typical Section
(Dinez, 1980)

underdrain particles, and be removed from stormwater discharge. Other possible benefits include reducing the rate and volume of runoff (if used in place of impervious pavements), reducing erosion (if used to pave previously unpaved areas), and maintaining soil moisture for use by local vegetation, and possibly for ground water recharge.

The initial cost of porous asphalt pavement has been estimated to be about 50 percent greater than for impervious asphalt pavements (Diniz 1980), although much of this high estimate may reflect unfamiliarity with construction requirements since materials and techniques do not appear inherently more expensive. Probably of greater economic concern is the durability of porous asphalt pavement, and the need for repair and maintenance.

Conventional asphalt pavement is designed to be impermeable, minimizing water penetration to prevent water caused damage. In areas which routinely experience freezing weather, porous asphalt may be expected to deteriorate rather quickly. In warmer climates, its comparative durability has not been adequately tested. Furthermore, it is unclear how well permeability would be maintained, or if deposits of oil and grease, and dirt, would plug surface pores and render pavement relatively impermeable within a short time.

Perhaps the largest obstacle to using porous asphalt pavements is that conventional engineering design standards call for the use of non-permeable materials. Thus, to allow its use in roadways, major changes would need to be made in construction standards. Construction and maintenance cost increases would have to be reviewed as a function of water quality and other downstream benefits, considerations that are not typically considered in engineering design of roadways.

Given these relatively large obstacles, porous pavements probably have their most immediate potential application in parking areas. Responsibility for installing and maintaining these surfaces can be assigned to the developer/owner, as a cost of preventing pollution as part of their business. While the water quality benefits accrued from controlling individual parking areas may appear small, the modeling results clearly show that in composite, controlling parking areas can make a significant reduction of overall loading from a watershed.

Greenbelts. Oil and grease can be degraded through simple biological processes. Intercepting stormwater from areas of high pollutant concentrations with greenbelts would provide opportunity for capturing and degrading oil and grease before entering a storm sewer. While the efficacy of greenbelts in treating stormwater is uncertain, since research has not been reported documenting either application or performance, analogies to soil cultivation of oily wastes are indicative of their potential as an effective treatment tool.

A simple system can be envisioned where runoff is channeled from a parking area or roadway onto a greenbelt. The greenbelt would contain an upper layer of porous topsoil supporting vegetation. Underlying the topsoil would be a gravel layer, serving to drain the topsoil, and as a reservoir from which stormwater could percolate into surrounding soils. The greenbelt could be curbed, with an inlet to a storm drain provided to handle large flows when the system infiltration and storage capacity would be exceeded.

The major cost of this type of system would be for land purchase. The price of the land, and the amount of area required to effectively contain most storm flows, would be very site dependent. Construction costs would be relatively low in developing areas, but more expensive in developed areas where existing asphalt or concrete would have to be removed. Maintenance costs might also be higher than for a conventional paved area with a storm drain, although they should not be substantial.

Benefits incidental to water quality improvement might mitigate much of the additional costs. Greenbelts would be visually attractive, and could offer such amenities as benches and shade trees. During storm events, when they were inundated, their use by pedestrians would be minimal. During pleasant weather, they could receive considerable use as rest areas, facilities to have lunch, or to wait for a bus. They would also reduce peak storm flows, reducing the need for downstream flood control facilities.

Water quality benefits would be expected for a number of pollutants. Much of the particulate and heavy metal load would be captured by the system. Oil and grease should adsorb to soil and underlying gravel drainage, and degrade over time. It is difficult to predict oil and grease decomposition rates, since work on soil treatment has focussed on cultivation and response following spills, with loadings much higher than anticipated from urban runoff (Kincannon 1972, Burk 1976). Furthermore, conditions such as availability of dissolved oxygen and nutrients will strongly influence degradation rates. However, the envisioned system should develop a microbial flora capable of using oil and grease as a food source, causing degradation. Some on-site maintenance may be required to ensure that anaerobic conditions do not develop in the soil. The use of fertilizers may also be required.

Adsorbents in Sewer Inlets. Catch basins, a common feature of storm sewers, are designed to capture large particles and prevent them from restricting flows through the drainage system. Historically, catch basins have been used primarily where the sewer gradients were insufficient to result in self-cleaning velocities. Accumulated solids are collected and removed periodically, a less expensive process than removing deposits from sewer lines.

Oil and grease will not ordinarily be captured in catch basins in great quantity. While there is some tendency for oil and grease to sorb to particles, which will settle in a catch basin, much of this material will exist in a dissolved or colloidal state ((Eganhouse and Kaplan 1981, Stenstrom et al. 1984). We propose that adsorbents, installed in sewer inlets, may provide a mechanism for removing much of the oil and grease load from runoff before it enters the storm drain.

This system would work by having runoff flow through a bed of adsorbent material before discharging into the storm sewer. Periodically, the adsorbent would be replaced, with the spent adsorbent either disposed of, or (preferably), regenerated through desorption and recovery of oil and grease.

A primary consideration would be the design of an appropriate hydraulic structure that would allow sufficient contact between stormwater and the adsorbent, without restricting flow sufficiently to cause flooding. One possible design for this structure is shown on Figure 2. Stormwater would impact on a perforated splash plate, dispersing flow over the sorbent surface and reducing water velocity to minimize sorbent packing. Stormwater would percolate through the adsorbent, and enter the storm drain. A removable fine mesh bag would contain the adsorbent; replacement of the adsorbent could be accomplished by removing the exhausted bag and replacing it with one containing fresh adsorbent. An overflow outlet device would be connected directly to the storm drain, so during periods of flow higher than the rate of infiltration through the adsorbent, flooding would not occur.

Consideration also would have to be given to the ability of various sorbent materials to remove oil and grease from runoff. While the use of sorbents to remove oil and grease is common, applications usually address spills, where relatively high concentrations are present (Gumtz 1973).

Calculations based on typical oil and grease loading and adsorptive/density properties of adsorbents reveals a good potential for the proposed system. One manufacturer reports that its product is able to sorb up to 12 times its weight in oil (King 1982). The density of the material as shipped is approximately 53 kg/m$^3$. The 0.2 m$^3$ volume of adsorbent specified in the prototype design would have the absolute potential (at 100 percent efficiency) of capturing about 127 kg of oil and grease. In the 2700 m$^2$ commercial parking lot in Richmond, with an average oil and grease loading rate of 0.24 g/m$^2$-cm(rainfall) and annual average rainfall of 50 cm, annual oil and grease discharge would be about 32 kg. Thus, annual replacement of the adsorbent would be sufficient, with a potential for extending replacement to intervals of several year.

Furthermore, the volume of flow can be estimated that would pass through the adsorbent bed (compared to flow bypassing the system due to intense precipitation). Assuming a filter flow rate of $2.7 \times 10^{-3}$ m$^3$/sec-m$^2$, and an adsorbent filter with surface area of 0.7 m$^2$, runoff could filter through the structure at a maximum rate of about $1.9 \times 10^{-3}$ m$^3$/s. Examination of the rainfall pattern from this site reveals that flows exceeding $1.9 \times 10^{-3}$ m$^3$/s would account for about 26 percent of the total flow.

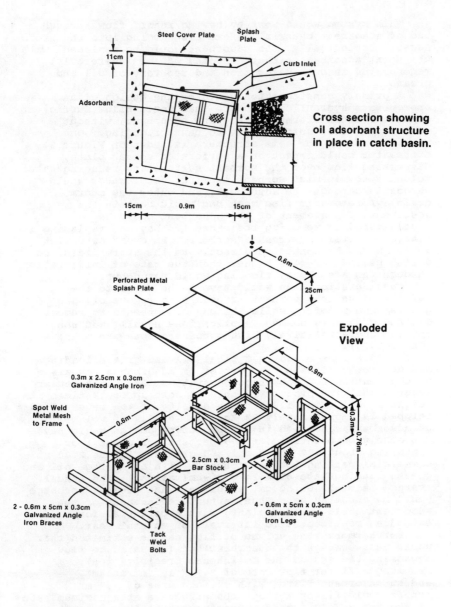

Figure 2. Prototype Oil Adsorbant Retention Structure

Thus, the system would treat approximately 74 percent of the total runoff. The percentage of oil and grease treated by this system might exceed 74 percent, however, because during the initial stages of a storm while flows are still low, concentrations of oil and grease may be in excess of average due to a "first flush" effect.

Surface Cleaning Using Wet Scrubbing. Street and parking lot sweeping is routinely performed in many areas. However traditional sweeping practices appear to be fairly ineffective in removing pollutants and protecting runoff water quality. Several studies done for the U.S. EPA National Urban Runoff Program have shown that sweeping is not cost-effective in removing pollutants from runoff (Heaney 1986). For example, Novotny (1985) reported that pollutant loading and concentration in runoff will not vary significantly from swept and unswept sites in a watershed with less than 80 percent imperviousness. Pitt (1979) estimated the cost effectiveness of typical street cleaning in removing pollutants, although not including oil and grease in these measurements.

Traditional sweeping practices appear particularly ineffective in removing oil and grease. Most oil and grease is found on very fine particles in runoff (Shaheen 1975; Hunter et al. 1979). Sartor et al. (1974) reported that traditional sweeping practices leave behind most of this fine material (85 percent of the material finer than 43 microns and 52 percent of the material finer than 246 microns), thus leaving the particles containing the majority of the deposited oil and grease.

While traditional sweeping does not appear to be effective, substantial oil and grease removal may be accomplished by concentrating efforts to those locations receiving heavy deposits. Innovative sweeping techniques would have to be employed. A system with substantial potential would be wet sweeping, using specially designed street sweepers. A spray of water and biodegradable soaps or detergents would be applied to solubilize oil and grease on pavement surfaces. Through a combination of sweeping and vacuuming, the liquid would be recovered, now containing pollutants removed from the surface. A sophisticated version of this system would include a filtration system on the sweeping truck to recycle the water. This proposed system is a hybrid of existing systems, and no such device currently exists. Prototype machines would need to be developed and evaluated to determine optimal performance characteristics, and the cost efficiency of the process. The effects of wet sweeping would also have to be evaluated on the basis of pavement longevity.

## CONCLUSIONS

Awareness of environmental degradation from urban runoff is growing, and regulatory activities are anticipated to force increased concern about this pollution source. Although urban stormwater eventually will be permitted through the National Pollution Discharge Elimination System (NPDES) at point sources of discharge, opportunities for quality control appears to be greater near the multiple points of heavy deposition rather than the point of discharge into receiving waters. Predictions from modeling in the San Francisco Bay area show that controlling a relatively small fraction of the watershed will have a major impact on overall stormwater quality. Thus, imposition of controls near the points of deposition appear to offer a practical means of controlling overall quality of watershed stormwater discharge.

Four source control measures have been described which have the potential to remove substantial quantities of oil and grease before entering the storm sewer system. None of these measures have been tested, so prototype systems would have to be developed and monitored in pilot projects before widespread implementation could be considered. The largest obstacle to their development is probably the lack of assurance of a market, given the lack of clear direction in past regulatory programs. Technological requirements appear to be relatively simple, but non-trivial. For example, adsorbents would have to be tested to determine the optimal system to adsorb oil and grease from runoff flowing into a catch basin, with techniques developed to regenerate the exhausted adsorbent without creating a waste disposal problem.

Evaluation of these source controls also needs to be compared to "end-of-pipe" treatment. Detention systems (Whipple 1983), groundwater recharge basins (Hannon 1980), and wetlands (Silverman and Chan 1988) have all been suggested as techniques capable of economical treatment of urban stormwater. None of these methods have been adequately tested, and a comparison of upstream and downstream cost effectiveness of various controls will depend in large part on site specific conditions. Source controls have the advantage of requiring treatment of a relatively small fraction of the total watershed flow to result in substantial pollutant removal. However, in some cases, "end-of-pipe" treatment may be preferred because of advantages in land costs, the need to recharge groundwater aquifers, the desire to enhance wetland habitats, or to provide other benefits from the application of freshwater.

While these source controls are not projected to be extremely expensive, new funding sources would have to be found. One of the advantages of source control is that responsibility for construction and maintenance may be given to the property developer or owner. A local government could impose as a condition for a conditional use permit for a commercial development that porous pavements be used and maintained throughout all parking areas. Any expenses in excess of conventional pavements would have to be considered as a cost of doing business in that region.

It would probably be much more difficult to install source controls in developed areas than developing areas. A major obstacle to implementing source control is that responsibility for protecting water quality is held by different departments or agencies than those responsible for pavement construction and maintenance. A state highway department's primary interest is to maintain the quality of the highways; suggestions to use other than standard engineering practices to construct a porous roadway that may result in maintenance problems will probably not be well received. Thus, even if the economic benefits in water quality enhancement (or related features such as reduction in downstream flood control requirements or aesthetic improvements) exceed additional construction or maintenance costs, a substantial economic burden would still be perceived. Thus, initial opportunities for implementation of source controls appear greatest in areas of new development.

While local governments have some ability to force use of source controls, there may be considerable resistance to such efforts. The impact of a typical parking or commercial area on overall watershed loading will be very small; only in composite will reductions in pollutant loading be substantial. Thus, it may be difficult to convince decision makers of the need to require these controls. However, by using this "soft" solution, in contrast to development of more traditional treatment works, substantial cost savings may be realized. Perhaps equally important, it is not economically feasible in most communities to build large treatment works; requiring source control diffuses the economic burden and provides local government with a practical management tool.

Use of these source controls would increase some costs. However, it is important to recognize that their utility must be evaluated on the basis of overall benefits, not just increased cost as compared to conventional practices. For example, while arguments may be anticipated that using porous pavements is undesirable because of increased costs of pavement maintenance, an analysis of the desirability of this use would be incomplete without consideration of water quality and other benefits. Similarly, a public works department would be anticipated to resist adding to their duties by requiring annual collection and replacement of adsorbents from catch basins. Adding water quality controls cannot be done without cost; control measure utility cannot be evaluated on a scale that maintains that they must function at no greater cost than current conventional practices.

While a few communities or developments may incorporate source controls in their design (for example, some developments use paving stones in parking areas for aesthetic reasons), large scale implementation will probably only be driven by regulatory direction. Hopefully, the control measures described here, plus any other systems envisioned to have a practical potential for controlling pollutants in urban runoff, will be ready for application before controls are required. Moreover, the ability and expense of controlling urban runoff quality should be well understood before final regulations are promulgated.

ACKNOWLEDGMENTS. This work was supported by the Association of Bay Area Governments. We thank Terry Bursztynsky and Yoram Litwin for their contributions to the evaluation of control measures.

REFERENCES.

Burk, J.C. 1976. A Four Year Analysis of Vegetation Following an Oil Spill in a Freshwater Marsh. Water Resources Research Center, University of Massachusetts, Amherst, Mass., OWRT Project A-042-Mass, Publication 71.

Connel, D.W., 1982. An Approximate Petroleum Hydrocarbon Budget for the Hudson Raritan Estuary - New York. Mar. Pol. Bul., 13(3):89-93.

Diniz, E.V. 1980. Porous Pavement, Phase 1 - Design and Operational Criteria. EPA-600/2-80-135, U.S. EPA, Cincinnati, Ohio.

Eganhouse, R.P and I.R. Kaplan, 1981. Extractable Organic Matter in Urban Stormwater Runoff. 1. Transport Dynamics and Mass Emission Rates. Environ. Sci. & Technol., 15(3):310-315.

Eganhouse, R.P., B.R.T. Simoneit and I.R. Kaplan, 1981. Extractable Organic Matter in Urban Stormwater Runoff. 2. Molecular Characterization. Environ. Sci. & Technol., 15(3):310-315.

Eganhouse, R.P. and I.R. Kaplan, 1982. Extractable Organic Matter in Municipal Wastewaters. 1. Petroleum Hydrocarbons: Temporal Variations and Mass Emission Rates to the Ocean. Environ. Sci. & Technol., 16(3):180-186.

Fam, S., M.K. Stenstrom and G. Silverman, 1986. Hydrocarbons in Urban Runoff. Journ. of Environ. Engin., 113(5):1032-1046.

Gumtz, G.D., 1973. Oil Recovery System Using Sorbent Material. EPA-670/2-73-068. U.S. EPA.

Hannon, J.W., 1980. Underground Disposal of Storm Water Runoff: Design Guidelines Manual. Report No. FHWA-TS-80-218, Federal Highway Administration, Office of Implementation, HRT-10, Washington, D.C.

Heaney, J.P., 1986. Research Needs in Urban Stormwater Pollution. Jour. of Wat .Res. Plann. & Manage. 112(1):36-47.

Hunter, J.V. et al., 1979. Contribution of Urban Runoff to Hydrocarbon Pollution. Jour. of the Wat. Poll. Contr. Fed., 51(8):2129-2138

Kincannon, C.B., 1972. Oily Waste Disposal by Soil Cultivation Processes. EPA-R2-72-110. U.S. EPA.

King, R. 1982. Personal communication, January 21, 1982. Technical Service Representative, 3M Company.

Novotny, V., H.M. Sung, R. Bannerman and K. Baum, 1985. Estimating Nonpoint Pollution from Small Urban Watersheds. Jour. of the Wat. Poll. Contr. Fed., 57(4):339-348.

Pitt, P., 1979. Demonstration of Nonpoint Pollution Abatement Through Improved Street Cleaning Practices. EPA-600/2-79-161. I/S. EPA.

Sartor, J.P. et al., 1974. Water Pollution Aspects of Street Surface Contaminants. Jour. of the Wat. Poll. Contr. Fed., 46:458

Shaheen, D.G., 1975. Contributions of Urban Roadway Usage to Water Pollution EPA-600/2-75-004. U.S. EPA.

Silverman, G.S. and E. Chan, 1988. Seasonal Freshwater Wetland Development in South San Francisco Bay. Restoring the Earth Conference, January 13-16, Berkeley, California (proceeding in press)

Silverman, G.S., M.K. Stenstrom and S. Fam, 1985. Evaluation of Hydrocarbons in Runoff to San Francisco Bay. Completion report to U.S. Environmental Protection Agency, under grant C060000-21. Association of Bay Area Governments, Oakland, Ca.

Stenstrom, M.K., G.S. Silverman and T.Z. Bursztynsky, 1982. Oil and Grease in Stormwater Runoff. Completion report to U.S. Environmental Protection Agency. Association of Bay Area Governments, Oakland, California.

Stenstrom, M.K., G.S. Silverman and T.A. Bursztynsky, 1984. Oil and Grease in Urban Stormwaters. Journal of the Environmental Engineering Division, ASCE, 10(1):58-72.

Whipple, W., Jr. and J.V. Hunter, 1979. Petroleum Hydrocarbons in Urban Runoff. Wat. Res. Bul., Amer. Wat. Res. Assoc., 15(4):1096-1105.

Whipple, W. et al., 1983. Stormwater Management in Urbanizing Areas, Prentice-Hall, Inc., Englewood Cliffs, New Jersey.

## Discussion of Dr. Silverman's Paper

Question:

I would first like to observe that your finding of about 15 mg/l of oil and grease is consistent with NURP study findings and what has been found in highway runoff. Now, to come to your receiving water--San Francisco Bay--you report effects including cancers and tumors in fish. Is that an oil and grease problem?

Answer:

No, not really. It is an organics problem, but the source may be pesticides entering from the Sacramento River.

Question:

Are there any toxics in oil and grease?

Answer:

Yes, there is a very good chance that they are sources of toxic reactions.

Question:

Is the issue really oily sheens on the water or toxicity in fact?

Answer:

Both.

Question:

Road runoff does contain oil and grease, but there is a problem in finding an upstream (rural) control area with which to contrast what we want to call the "urban" oil and grease load. What is the control to make this separation?

Answer:

Rather than "urban" versus "rural," the separation is simply by land use type. In undeveloped areas, we find oil and grease runoff concentrations to be 1-5 mg/l, and the sources are crankcase oil. In industrial areas, the concentrations are simply much higher.

Comment:

We have performed GC/MS analyses on extensive shellfish areas around the coast. Frankly, we have found peaks in both rural and urban areas that are characteristic of oil and grease constituents. But "nasty" things like benzo-a-pyrene do come from specific industrial areas. So I can confirm the land-use distinction as opposed to the more general "urban" versus "rural" source distinction.

Program to Reduce
Deicing Chemical Usage

Byron N. Lord*

**ABSTRACT**

Our Nation's transportation system has become the veins and arteries of the lifeblood of our national economy. Rapid, reliable movement of people, goods, and services to supply raw materials and the work force to produce and deliver are critical to the health and welfare of our society. This capability must be year-round. In our northern States, this requires the maintenance of clear open roads and streets throughout the winter. Over the years (since about 1948), our dependency on deicing chemicals has been increased to provide "bare pavement" for safe and efficient winter transportation. Concern about the effects of sodium chloride (salt) on our environment and water quality and on automobile and highway bridge deck corrosion also increased with this chemical usage. Beginning in the late 1960's and culminating today with an $8 million research effort in the Strategic Highway Research Program (SHRP), the highway community--industry, government, and academia--has undertaken measures to minimize our dependency on chemicals.

**INTRODUCTION**

The lifeblood of our national economy moves on our highways. The economic well-being of the United States is dependent upon our ability to maintain year-round access to farms, factories, ports, offices, schools, hospitals, etc., on our roads and highways. Snow and ice can cause major disruptions in nearly every State, with Hawaii perhaps the only exception. At present, more than $1 billion is spent and approximately 8 million tons of sodium chloride (salt) are supplied to maintain our roadways during winter snow and ice conditions. While it is difficult to accurately quantify these costs (Welch, 1976), it is even more difficult to determine the total cost to users in delay, corrosion, accidents, environmental quality, etc. (Brenner, 1976).

Beginning around 1965, in an effort to control these costs while continuing to maintain a safe open roadway, the highway community undertook research and development programs to look for improved technology. This search took

---

* Chief, Pavements Division, HNR-20, Federal Highway Administration, 6300 Georgetown Pike, McLean, VA 22101-2296.

many paths and many participated in it. This paper will attempt to give an overview of the approach taken, the improvements in technology implemented, and the direction the program is taking.

## BACKGROUND

Beginning in the late 1940's and early 1950's, the "bare pavement" policy was gradually adopted by highway agencies as a standard for pavement condition during the winter to provide safer highways. "Bare pavement" has become a useful concept of maintenance because it is a simple and self-evident guideline for highway crews. But it should describe the minimum of salt needed rather than a maximum application. Highway authorities were responding to a series of technological changes and rising public demands, including rapidly increasing numbers of motor vehicles. Dispersion of city populations into suburbs, higher travel speeds, and growing dependence upon cars for commuting and trucks for commerce increased the need for snow-free winter roads. Since salt was the most effective and economical deicing chemical available, highway authorities adopted it as their primary weapon against snow and ice.

The most commonly used salts are sodium chloride (NaCl) and calcium chloride (CaCl). For simplicity, these are referred to as "salt." The current annual use of highway salts is around 10 million tons of sodium salts and 0.3 million tons of calcium salts. Highways also receive about 11 million tons of abrasives. The eastern and north-central sectors of the country use more than 90% of all the salt. The use of salt has increased during the past 40 years. Salting rates are generally in the range of 200 to 500 pounds per mile of highway per application. Over the winter season, many roads and highways in the United States may annually receive more than 50 tons of salt per mile. However, prescribed application rates vary considerably from town to town and from State to State. This results partly from differences in ice control strategy and policy.

Concern about the effect of salt on the natural environment began to arise about 1958 with the evidence of damage to roadside sugar maples (a salt-intolerant species) in New England. Shortly thereafter, concern arose over contamination to drinking water from wells adjacent to or under unprotected salt storage areas. During this same period the effects of this increased use of salt were further evidenced by pitting and "rust out" of automobiles. The snow belt became known as the "rust belt." Due to changes in automobile corrosion protection technology, this term is probably unheard of by today's junior high school population and will soon be only a memory. In addition, as early as 1964, the highway community began to observe the effects of corrosion upon highway structures, particularly bridge decks. The solution of one problem, the need for increased winter mobility, had generated a plethora of others. It was time to begin the search for solutions.

The nature of snow and ice removal activities is extremely complex. Many factors interact to confound efforts to improve technology. Winter maintenance is an activity governed almost entirely by the whims of nature. While we have not yet learned how to control the weather, we are beginning to better understand it, monitor conditions, and predict what will occur. In addition, snow and ice are almost universally a problem. All but seven States (Alabama, Florida, Georgia, Hawaii, Louisiana, Mississippi, and South Carolina) can plan on snow or ice affecting some part of their system every year. In fact, with perhaps the exception of Hawaii, every State has some experience with snow and ice. To promptly respond to local conditions, winter maintenance is decentralized and snow and ice control programs are administered at the lowest local organizational element.

Sodium chloride is the most commonly used tool in the snow and ice control arsenal. It is an almost universally available material. It is generally effective in accomplishing its intended purpose, ice disbonding. In addition, it is generally very low-cost. At the mine, high purity rock salt may be purchased in bulk for approximately $15-$20 per ton. With the addition of transportation, salt cost is generally from $35-$70 per ton. While one would view salt's cost effectiveness and availability as highly desirable, these characteristics also serve as a significant factor in the evolution of alternative technology.

Without attempting to get bogged down in a complex economic analysis of the costs and benefits of deicing chemicals, and there have been several (Brenner, 1976; Welch, 1976; Nottingham, 1983; TRB, 1979), there is an overriding economic constraint in implementing new technology. Alternatives must replace existing technology within very limited bounds of conformance to existing resource constraints and capabilities. Nearly $1 billion annually is expended in winter maintenance out of approximately $15 billion for all highway maintenance. It should also be pointed out that no Federal funding is available for winter maintenance. Local governments must provide for all increases in costs incurred. Any significant increase in the cost of controlling snow and ice will detract from available resources which are already seriously constrained. Therefore new technology should at least be comparable to, or even more desirably, reduce costs.

## HISTORIC APPROACH

The search for new technology took several directions. The automobile industry has implemented major technology changes to protect or prevent corrosion. The highway community is continuing pursuit of measures to protect or prevent corrosion of bridge decks and structures. Changes in the design of highways have been implemented to intercept salt-laden runoff to protect drinking water supplies.

These are all key components in the approach to provide safe, economical, and effective measures for winter operation of our transportation network. However, the focus of this paper is control of the source of the problem ... but is this problem salt? On one hand, it was the increased use of salt which resulted in the problems to environmental quality, corrosion, etc. However, the basic problem is the very strong bond of ice with the highway pavement. If it did not snow or freeze, we would not have to concern ourselves with related problems. Unfortunately we cannot all move to Hawaii, and we have not mastered control of the weather.

Early in the 1960's, the search for improved deicing technology took off down the paths of alternative deicing chemicals, reduced chemical use, improved operational practices, pavement heating, pavement modification, and mechanical approaches. Each of these approaches had varying success.

## ALTERNATIVE DEICING CHEMICALS

The search for alternatives to salt has been pursued by many. Research has identified salt substitutes that to a limited degree have demonstrated potential. To be an acceptable, implementable, and affordable substitute for salt, the alternative must have effective melting range similar to salt, lack detrimental effects, and be cost-comparable. However, there has not been sufficient success to warrant application other than on an experimental basis or under highly unique conditions. These chemicals include formamide, urea, urea-formamide mixture, tetrapotassism phosphate (TKPP), ethylene glycol, and ammonium acetate (Dunn, 1980).

At present the only chemical alternative identified which still merits pursuing is calcium magnesium acetate (CMA). CMA is made from delometic limestone treated with acetic acid. A substantial hurdle to the implementation of CMA is its cost, which is estimated to exceed salt by a factor of 10 to 20. Efforts have been made to find more effective production technology, with some limited success. However, cost remains high. Significant research has been conducted to evaluate CMA's effects on the environment, corrosion potential, as well as its deicing ability. While CMA does not overcome all the undesirable characteristics of salt, it is comparable in deicing (however, more material must be applied to effect the same deicing); has less potential to affect the environment; and is not as corrosive. At present, field testing to determine operational characteristics are being completed (Chollar, 1988).

There is considerable interest in CMA. Many States are following its progress closely. However, at present, commercial supplies are limited. While its potential to replace salt is not yet established, the significant cost differential, unless overcome, will probably restrict future use of CMA to special areas where the continued use of salt is undesirable.

## REDUCED CHEMICAL USE

Measures to reduce the chemical usage of deicers have resulted in one of the more successful approaches. These include measures to improve the performance of the deicer by providing methods to get the chemical to the proper location in the proper amounts, and keep it there. "Prewetting" of salt, normally with liquid calcium chloride, has resulted in quicker melting (Bozarth, 1973) and less salt loss due to bounce and scatter. The use of salt brine applicators will achieve similar results. Required modifications to equipment and practice have inhibited implementation of this technology. However, a few States with suitable sources of "waste brines" from oil and gas development are pursuing this technology (Eck, 1987).

Perhaps one of the most effective measures for reducing chemical application has been the use of the calibrated spreader (Besselievre, 1976). When properly calibrated and used in conjunction with optimal application rates, significant reductions in salt can be achieved. One of the big plusses for this technology is its close working within the framework of existing organizational constraints. Manufacturers have developed technology compatible with existing spreaders and capable of synchronized operation from within the cab. As with any piece of equipment, maintenance is essential to effectiveness.

Another key component of the spreader technology is the "optimum application rate" (Minsk, 1982). In fact, without management control, standards of practice, and employee training, the best of equipment is of limited value (Cohn, 1974). It is here where the "plow meets the road."

## IMPROVED OPERATION PRACTICES

The key component in an effective source reduction program are the personnel required to implement the program. One of the many leaders in effective use of chemicals has been the Salt Institute. Their "sensible salting" program (Salt Institute, 1977), initiated in 1967, has been in many cases the impetus for change. "Sensible salting" emphasizes optimizing the effectiveness of every aspect of the snow and ice control program, while maintaining the safest roads possible and protecting the environment. The "sensible salting" program encompasses an integrated program of planning, proper storage, maintenance of equipment, employee training, spreader calibration, proper salt application, safety, and environmental awareness.

While all of these are key concerns, perhaps one of the most significant and most effective in source control of deicing chemicals has been the implementation of effective storage of chemicals (Richardson, July 1974). Throughout the United States, particularly among the State highway agencies but also local agencies, there have been active programs to bring salt "under cover" (Wyant, 1983).

On first examination, it may appear covered storage on impervious pads offers limited benefit. However, rain has the potential to reduce a salt pile at the rate of 1/4% per annual inch of precipitation. In an area receiving 40 inches per year (10/cm) (Richardson, 1974), an exposed pile can lose 5% of its volume. This is a highly concentrated point of loss having a significant potential for pollution to ground and surface waters. There are thousands of salt storage facilities across the United States, and many are now under cover. Plans are underway to bring most under cover. One of the big benefits in proper storage is significant cost savings. At $30 per ton, a 5% annual loss in the 8 million tons represented 400,000 tons or $12 million. This does not include the savings in improved environmental quality, corrosion of highway agency equipment, or replacement of wells for drinking water supplies.

## PAVEMENT HEATING

The primary objective of any deicing program is to disbond ice or snowpack from the pavement so that the pack can be pushed from the pavement. It is not cost-effective to attempt to melt the entire snow mass. The pavement heat source must supply sufficient heat to break the bond or prevent its formation. In addition, it must also provide sufficient heat to prevent the bond from reforming. Within the constraints of current technology, this requires raising the temperature of the pavement while overcoming heat loss to the surrounding ground and air. To maintain an adequate disbonding temperature, a major investment of heat is required.

There are basically two methods to apply this heat: in-pavement circulating fluid or electrical resistance. Fluid recirculation systems may use "heat pipe" technology which incorporates "earth heat," or they may be warmed by an external heat source which could include waste heat from nearby power plants or other sources. Geothermal springs have also been explored as heat sources. The pipes are laid a few inches below the pavement surface during construction and the pavement temperature is maintained slightly above the freezing point. These systems are costly to install. For those requiring external heat sources, operating costs exceed salt on the order of 15 to 30 times (Murray, 1972). Electrical pavement heaters are somewhat less expensive to install. However, operating costs are high. At present, pavement heating technology is not practiced except in limited locations due to the severe cost differential and the difficulty of adding it to existing systems.

## PAVEMENT MODIFICATION

Another alternative approach to disbonding is to modify the surface to reduce adhesion of the ice. Limited research into hydrophobic or icephobic coatings was conducted without real success. While the coatings exhibited promise, they were unable to meet the goal of a seasonal treatment (Ahlborn, 1976). The requirements for a surface treatment are strenuous. While it must weaken or prevent bonding, it must not decrease traction during no snow conditions. In addition, it must persist in an extremely harsh environment.

One of the most effective techniques for prevention of ice bonding has been pre-salting prior to the storm. This may be accomplished by spraying or flooding the surface or by building the treatment into the pavement surface. This built-in deicing treatment, "Verglimit," was developed in Switzerland in 1973 and has been in use in Europe for the last 14 years. Verglimit is derived from the French "limité le verglas," "end slippery ice." This process incorporates calcium chloride flakes to which a 5% sodium hydroxide is added. The treated flakes are then coated with polymerized linseed oil to protect the flakes from water. The flakes are then incorporated into an asphalt concrete mix which is used to overlay the pavement surface. This mix is laid and compacted using conventional technology. As the pavement surface wears, the treated flakes are exposed, providing a continuous source of deicer. The Verglimit performs two functions: it retards pavement icing and promotes ease of snow removal.

Verglimit has been evaluated in the United States for about the past 10 years. Recently completed tests in New Jersey (Rainiero, 1988) documented Verglimit's performance. While Verglimit did not eliminate the need for spreading of salt, applications were reduced from 40 to eight, compared to control locations. "Snow adhering to the surface was for the most part nonexistent," except during low traffic and highway snowfall. In spite of the satisfactory performance, it should be noted the costs for Verglimit are significantly higher than for conventional pavement. Material cost in 1986 was estimated at $101 per ton compared to approximately $60 for conventional overlay. This will probably limit future application of this treatment to high safety areas such as ramps, severe curves, or bridge sections.

## MECHANICAL APPROACHES

Current plow technology is designed to remove the disbonded snow and ice from the pavement. Research into development of mechanical measures to remove snow and ice has met with little success (TRB, 1979). Due to the tenacious nature of the pavement-ice bond, mechanical measures to break the bond frequently break the pavement.

## SUCCESS TO DATE

It is difficult to measure the success of these programs. Certainly, significant achievements have been made. The highway agencies' diligent program of upgrading storage facilities has resulted in providing covered storage on impervious pads to most of the major salt storage areas. It is impossible to say how many remain uncovered. However, progress has been substantial.

Major progress has also been made in employee training and awareness. The "some's good; more's better" approach to snow and ice control has undergone significant education. Personnel are well aware of the potential environmental costs in addition to chemical wasteage. Through State highway agencies, Rural Technical Assistance Programs, and other programs such as the Salt Institute's "Sensible Salting," seminars and training programs are held throughout the United States every fall to review proper application, calibrate equipment, and promote safe, effective snow and ice control.

Information completed by the Salt Institute showed a rising trend in salt usage nationwide, peaking in approximately 12 million tons in the winter of 1978-1979, but dropping to 6.9 million tons in 1981-1982 and 8 million tons in 1982-1983 (Salt Institute, 1978, 1982, 1983). Current information places national consumption in the area of 10 million tons annually. While these are gross figures and are subject to reporting accuracy and variability in climate (a severe winter could result in a significant increase), it appears salt usage has decreased. This is due in large measure to the programs outlined above.

## FUTURE

In 1984, "America's Highways, Accelerating the Search for Innovation" was published. This report was the result of the Strategic Transportation Research Study (STRS) by the Transportation Research Board (TRB) commissioned by the Federal Highway Administration (FHWA). This research was initiated in response to concern over the condition of our national highway transportation system and the substantial investment required to preserve it. The STRS program recognized the historic approach to research had focused on incremental solutions to current local programs. The STRS objective was innovation and major technological advancement (TRB, 1984). Nine questions were asked in an effort to focus on this objective:

1. Will the research yield big payoff?
2. Is it currently neglected?
3. Will it address important research previously hampered by institutional or organizational barriers?
4. Can research findings be used?
5. Does the research require a large-scale effort?
6. Does it require an integrated or national approach?
7. Does the research respond to new and potential changes in national policy?
8. Does the research use or respond to other technological changes?
9. Will the research affect safety or the environment significantly?

On the basis of this selection strategy, six priority areas were identified. Included in this was "Chemical Control of Snow and Ice on Highways." While it was not identified as an area neglected in terms of the range of activities underway, it was determined that work was splintered, significantly underfunded, and in need of a major national research effort to focus the attack.

As a result of the STRS report, Strategic Highway Research Program (SHRP) was organized to address the needs identified. SHRP is a $150-million, 5-year, highly focused, results-oriented research program. It is funded under the Surface Transportation and Urban Relocation Assistance Act of 1987. SHRP is supported through a mutual agreement with FHWA, American Association of State Highway and Transportation Officials, and the National Research Council. SHRP is administered as an independent unit of the National Research Council. It will address six technical areas: Pavement Performance, Asphalt, Cement and Concrete, Concrete Bridge Protection, Maintenance Cost Effectiveness, and Control of Snow and Ice (SHRP, 1987).

The objectives of the "Snow and Ice Control" studies are to provide more cost-effective ways to prevent or remove the buildup of snow and ice on highways and bridges during winter conditions; reduce the deterioration of bridges, pavements, and vehicles; and mitigate adverse environmental consequences of the use of snow and ice control chemicals.

Five projects totaling $8 million have been identified as follows:

1. <u>Ice-Pavement Bond Prevention</u>

Ice and refrozen compacted snow form an exceedingly tenacious bond to asphalt and portland cement pavements. The nature of this bond and the mechanisms by which it

develops are not clearly understood. Research on the
fundamental chemical and physical structure and mechanics of
bond formation may provide a basis for developing techniques
and devices for interfering with or preventing bond
formation without adversely affecting pavement properties
such as friction. This project will explore chemical
modification of the surface by incorporating chemicals into
the pavement itself and through physical methods of surface
modification, such as icephobic materials. Funding
estimates project a budget of $2.1 million for this research.

## 2. Ice-Pavement Bond Destruction

Most present techniques depend on destroying the bond once
it has formed; therefore there appears to be greater
probability of success in developing new techniques,
materials, and equipment to make significant improvements in
ice removal. Key research tasks include fundamental studies
of ice-substrate bond; physical surface modification (such
as deformable or sacrificial surfaces); non-contact
disbonding (e.g., microwave and laser radiation); contact
disbonding (e.g., mechanical cutting, breaking, listing);
evaluation methodologies for deicing chemicals; and
improvement of sodium chloride as a deicer. Funding
estimates project a budget of $3.25 million for this
research. The fundamental studies of ice-pavement bonding
in these two programs will be conducted concurrently by two
researchers sharing progress in hopes of a technological
breakthrough.

## 3. Improved Displacement Plows

U.S. plow designs have evolved empirically, with scant
attention paid to the physical properties of the material
being handled and with little consideration given to the
aerodynamic and hydrodynamic principles involved in the flow
of fluidized snow. As a consequence, the energy expended in
displacing snow is disproportionate to the work performed,
and low cast distance requires unnecessary rehandling of
snow.

The plow research will establish design criteria and prepare
design details and specifications for displacement plow
systems based on aerodynamic/hydrodynamic principles,
material handling characteristics of snow, and ice cutting
machines. The project will lead to standard designs for
plows for different types of snow and climatic conditions.
This program is funded at a projected budget level of $1.2
million.

## 4. Control of Blowing Snow

Snow fences and other means of reducing the amount of snow
reaching a roadway help cut the costs of snow clearing,
reduce the formation of compacted snow, and reduce the need

for chemicals. The study will develop criteria for properly designed snow fences installed in the proper location to control wind-blown snow in many situations. Considerable progress has been made in the design of snow fences over the last few years, yet more remains to be done to increase their effectiveness in a diversity of conditions. It may be possible to improve the effectiveness of snow fences through research on free-turbulent mixing in the vicinity of barriers to allow prediction of surface shear stress. Proper design of roads, overpasses, guardrails, and vegetation to prevent many severe drifting situations also will be studied. This study will be funded at $0.25 million.

## 5. Management

Snow and ice control is the single most costly maintenance function performed by many northern States and cities. Staffing levels and equipment specifications are very often based on winter maintenance functions and affect all other seasonal activities. This study will identify and evaluate alternative means of sensing and communicating road and bridge conditions to the maintenance organization to reduce response time of maintenance crews and to provide maximum public awareness of road conditions. This research is funded at $1.2 million.

SHRP became operational on April 2, 1987, with the passage of the Surface and Transportation and Uniform Relocation Assistance Act of 1987. Since that time SHRP has organized, staffed, solicited for contractors, and has research underway on the two fundamental studies of ice bonding. The SHRP program will lead the search for the future in continued reduction in deicing chemical usage.

The highway community has made significant progress in reducing chemical usage. There is good potential for further reductions by continued research, development, and implementation. Perhaps some day technology will release our highways from nature's bonds of ice.

## REFERENCES

Ahlbom, G., & Poechlmann, H., "Development of a Hydrophobic Substance to Mitigate Pavement Ice Adhesion," U.S. Environmental Protection Agency, Cincinnati, Ohio, December 1976.

Besselievre, W., "Automatic Controllers for Hydraulically Powered Deicing Chemical Spreaders," Federal Highway Administration, FHWA-RD-76-505, Washington, D.C., August 1976.

Bozarth, F., & Huisman, C., "Implementation Package for Use of Liquid Calcium Chloride to Improve Deicing and Snow Removal Operations," Federal Highway Administration, FHWA-IP-73-2, Washington, D.C., April 1973.

Brenner, R., & Moshman, J., Benefits of Costs in the Use of Salts to Deice Highways," The Institute for Safety Analysis, Washington, D.C., November 1976.

Chollar, Brian, "Field Evaluation of Calcium Magnesium Acetate During the Winter of 1986-1987," Public Roads, Federal Highway Administration, Washington, D.C., June 1988.

Cohn, M., & Fleming R., "Managing Snow Removal and Ice Control Programs," American Public Works Association Special Report No. 42, Chicago, Illinois, 1974.

Dunn, S., & Schenk, R., "Alternative Highway Deicing Chemicals," Federal Highway Administration, FHWA-RD-79-108, Washington, D.C., March 1980.

Eck, R., & Sack, W., "Determining Feasibility of West Virginia Oil and Gas Field Brines as Highway Deicing Agents," West Virginia Department of Highways, Morgantown, West Virginia, January 1987.

Jones, P., "Evaluation of Ontario Municipal Experiments to Reduce the Use of Road Salts," Institute for Environmental Studies, University of Toronto, Toronto, Canada, October 1980.

Long, D., & Baldwin, J., "Snow and Ice Removal from Pavement Using Stored Earth Energy," Federal Highway Administration, FHWA-TS-80-227, Washington, D.C., December 1972.

Minsk, L. David, "Optimizing Deicing Chemical Application Rates," Federal Highway Administration, Washington, D.C., August 1982.

Murray, D., "A Search: New Technology for Pavement Snow and Ice Control," U.S. Environmental Protection Agency, Washington, D.C., December 1972.

Nottingham, S., et al., "Costs to the Public Due to the Use of Corrosive Deicing Chemicals and a Comparison to Alternative Winter Road Maintenance Procedures," Alaska Department of Transportation and Public Facilities, Fairbanks, Alaska, December 1983.

Pravda, M.F., et al., "Augmentation of Earth Heating for Purposes of Roadway Deicing," Federal Highway Administration, FHWA-RD-79-80, Washington, D.C., February 1980.

Rainiero, J., "Investigation of Ice Retardant Pavement, Verglimit," Transportation Research Board, Washington, D.C., January 1988.

Richardson, D.L., et al., "Manual for Deicing Chemicals: Storage and Handling," U.S. Environmental Protection Agency,

Cincinnati, Ohio, July 1974.

Richardson, D., et al., "Manual for Deicing Chemicals: Application Practices," U.S. Environmental Protection Agency, Cincinnati, Ohio, December 1974.

Salt Institute, "The Snowfighter's Handbook," Salt Institute, Alexandria, Virginia, 1977.

Salt Institute, "Salt Facts," Salt Institute, Alexandria, Virginia, April 1983.

Salt Institute, <u>Survey of Salt, Calcium Chloride, and Abrasive Use in the United States and Canada</u>, Salt Institute, Washington, D.C., 1973-1974, 1978-1979, 1982-1983, 1984-1985.

<u>Strategic Highway Research Program: Annual Report</u>, National Research Council, Washington, D.C., October 1, 1987.

Strategic Highway Research Program, <u>Focus</u>, No. 21, National Research Council, Washington, D.C., June 1, 1987.

Strategic Highway Research Program, <u>Research Plans</u>, National Research Council, Washington, D.C., May 1986.

Transportation Research Board, "America's Highways, Accelerating the Search for Innovation," Special Report 202, National Research Council, Washington, D.C., 1984.

Transportation Research Board, "Minimizing Deicing Chemical Use," National Cooperative Highway Research Program Synthesis of Practice 24, National Research Council, Washington, D.C., 1974.

Transportation Research Board, "Snow Removal and Ice Control Research," Special Report 185, National Academy of Sciences, Washington, D.C., 1979.

Welch, B., et al., "Economic Impact of Highways Snow and Ice Control," Federal Highway Administration, FHWA-RD-77-20, Washington, D.C., September 1976.

Wyant, D., & Long, R., "Proper Storage of Deicing Salts," Virginia Highway and Transportation Research Council, Charlottesville, Virginia, March 1983.

# Discussion of Mr. Lord's Paper

Question:

You report apparent reductions in the amount of de-icing salt used nationally. Did you determine why they occurred?

Answer:

Not specifically. It is hard for me to say that the reductions did not occur simply because we had a series of milder winters. On the other hand, there were increases in road miles to treat in the same period. We are confident, however, that applied salt tonnage reductions have occurred.

Question:

Is ice-pavement bonding the whole issue? Isn't slickness of the original road surface partly responsible for the bonding that takes place and hence the amount of salt required?

Answer:

Yes, it is. But we have to get down to bare pavement, and the bonding layer on any roadway, which is about the same thickenss anywhere, is the tough nut to crack.

Question:

Have you considered urea?

Answer:

We looked at it. It's more expensive , and there are many environmental negatives about urea compared to its alternatives.

Question:

Did you try a salt and sand mixture?

Answer:

Yes; but it provides more traction, not sufficient bonding melt.

Question:

What is the temperature at which the salt is effective?

Answer:

$18°$ F.

Question:

Doesn't heat content of the roadway depend on traffic flow?

Answer:

Not really; not when the ambient conditions are at freezing or well below.

Question:

Have you tried chromates?

Answer:

The adding of anticorrosives wasn't all that effective at de-icing, and it was more expensive than its alternatives.

# SOURCE TRACING OF TOXICANTS IN STORM DRAINS

Thomas P. Hubbard and Timothy E. Sample*

## Abstract

Sources of toxic chemicals in storm drains are often difficult to locate and even more difficult to control. Receiving environment chemistry, storm drainage maps, land use information, and sampling of sediments in storm drains, can be used to identify significant sources of heavy metals and organic toxicants from commercial and industrial facilities. Once runoff contaminated by past and present practices is controlled, in-line sediments can be removed from the drains to prevent further degradation of the receiving environment.

## Introduction

Just south of downtown Seattle, Washington, the Duwamish River discharges into Puget Sound. Historically, local industries and cities have used it for the disposal of wastes and wastewater. The Duwamish also receives runoff from dairy farms, burgeoning suburbs, miles of interstate highway, and the industrial heartland of the Pacific Northwest. EPA and the Washington Department of Ecology (WDOE) designated the lower Duwamish as one of the four worst water quality problems in the state (EPA et al. 1980). Despite its problems, the river remains an important regional resource: anadromous salmon and steelhead runs are valued at more than $10 million per year (Harper-Owes 1983). The river is also used for commercial navigation, public recreation and wildlife habitat.

Pollution control efforts, over the past 25 years (especially in the reduction of point sources), have substantially improved water and sediment quality in the Duwamish River (Harper-Owes 1983 and Sample 1987), but it was recognized that additional control measures were necessary to protect the river and its resources. EPA and WDOE awarded the Municipality of Metropolitan Seattle (Metro) a Section 208 (nonpoint source planning) grant to study impacts to beneficial uses of the river and to develop the Duwamish Clean Water Plan (Metro 1983).

---

* Water Quality Planners, Municipality of Metropolitan Seattle (Metro), 821 2nd Avenue, Seattle, WA 98104

Problems have been identified in the water, sediment and biota of the Duwamish. In the water column, concentrations of copper exceeded the EPA acute criterion for marine life, and lead exceeded the chronic criterion. In bottom sediments, concentrations of several metals (copper, lead, arsenic, zinc and mercury) and organic toxicants (PCBs, PAHs and pesticides) were elevated many times greater than Puget Sound background conditions. Biological abnormalities were frequently observed in the river's bottomfish (Malins et al., 1982, 1984).

Mass balances comparing existing conditions and known sources indicated that 80 to 99% of the toxicants could not be attributed to permitted sources. Longitudinal plotting of sediment chemistry indicated large uncontrolled sources near the mouth of the river (Harper-Owes 1983). Because high concentrations of toxicants were found in the water column and the top 2 cm of the sediment despite the absence of any large point source discharges, present sources of contamination, possibly storm drains from commercial and industrial areas, were suspected sources of toxicants. One of the eleven recommendations of the Duwamish Clean Water Plan was "control of toxicants in the West Waterway" near the mouth of the river.

## Method and Results

To implement the plan's recommendation for control of toxicants, EPA and WDOE awarded a Section 205(j) (water quality management planning) grant to Metro for the Duwamish Industrial Nonpoint Source Investigations in 1983. Stormwater and sediment of major storm drains discharging to the river were sampled, and commercial and industrial sites were visited to determine if they were contributing heavy metals and organic toxicants to the river. Sampling of two drainage systems, followed by site visits, identified three significant sources of toxicants. Once sources were controlled, remedial actions removed the contaminated sediment from the drain to prevent further degradation of the river.

Studies in California, Alaska and Illinois (Meiggs 1980) have shown that sediments could be used to trace sources of toxicants. A study in Michigan (Schmidt and Spencer 1986) reported that improper discharge of stormwater from commercial and industrial facilities was a major contributor of toxicants to an urban river.

## Lead in the SW Lander St. Storm Drain

Studies in the late 1970s and early 1980s found elevated concentrations of lead in the water, sediment and biota of the lower Duwamish River. In the water column, lead exceeded EPA's chronic criterion for marine life during most of the year. In sediments of the river, concentrations of lead were 22 times greater than Puget Sound background conditions (Harper-Owes 1983). Duwamish mussels had 30 times (Schell and Barnes 1974) and sea gulls had nine times more lead in their tissues than animals from rural areas of Puget Sound (Riley et al., 1983).

Mass balances, comparing existing conditions and known sources, showed that 86% of the lead in the river could not be attributed to permitted sources. Sediment chemistry from several sampling efforts were plotted on a profile of the river (Figure 1). Concentrations of most toxicants were not evenly distributed; there was a definite peak in the West Waterway 200-400m upstream from the mouth of the river where concentrations averaged 750 ppm (dry weight), the highest in the entire system (Harper-Owes 1983). Analysis of sediment data from the West Waterway (Figure 2) identified a probable source on the east bank of the river. Concentrations showed a definite gradient: those closer to the SW Lander St. storm drain were higher and immediately below the drain's outfall, lead concentrations reached 12,990 ppm (Farris 1980).

A review of City of Seattle storm drainage maps and land use information of the area identified a smelter, which recovered lead from crushed batteries. The smelter was a block from the river in an area drained by the SW Lander St. storm drain. There are two storm drains with outfalls at SW Lander St. One is a small, privately-owned drain, which services an oil storage facility, and the other is a larger system owned by the City of Seattle, which drains approximately two square blocks of Harbor Island including the lead smelter. Other land uses in the drainage area are petroleum tank farms, small warehouses, and ship repair facilities. Air quality data collected by the Puget Sound Air Pollution Control Agency (PSAPCA) found 182,000 ppm (18%) lead in soil samples collected on unpaved parking lots around the smelter (PSAPCA 1980). Fugitive dust from the smelter was suspected as a major source of lead contamination in the SW Lander St. storm drain and the Duwamish River.

Sediment and composite stormwater samples were collected at four manholes and analyzed for heavy metals. Stormwater samples were difficult to collect because the drains were tidally-influenced and during Seattle's

# SOURCE TRACING OF TOXICANTS

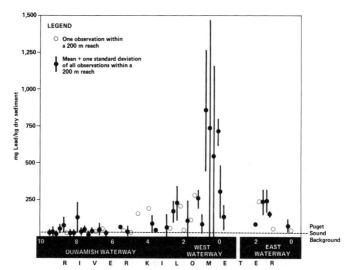

Longitudinal Variation in Sediment Lead Concentration
in the Duwamish Estuary, 1972 - 1982 (n=133 observations)

(from Harper-Owes 1983)

**FIGURE 1**

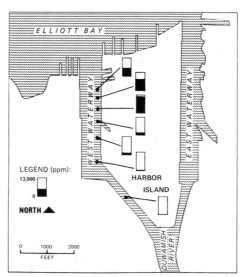

Lead in Duwamish River Sediments in PPM (3/80)
**FIGURE 2**

winter rainy season, minus tides occur only late at night. Samples were analyzed according to EPA methods for analyzing wastewater and solid wastes (EPA 1979 and 1982). Water and sediment lead concentrations were much higher in the municipal system (Figure 3). Composite stormwater samples had 1.2-2.3 ppm lead, ten times greater than stormwater in suburban areas near Seattle (Galvin and Moore 1982), and exceeded the EPA acute criterion for marine life (0.14 ppm.). Storm drain sediment concentrations of lead were 247,000-358,000 ppm (25-36%), 500 to 700 times greater than typical street dust in the Duwamish basin (460 ppm lead) (Galvin and Moore 1982).

Using the storm drain data, a Metro site visit team inspected all of the facilities which could contribute lead (and other toxicants) to the SW Lander St drain. With the exception of the lead smelter, none of the businesses used lead. Samples of sediment from catch basins on the smelter site were not collected, but the site visit team found dark red sediment, similar in color to the smelter's bag house dust covering much of the site including several catch basins on the facility. The smelter's management and the Metro team agreed that additional storm drain samples were unnecessary and that remedial actions should be taken.

Source controls of fugitive dust mandated by PSAPCA reduced lead emissions in the air from 4 ug/m$^3$ in 1974 to less than 1 ug/m$^3$ in 1981. The lead control plan (PSAPCA 1980) recommended additional source controls. In 1983, the City of Seattle paved parking lots in the area around the smelter to reduce resuspension of lead-contaminated dust disturbed by vehicles driving on unpaved lots. In the same year, the smelter was closed permanently. With the closure of the smelter and paving the parking lots, the major source of lead in this area had been eliminated. A cooperative cleanup effort was initiated to remove the contaminated sediment from the drain before it could be carried to the river during winter storms.

City of Seattle Sewer Utility crews constructed a temporary sandbag weir in the drain at the last manhole before the river. Using high pressure hoses, the drain was flushed. Most of the sediment was trapped behind the weir, and the wash water plus suspended sediment was discharged to Metro's sanitary sewers via the smelter's pretreatment system. Twenty cubic yards of sediment were collected from the drain. The sediment was shipped to a smelter in Oregon and it was smelted to recover the lead.

The SW Lander St. monitoring had shown that sediment chemistry was effective in identifying sources of pollution, and sediment grabs were easier than composite stormwater samples to collect--especially in drains which

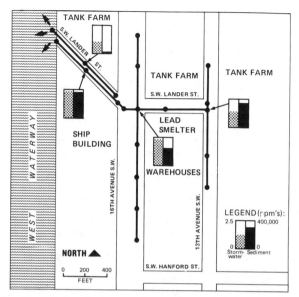

**Lead in S.W. Lander Street: Stormwater and Sediment (3/84)**
FIGURE 3

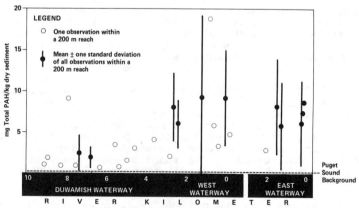

**Longitudinal Variation in Sediment PAH Concentration in the Duwamish Estuary, 1978 - 1982 (n=40 observations)**
(from Harper-Owes 1983)

FIGURE 4

are tidally-influenced. In this system, there was a single source which had recently ceased operations, and the sediment could be reused, and therefore not designated as a hazardous waste.

## Metals and Toxic Organics in the SW Florida St. Drain

Across the river at SW Florida St., there were multiple sources of toxicants, continuous operations, hazardous waste concerns and an EPA criminal investigation underway--a situation very different from the relatively straightforward SW Lander St. investigation.

Similar to lead, elevated concentrations of other heavy metals and organic toxicants were found in Duwamish water, sediment and biota (Harper-Owes 1983). Plotting of polycyclic aromatic hydrocarbons (PAHs) in bottom sediments (Figure 4) showed that the West Waterway had some of the highest concentrations of PAHs and heavy metals (copper, lead, arsenic and zinc) in the entire river (Figure 5). It was also one of the few places where polychlorinated biphenyls (PCBs) were found in detectable concentrations. Most of the PCBs were Aroclor 1260, which was primarily used in electrical transformers. A review of the data revealed definite gradients of concentrations indicating probable sources on the west bank of the West Waterway. Again, there were no known point sources in the immediate vicinity of the largest concentrations, but a municipal storm drain discharged into the river at SW Florida St.

By consulting city drainage maps and land use information, possible sources of toxicants were identified. A wood treatment facility on the west bank of the West Waterway was suspected as a source of PAHs and heavy metals because creosote and copper arsenate are used in wood treatment. The drainage area also had a major shipyard, port facilities, and a scrap metal recycler. The SW Florida St. system has two major trunks and also receives effluent from a small combined sewer overflow (CSO), which overflows less than once a year. Based on the SW Lander St. experience, stormwater samples were not collected. Grab samples of sediment were collected from four storm drain manholes and the CSO. One station was placed upstream of the wood treatment facility and the other was downstream. The samples were analyzed for heavy metals and priority pollutant organics (EPA 1979 and 1982).

The sediment in the CSO and trunk under 26th St. SW had very low concentrations of metals and organics, but the SW Florida St. line had 57-161 ppm total PAHs and high concentrations of copper and arsenic, especially

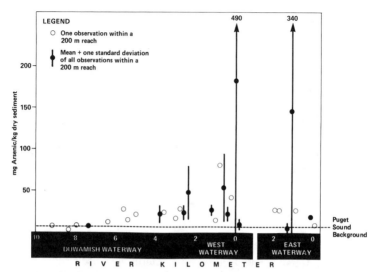

**Longitudinal Variation in Sediment Arsenic Concentration in the Duwamish Estuary, 1972 - 1982 (n=54 observations)**

(from Harper-Owes 1983)

**FIGURE 5**

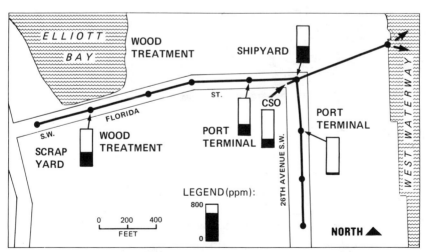

Arsenic in S.W. Florida Street: Storm Drain Sediment (4/84)

**FIGURE 6**

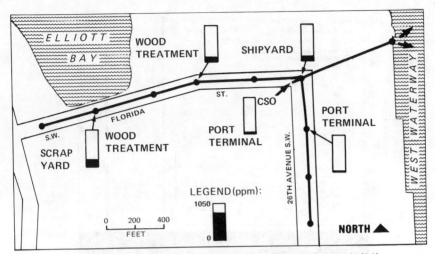

PCBs (total) in S.W. Florida Street: Storm Drain Sediment (4/84)

**FIGURE 7**

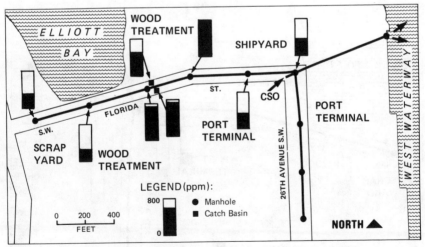

Arsenic in S.W. Florida Street: Storm Drain Sediment (10/84)

**FIGURE 8**

around the wood treatment facility (Figure 6). Elevated concentrations of PCBs, exceeding EPA's hazardous waste designation (50 ppm), were found, and the gradient of concentrations indicated a source upstream of the wood treatment facility. Similar to river sediments near the mouth of the drain, most of the PCBs were Aroclor 1260.

To confirm the high PCB concentrations and to collect sediment from the wood treatment facility's catch basins, the SW Florida St. line was resampled six months later. Based on the data from the previous sampling, in-line sediment was collected at every manhole of the SW Florida St. portion of the storm drain system (Figure 6) and from two catch basins adjacent to the wood treater. No sediment was collected from the CSO and the 26th SW line. High concentrations of PAHs, copper, and arsenic were found at the wood treatment facility, and the PCB data pointed to a source at the head of the line where the scrap yard's stormwater discharged to the system.

Because of EPA criminal investigations of the wood treatment facility, that site was not visited, but discussions with the scrap yard confirmed that they handled electrical transformers from several utilities and waste oil had been spilled on the site. The wood treater pleaded guilty to criminal charges under the federal Resource Conservation and Recovery Act (RCRA) and Clean Water Act and agreed to cease illegal discharges of wastewater to the storm drain. The scrap yard had ceased handling PCB transformers.

Cleanup of this drain was much more complicated because two industries were involved. The sediment was probably hazardous waste and not reusable, and city engineering department crews were not experienced in handling hazardous materials. Furthermore, there was no convenient pretreatment system for disposal of wastewater generated by the cleanup.

City crews installed a drag machine (buckets pulled by cables), which scraped the oily sediment out of the line, and vactor trucks were used to clean the catch basins. High pressure hoses were used to dislodge the remainder. An improvised settling pond was constructed to pretreat the water before it was discharged to the sanitary sewer. Twenty cubic yards of contaminated sediment were held in lagoons and later disposed in a sanitary landfill. The responsible industries agreed to split the costs of disposal.

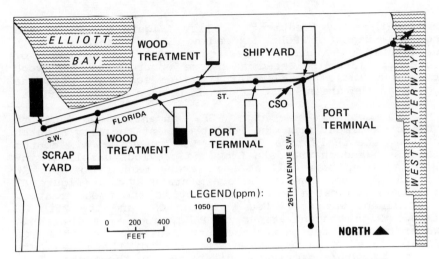

PCBs (total) in S.W. Florida Street: Storm Drain Sediment (10/84)

**FIGURE 9**

Summary

1. The distribution of toxicant concentrations in receiving environment sediments can indicate likely areas for possible source tracing.

2. Land use information and storm drainage maps can assist in designing monitoring programs to identify sources of toxicants.

3. Storm drain sediments can be effective integrators for tracing sources. A few samples of sediments in major tributaries can focus intensive monitoring and site visits and are much more efficient than stormwater samples collected during storm events.

4. Stormwater can transport toxicants from commercial and industrial areas into publicly-owned storm drains, which will eventually discharge into the environment.

5. After implementation of source controls, the drains can be cleaned to prevent further environmental degradation.

## Acknowledgements

Special thanks to Robert Swartz, Dave Galvin and Lori Nye, Metro, Water Resources Section, and Betty Hageman, Metro, Graphics Section, in preparing this manuscript.

Financial support for this study was supplied by EPA and the Washington Department of Ecology through a 205(j) grant.

## References

EPA (1979). <u>Methods for the Chemical Analysis of Water and Wastes</u>. EPA Document 600/4-79-020. Washington, DC: US Environmental Protection Agency.

EPA (1982). <u>Test Methods for Solid Waste. Physical/ Chemical Methods</u>. EPA Document SW-846. Washington, DC: US Environmental Protection Agency.

EPA, Washington Department of Ecology, and Washington Department of Health and Social Services (1981). <u>Fiscal Year 1981 State/EPA Agreement</u>. Seattle and Olympia: EPA and the State of Washington.

Farris, G. (1980). Letter to Dave Nunnallee, Washington Department of Ecology (3/24/80).

Galvin, D., and R. Moore (1982). <u>Toxicants in Urban Runoff. Metro Toxicant Program Report #2</u>. Seattle: Municipality of Metropolitan Seattle (Metro).

Harper-Owes (1983). <u>Water Quality Assessment of the Duwamish Estuary, Washington</u>. Seattle: Municipality of Metropolitan Seattle (Metro). 189pp.

Malins, D., McCain, B., Brown, D., Sparks, A., Hodgins, H., and Chan, S. (1982). *Chemical Contaminants and Abnormalities in Fish and Invertebrates from Puget Sound*. NOAA Technical Memorandum OMPA-19. Rockville, Maryland: US Department of Commerce.

Malins, D., McCain, B., Brown, D., Chan, S., Myers, M., Landahl, J., Prohaska, P., Friedman, A., Rhodes, L., Burrows, D., Gronlund, W., and Hodgins, H. (1984). *Chemical Pollutants in Sediments and Diseases of Bottom-dwelling Fish in Puget Sound, Washington*. Environ. Sci. Technol. 18:705-713.

Meiggs, T. (1980). "The Use of Sediment Analysis in Forensic Investigations and Procedure Requirements for Such Studies", in *Contaminants and Sediments*. Vol 1. pp. 297-308. Ann Arbor, Michigan: Ann Arbor Science.

Metro (1983). *Duwamish Clean Water Plan*. Seattle: Municipality of Metropolitan Seattle (Metro).

Puget Sound Air Pollution Control Agency (PSAPCA) (1980). *Airborne Lead: A Plan for Control*. Seattle: PSAPCA. 53pp.

Riley, R., Crecelius, E., Fitzner, R., Thomas, B., Gurtisen, J., and Bloom, N. (1983). *Organic and Inorganic Toxicants in Sediment and Marine Birds from Puget Sound*. NOAA Technical Memorandum NOS OMS 1. Rockville, Maryland: US Department of Commerce.

Sample, T. (1987). "Rediscovery of the Lower Duwamish River Estuary--Solutions to Pollution by Point and Nonpoint Source Controls" in *Coastal Zone '87. Proceedings of the Fifth Symposium on Coastal and Ocean Management* edited by O. Magoon, H. Converse, D. Miner, L. Tobin, D. Clark and G. Domurat. New York: American Society of Civil Engineers. Vol. 2: 2134-2140.

Schell, W., and Barnes, R. (1974). "Lead and Mercury in the Aquatic Environment of Western Washington" in *Aqueous-Environmental Chemistry of Metals* edited by A. Rubin. Ann Arbor, Mich.: Ann Arbor Science Publ. pp. 129-165.

Schmidt, S., and Spencer, D. (1986). "The Magnitude of Improper Waste Discharges in Urban Stormwater Systems". Journal WPCF 58:744-748.

## Discussion of Mr. Hubbard's Paper

Question:

Why did your own city staff members get into those suits and masks and go down into the sewers? Why didn't you use hazardous waste clean-up contractors?

Answer:

The City owned the drains at issue. We felt responsible.

Question:

You said a storm water utility is being formed now, partly, I take it, in response to this issue. Why is that?

Answer:

In part, I think it was us. When I called the Engineering Department to tell them I had found **another** storm drain with a problem, they began to get excited. Plus, up-scale Belleview is across the way, and they have a leading utility. And the specter of the NPDES permit program caused the local people to want to seize control of this issue.

Question:

Have you identified or do you plan to identify the effects in the Sound of natural sources of lead or of the lead that was flushed from these sewers into the Sound in years gone by?

Answer:

Such studies are not planned now, but we're talking about them.

Question:

What happens to the dredgings?

Answer:

That occurs upstream on the Duwamish from the area I discussed here. There are sediment traps upstream, too.

Question:

Do you still have people eating the fish they catch out there? (... from the lower Duwamish estuary?)

Answer:

We've signed the areas in five languanges, warning of the

frequencies of discharge events and of fish and shellfish parts to be avoided.

# Long-Term Effects of Urban Stormwater on Wetlands

Richard R. Horner, Member, ASCE[*]

## Organization of a Long-Term Research Effort

Wetlands are lands transitional between terrestrial and aquatic systems where the water table is usually near, at, or above the surface, and hydric soils and hydrophytic vegetation predominate (Cowardin et al., 1979). They are recognized as potential providers of numerous ecological and societal functions of value. These functions include hydrologic and water quality regulation, primary production, consumer support, and various social amenities.

With the growing recognition of these functions of wetlands has come increasing interest in both exploiting certain functions for environmental management and protecting overall functional integrity. The greatest interest nationwide in employing wetlands in management has concerned providing advanced treatment for sewage effluents, and a number of relatively small systems are doing so. Now getting more emphasis is using freshwater wetlands to route and store urban stormwater for discharge quantity control. Also developing rapidly is interest in combining this purpose with capturing pollutants in nonpoint source runoff. While surface water managers and developers have promoted planned use of these functions of wetlands in developing areas, resource managers have expressed concern about the potential effects of such actions on the remaining functions.

While substantial losses have occurred, the Puget Trough region of Washington State retains numerous and diverse wetland resources. Many of these wetlands lie within the zones of current and projected development around Puget Sound cities. Particularly at issue are the widely distributed palustrine wetlands in the uplands of central and southern Puget Trough counties. Some of these wetlands are being considered for designation as stormwater retention/detention ponds, and may be modified morphologically and hydrologically for that purpose. A larger number would receive an altered pattern and quality of runoff, in a more incidental fashion, if their catchments are developed.

---

[*]King County Resource Planning, 506 Second Avenue, Seattle, WA 98104; and University of Washington, FX-10, Seattle, WA 98195.

In 1986 scientists and managers associated with local, state, and federal agencies that have wetland protection and stormwater management responsibilities in the region joined to consider the need for research on the relationships between urban stormwater and wetlands. This group served and still serves as a technical advisory committee for what became known as the Puget Sound Wetlands and Stormwater Management Research Program. King County is staffing and coordinating the program, drawing on funding and other support provided by a number of the participants. The broad questions initially stated by this group were:

1. What impacts on wetland ecosystems and their functions could be associated with their use for storing urban stormwater?

2. What are the potential water quality benefits to downstream receiving waters of draining urban stormwater through wetlands?

Literature review

The initial thrust of the program was a literature review concerning these broad questions. Evaluation of the results of this survey led to the following conclusions (Stockdale, 1986a, b; Stockdale and Horner, 1987), which generally defined what we knew about urban stormwater and wetlands at the outset of the program:

1. A number of physical, chemical, and biological mechanisms can operate in wetlands to remove and hold pollutants entering with influent water.

2. Wetlands have been investigated extensively as sites for polishing municipal wastewater treatment plant effluents, and the resulting experience should be generally applicable to stormwater applications. However, little investigation of this type has occurred in the Pacific Northwest.

3. Stormwater applications have been developed much less, relative to sewage treatment, but are now receiving more attention.

4. A large amount of data is available on treatment efficiencies afforded by wetlands (almost all in polishing sewage effluents). These data generally indicate relatively efficient removal and retention for most classes of pollutants, but seasonal nutrient export has been documented.

5. Some techniques have been suggested, but not widely tested, for managing natural wetlands to improve removal and retention of pollutants.

6. Design strategies have been developed for artificial wetlands to serve the treatment function, and substantial experience has been gained with such systems, especially in Europe.

In preparation for designing a research program, the literature review also took note of major voids in knowledge concerning wetlands and stormwater, as follows:

1. Very little study has been given to the impacts on overall ecosystem functioning of routing point or nonpoint source effluents through wetlands. Especially neglected have been long-term impacts.

2. The respective roles of contaminants and hydrologic change in affecting wetland ecosystems are unknown.

3. Many specific questions of impact were suggested by the inquiry, especially concerning the transport, fate, and effects of toxicants on the wetland system and on adjacent surface water and groundwater.

4. There is very little Pacific Northwest confirmation of results from elsewhere on the water quality benefits of wetland treatment.

## Management needs survey

With the literature review complete, it was desired to develop a foundation for the research program by establishing the specific needs of resource and stormwater managers. Accordingly, a formal management needs survey was conducted among the advisory committee membership. Starting with a list of 18 needs suggested by members, a process was applied to rank the alternatives according to agreed-upon criteria. The process used was developed by Mar et al. (1985) and Horner et al. (1986) and involves computerized manipulation of matrices of relative pairwise weightings of alternatives for each criterion, and of the perceived relative importance of the criteria. The result was a ranked list of the alternative management needs. Those ranking highest, which became the basis for research program design, were:

1. Means of assessing short- and long-term impacts of urban stormwater on functioning of the palustrine wetlands potentially most affected by projected development in the Puget Sound region;

2. Criteria for evaluating proposed stormwater discharges by palustrine wetland type;

3. Improved understanding of hydrologic features and functioning of wetlands as a basis for developing management guidelines pertaining to hydrologic change;

4. Improved understanding of factors critical to achieving urban stormwater quality improvement in wetlands and how to manage that process; and

5. Determination of allowable flood storage capacities that avoid impairment of overall wetland functioning.

Reflective of both the literature review and management needs survey results, it was recognized that the program could make its principal contribution by comprehensively considering long-term effects of urban stormwater discharges on wetlands. At the same time, it could help to fill certain gaps in the knowledge of wetland water quality functioning, as well as test conclusions on wetland treatment developed in other settings for application to Puget Sound area urban stormwater. The program is a response to recent calls (e. g., Wolfe et al., 1987) for long-term research directed toward hypothesis testing for improving resource management and regulation.

General research program design

Responding to the identified needs, a research program consisting of four components, as follows, was designed and implemented:

1. A synoptic survey of the characteristics of palustrine wetlands that have and have not been affected by urban stormwater runoff in the past;

2. A long-term investigation of the functional impacts of urban stormwater discharges;

3. A study of water quality benefits to downstream receiving waters of draining urban stormwater through wetlands; and

4. Laboratory-scale or short-term field experiments to answer specific research quaestions.

The synoptic survey was performed during the spring and summer of 1987 and will be reported in this paper. The remaining tasks are being initiated this year and will be described briefly in the concluding section of the paper.

Synoptic Survey

Methods

Approach and site selection. The survey was intended to compare and contrast wetlands that have been affected by urban runoff for an extended period versus similar systems that drain catchments not developed for urban use, agriculture, or other intensive human activities. Its purposes were to perform pilot work for succeeding studies and to identify appropriate preliminary management guidelines that might be tested in that later work. The survey involved observations and sampling on a single day during the growing season, concentrating on ecosystem components that were expected to exhibit any accumulated effects of urban stormwater.

The survey covered 73 palustrine wetlands in King County, 46 that receive urban runoff and 27 that do not. A greater number of urban sites was selected in order to represent some variety in developed land use. The wetlands represented palustrine open water (POW), emergent (PEM), scrub-shrub (PSS), and forested classes (PFO) (Cowardin et al., 1979).

Observations, sampling, and analyses. A program of observations, sampling and on-site analyses was carried out in all survey wetlands. Samples were taken from 31 of the sites to the Municipality of Metropolitan Seattle Water Quality Laboratory for analysis of certain quantities. This allocation provided a very large data base for statistical analyses on field-measured quantities and a smaller, but still substantial, set of data on the quantities that are more expensive to analyze.

Table 1 summarizes the monitoring carried out in the survey program. Horner et al. (1988) provide details on sample collection and analytical methods. In general, analytical effort concentrated on the soils and plants, which were expected to demonstrate any accumulated effects to a greater degree than more transient elements, such as water.

Data analysis. Initial data analyses were concerned with determining how wetlands affected by urban runoff differ from unaffected wetlands. Therefore, the significance of differences between the two groups was tested statistically for each set of measurements, using a two-sample t-test, Mann-Whitney-Wilcoxon nonparametric test, or chi-square test (Zar, 1984), as appropriate. These analyses were performed using SAS software (SAS Institute, Inc., 1985).

Each hypothesis test was displayed graphically by plotting the interval of nonsignificant difference (IND) (Conquest, 1986). If the mean of one population falls within the IND of the other, the two populations are not significantly different for the selected significance level (alpha), and vice versa. The basic significance level used in the tests was 0.05, although 0.10 was also investigated. Associations among certain variables were investigated with the use of correlation and multiple linear regression analyses (Zar, 1984; SAS Institute, Inc., 1985).

Certain biological data were analyzed using other procedures. Aquatic invertebrate data were subjected to discriminant and stepwise discriminant analyses to identify

the taxa that best discriminate wetlands affected and
unaffected by urban runoff (Morrison, 1976). Also,
diversity indices (Shannon and Weaver, 1963) were computed
for the adults and larvae collected in each wetland. Plant
cover-abundance data were analyzed with the use of the
ordination technique TWINSPAN (Gauch, 1982) to determine
whether plant community characteristics distinguish
affected and unaffected wetlands.

Table 1. Summary of Synoptic Survey Monitoring

| Category | Measurement |
|---|---|
| Qualitative observations | Miscellaneous characteristics |
| Soils | Texture |
| | Organic content |
| | pH |
| | Oxidation-reduction potential |
| | Metals (Pb, Cd, Zn, Cu) |
| | Total phosphorus |
| | Total nitrogen |
| | Microtox |
| | Carbon dioxide evolution |
| | Adenosine triphosphate |
| | Microfauna |
| Water column | pH |
| | Fecal coliforms |
| | Enterococci |
| Plants | Cover-abundance |
| | Metals (Pb, Cd, Zn, Cu) |
| Animals | Aquatic invertebrate counts |
| | Sighting of others |

## Results

*General*. Urbanized watersheds represented a variety of
land use types, ranging up to 100 percent low-density
residential (single family), high-density residential
(multi-family), or commercial plus roads. Various mixes of
those types also occurred. The runoff contributing areas
of the nonurbanized wetlands never had more than 10 percent
low-density residential land use (most had none), and no
high-density residential or commercial land uses occurred
in this group.

Chi-square tests were employed to compare the two classes
of wetlands on the basis of numerous qualitative
characteristics that had been noted systematically.
Wetlands in urbanized watersheds exhibited highly

significantly ($P < 0.01$) more frequent instances of human intrusion, debris, and sediment deposition. Significantly ($0.01 < P < 0.05$) more frequent in this group were examples of channelization, placement of fill, scum or foam, and odor. Also in more evidence in the affected wetlands were attempts to drain the wetland and visible oil, but these differences were marginally significant statistically ($0.05 < P < 0.10$). Not significantly different ($P > 0.10$) were signs of water level fluctuation, scour, water coloration, human effects on vegetation, and the presence of unhealthy plants and dead matter.

Soils. Among the physical and chemical characteristics tested in wetland soils, the following exhibited no significant differences in any zone at alpha = 0.05 (with only one isolated exception at 0.10):

| | |
|---|---|
| Texture | Total phosphorus |
| Organic content | Total nitrogen |
| pH | |

While the mean oxidation-reduction (redox) potentials in wetlands affected by urban runoff were generally higher than in the unaffected sites, meaning the soils were more highly oxygenated in the affected cases, the difference was significant at alpha = 0.05 only near the inlet. As would be expected, mean redox potentials were lowest (< 100 mv) in the deepest water (POW zones). They were below the level at which oxygen is generally fully depleted (approximately 250 mv) in the inlet, open water, and emergent zones, but not in the scrub-shrub or forested zones.

Figure 1 presents IND plots for the metals in soils. The inlet and emergent (PEM) zones exhibited significantly higher Pb. If the 0.10 alpha level is used, Cd and Zn differences would also become significant in the PEM zone. If the comparisons are based on soil wet weights instead of dry weights, recognizing that the wetland soils are usually saturated, some additional significant differences appear. Regardless of the basis, copper exhibited no significant differences in any zone, and no means differed significantly in the scrub-shrub zone (PSS).

The soil metals data were further investigated by examining the highest 10 percent of recorded values, representing 25 cases, of which 22 occurred in the inlet, POW, or PEM zones of wetlands affected by urban runoff. Of these 22 cases, nine occurred in catchments that have at least 95 percent high-density residential (HDR) or commercial land uses, or a combination of the two. An additional 11 cases (seven in inlet zones) were in watersheds with 50 percent or more low-density residential (LDR), or some combination of LDR with HDR and/or commercial. This analysis was formalized with stepwise multiple linear regressions of soil metal concentrations on land use percentages. Of 16 analyses attempted, 10 resulted in significant (P < 0.07) regression coefficients, with coefficients of determination ($R^2$) ranging from 0.12 to 0.81. In seven of these 10 cases the significant regression coefficients were associated with percent HDR, commercial, or both.

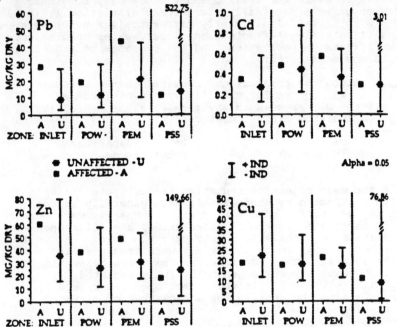

Figure 1. Intervals of Nonsignificant Difference (IND) for Four Metals

Figure 2 displays IND for the Microtox test series conducted for 15 minutes. Affected wetland open water and emergent zone soils exhibited significantly greater inhibition of bioluminescent bacteria output at alpha = 0.10. The difference was also significant in the open water zone in a five-minute exposure (not graphed). Inlet and scrub-shrub (not graphed) zones exhibited no significant differences.

The lowest 10 percent of Microtox values (signifying the greatest inhibition of bioluminescence) represented 14 cases, of which 12 were in urban wetlands. Five of these 14 have 75 percent or more HDR or commercial land uses in their watersheds, and an additional six have at least 50 percent LDR or LDR in combination with HDR. Stepwise multiple linear regressions involving the Microtox data and land use percentages resulted in only two statistically significant (P < 0.05) equations in eight attempts. Both indicated that potential toxicity was inversely related to the proportion of forested land use.

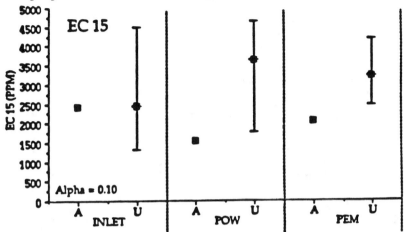

Figure 2. IND for 15-Minute Microtox Test Series (EC = effective soil concentration that caused a 50 % bioluminescent output reduction compared to controls)

Correlation analyses by wetland zone found few significant correlations between soil metal concentrations and Microtox results. Only Pb in open water areas and Zn in scrub-shrub zones were significantly correlated with inhibited bioluminescence.

Among the three indicators of soil microbial activity tested (ATP, carbon dioxide evolution, and microfauna counts), only mean carbon dioxide production in emergent zones on one of two sampling occasions differed significantly. There is no apparent biological reason for this singular result.

Therefore, the general physical, chemical, and microbiological nature of the substrates in wetlands affected and unaffected by urban runoff did not appear to differ in many respects. However, a number of instances of significantly elevated metal concentrations were noted, and were associated largely with the more intensive land uses. These differences between metals in affected and unaffected wetlands contrast with the results of comparing soil nutrients, where no significant differences occurred. Nutrients cycle relatively rapidly in wetland soils, while metals tend to be more cumulative over time (Horner, 1986), a tendency that seems to be evident in the wetlands surveyed. The presence of elevated metals is one possible explanation for the potential toxicity sometimes noted in

Microtox results. However, the associations are not entirely consistent among the zones; and few significant correlations related metal concentrations and inhibited bioluminescence. Organic toxicants, which were not measured, are another possible explanation for the potential toxicity, which was not a chance occurrence according to the statistical tests.

<u>Water column</u>. Water column analyses were limited because of the belief that more permanent components of the systems would better reveal differences in a survey. Mean pH was higher in the affected (6.38) compared to the unaffected (6.10) wetlands, a difference significant at the 0.07 but not the 0.05 level.

Figure 3 depicts IND for the water column bacteriolgical data. Means of both fecal coliforms and enterococci were significantly higher in the affected cases. The fecal coliform mean was above the Washington State standards for Class AA and A waters (geometric means of 50 and 100 organisms/100 mL, respectively), but below the 200/100 mL standard for Class B waters. The enterococci mean was below the proposed EPA criterion (Dufour, 1984) of 33/100 mL. Log-transformed fecal coliforms and enterococci were significantly correlated (Pearson correlation coefficient = 0.73, nonparametric Spearman correlation coefficient = 0.65).

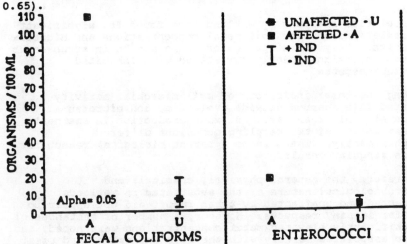

Figure 3. IND for Fecal Coliforms and Enterococci

Review of the data set revealed that seven of the 19 affected wetlands tested had fecal coliform counts greater than 200/100 mL (five also had enterococci above 33/100 mL). Six of these seven catchments have HDR or commercial land use, or both, but the percentages varied considerably.

Four had coliform counts above 400/100 mL, with a high of 4633/100 mL (enterococci ranged 80-230 in these wetlands, the highest values measured). One of two stepwise multiple linear regression analyses relating these measurements to land use was significant statistically ($P < 0.05$), but $R^2$ was only 0.27. In this equation fecal coliform count was positively associated with commercial land use and negatively with open (unforested) undeveloped land.

Plants and aquatic invertebrates. Plant cover-abundance was analyzed according to the ordination technique by wetland zones. The clearest separation between wetlands affected and unaffected by urban runoff occurred in the emergent zones, where the affected sites exhibited a prevalence of Phalaris arundinacea (reed canary grass), a recognized opportunist that tends to monoculture. The unaffected sites had a greater diversity, with Carex species (sedges) prominent in the composition.

Comparison of mean metal concentrations in tissues of four plant species found in affected and unaffected wetlands exhibited no statistically significant differences at $P = 0.10$. However, a number of values were below the detection limit, especially for Pb, and a proper hypothesis test could not be conducted in several cases. In future work a more sensitive analytical technique will be used to avoid this problem. There were no significant correlations between sediment and plant tissue metals concentrations in the various wetland zones. However, eight of 16 stepwise multiple linear regression analyses on land use percentages produced significant ($P < 0.05$) coefficients; all were associated with HDR, commercial, or roads, or some combination of two of these land uses. $R^2$ ranged from 0.17 to 0.98.

For adult and larval invertebrates, discriminating models were calibrated to classify wetlands in groups affected and unaffected by urban runoff. Among the larvae, the best discrimination was provided by the caddisfly genus Anabolia, the families Chironomidae and Heptageniidae, and the subfamily Lestidae (a damselfly), all more prevalent in unaffected wetlands; and by Dicosmoecus, another caddisfly, and two Hemiptera families, all more prevalent at affected sites. These models now require verification with an independent data set. Mean Shannon-Weaver diversity indices for adults were 0.74 and 0.62 in affected and unaffected wetlands, respectively. The equivalent indices for larvae were 0.53 and 0.64, respectively. Neither difference was statistically significant.

## Conclusions

The wetlands generally exhibited substantial within-group variability. This dispersion detracted from the ability to demonstrate differences statistically. To compensate, at least two possible significance levels were always investigated. When the null hypothesis was accepted, the P value was usually well above 0.10. Therefore, there is a fairly high degree of statistical confidence in the hypothesis test results.

The results present a somewhat mixed picture with respect to the differences between wetlands that have and have not received urban runoff over a long period. The most distinct differences appeared in comparisons of qualitative features, most of which are not the direct result of stormwater runoff but of the overall urban presence, or lack thereof. The consequences of these differences seem to be primarily aesthetic, rather than indicative by themselves of functional impairment.

The strongest quantitative distinction between the two classes of wetlands was in water column bacteriological indicators, followed by soil metal concentrations and the potential toxicity of the soil material. It does appear from these results that runoff from high-density residential and commercial areas has affected the wetlands surveyed differently than drainage from low-density residential catchments. Effects associated with the more intensively developed watersheds were both stronger and more widespread through the wetland zones. However, the actual biological consequences of these effects have not been established.

Only a few general management-related guidelines can be developed on the basis of these results. First, observation of extensive intrusion in and aesthetic degradation of urban wetlands, over and above any stormwater impacts, suggests that agencies that have custody of such wetlands should take steps to prevent disturbances of the type noted. Effective steps would protect the image of well managed stormwater control operations in wetlands from being tarnished by unrelated degradation.

Observation of much more sediment deposition in affected wetlands, coupled with the measurement of some elevated metal concentrations, probably introduced through solid-phase transport, suggest the consideration of presettling before any urban stormwater discharge to a natural wetland, especially from a densely developed catchment. Even relatively brief presettling could remove the largest particles and minimize the need to disturb the wetland by dredging. In series arrangement with a presettling pond, the wetland would still offer the majority of the discharge quantity control, as well as the extended residence time needed for settling finer particles and promoting chemical and biological pollutant removal mechanisms.

Documentation of the dominance in the emergent zones of some affected wetlands by an opportunistic species that tends to establish monocultures suggests an area for monitoring attention. It is hypothesized that succession to such patterns is caused more by modified hydrology than by the delivery of pollutants. This is a hypothesis that could not be tested in the one-time surveys, but will receive definite attention in future research. In the meantime, it is recommended that large deviations from predeveloped hydrologic patterns (depths, frequencies, and durations of inundation) be avoided.

Urban wetlands were found to have much higher bacterial counts than their unaffected counterparts, especially when they drain relatively intensively developed catchments. However, in the majority of cases these counts did not, at least in a single measurement, exceed existing or proposed standards. Moreover, humans rarely use wetlands for contact recreation or as a drinking water source. Therefore, the significance of this result depends mainly on the release of bacteria from the wetland and the use of downstream waters. This survey was unable to document transport, but that appears to be a fruitful subject for succeeding work. If wetlands are effective traps for bacteria in urban runoff, they could serve a significant role in reducing transport of those organisms to waters where they have a considerable impact (e. g., shellfish beds). A similar concern pertains to other pollutants that may be present in the influent or released from the soils.

## Future Work

In 1988 we are beginning work on all three remaining components of the program: the long-term study of urban runoff impacts, the investigation of water quality benefits that wetland treatment might provide, and a laboratory study. The long-term study is being carried out in 14 wetlands (with several more to be added), half that never become affected by urban runoff and half that become so affected after a period during which baseline monitoring will be done. The study will proceed for five years, although not necessarily five successive years. In one or more later years, annual coverage may be held in abeyance in order to ensure that resources are sufficient to document any slowly developing long-term effects. Monitoring will encompass water level fluctuation; water quality; the physical structure and chemistry of soils; soil microbial activity; plant community composition, productivity, and tissue metals concentrations; aquatic invertebrate identification and enumeration; bird censusing; and mammal and herpetofaunal surveys.

In the water quality benefit study, influent and effluent stations were placed in two wetlands, one that receives urban runoff and one that does not. Continuous flow gaging is being performed at these stations, and composite water samples will be taken during approximately 10 storms and during dry weather periods as well. Each wetland has a precipitation station to provide both quantity measurements and precipitation samples for analysis. In order to study the participation of groundwater in the hydrologic systems of these wetlands, a series of shallow piezometers was placed around them in order to measure the position of the groundwater surface and sample for water quality analyses. Analyses of these various samples will be similar to those listed for the long-term program.

The laboratory investigation that is being started in 1988 involves the hydrologic manipulation of wetland plant and soil microcosms. This study directly addresses the hypothesis stated above concerning the influence of hydrology on plant community development.

It is expected that this comprehensive program of studies will yield not only the desired guidance to manage wetlands that are under urban influence, but also substantial information to expand the state of knowledge of overall wetland functioning.

## Acknowledgments

The author is grateful for support for this work by the Washington Department of Ecology, through Coastal Zone Management Act funding, the Municipality of Metropolitan Seattle (Metro), and King County. Sample collection and analytical services were provided by the Metro Water Quality Laboratory. Statistical analyses were guided by Loveday L. Conquest, and F. Brandt Gutermuth and Lorin E. Reinelt performed computer work.

## References

Conquest, L., "Use of Intervals of Nonsignificant Difference (IND) in Graphical Procedures," U. S. Food and Drug Administration, Washington, D. C., 1986.

Cowardin, L, M., Carter, V., Golet, F. C., and LaRoe, E. T., "Classification of Wetlands and Deep Water Habitats of the United States," FWS/OBS-79/31, U. S. Fish and Wildlife Service, Washington, D. C., 1979.

Dufour, A. P., "Health Effects for Fresh Recreational Waters," EPA-600/1-84-004, U. S. Environmental Protection Agency, Cincinnati, OH, 1984.

Gauch, H. G., *Multivariate Analysis in Community Ecology*, Cambridge University Press, Cambridge, England, 1982.

Horner, R. R., "A Review of Wetland Water Quality Functions," *Wetland Functions, Rehabilitation and Creation* in the Pacific Northwest: The State of Our Understanding, R. Strickland, ed., Washington State Department of Ecology, Olympia, WA, 1986, pp. 33-50.

Horner, R. R., Mar, B. W., Reinelt, L. E., Richey, J. S., and Lee, J. M., "Design of Monitoring Programs for Determination of Ecological Change Resulting from Nonpoint Source Water Pollution in Washington State," Washington State Department of Ecology, Olympia, WA, 1986.

Horner, R. R., Gutermuth, F. B., Conquest, L. L., and Johnson, A. W., "Urban Stormwater and Puget Trough Wetlands," Proceedings of the First Annual Puget Sound Research Meeting, Puget Sound Water Quality Authority, Seattle, WA, 1988, pp. 723-746.

Mar, B. W., Lettenmaier, D. P., Horner, R. R., Richey, J. S., Palmer, R. N., Millard, S. P., MacKenzie, M. C., Vega-Gonzalez, S., and Lund, J. R., "Sampling Design for Aquatic Ecological Monitoring," Vol. 1-5, EPRI EA-4302. Electric Power Research Institute, Palo Alto, CA, 1985.

Morrison, D. F., Multivariate Statistical Methods, McGraw-Hill Book Company, New York, NY, 1976.

SAS Institute, Inc., SAS Users' Guide: Statistics, Version 5 Edition, SAS Institute, Inc., Cary, NC., 1985.

Shannon, C. E. and Weaver, W., The Mathematical Theory of Communication, University of Illinois Press, Urbana, IL, 1963.

Stockdale, E. C., "The Use of Wetlands for Stormwater Management and Nonpoint Pollution Control: A Review of the Literature," Washington State Department of Ecology, Olympia, WA, 1986a.

Stockdale, E. C., "Viability of Freshwater Wetlands for Urban Surface Water Management and Nonpoint Pollution Control: An Annotated Bibliography," Washington State Department of Ecology, Olympia, WA, 1986b.

Stockdale, E. C. and Horner, R. R., "Prospects for Wetlands Use in Stormwater Management," Proceedings of the Coastal Zone 87 Conference, Seattle, WA, 1987.

Wolfe, D. A., Champ, M. A., Flemer, D. A., and Mearns, A. J., "Long-Term Biological Data Sets: Their Role in Research, Monitoring, and Management of Estuarine and Coastal Marine Systems," Estuaries, Vol. 10, 1987, pp. 181-193.

Zar, J., Biostatistical Analysis, 2nd Ed., Prentice-Hall, Englewood Cliffs, NJ, 1984.

# Discussion of Dr. Horner's Paper

Question:

Rather than viewing a wetlands as a special place needing special protections, I am interested in how I can use a wetlands as a treatment device, even if I hurt it a little. Are there criteria around for how to use a wetlands for treatment?

Answer:

In the sanitary wastewater treatment field, the use of bulrushes, among other plants, has been studied extensively. There are at least 30 years of experience in that area. EPA even had a treatment guide out. So, yes, there are criteria out there for treatment with plant species, but from the advanced wastewater treatment area, not yet from the runoff area.

Question:

In your own wetlands that you studied, do you find that removals of urban runoff related constituents are controlled from the "bottom-up," as with soil and root uptake, or is the ultimate removal or control more from the "top-down," as in full plant utilization and herbivore grazing and assimilation?

Answer:

We don't have the investigative abilities in place to answer that now, but those area exactly the kinds of things these initial investigations I described are being expanded to find out.

# THE USE OF WETLANDS FOR URBAN STORMWATER MANAGEMENT

## ERIC H. LIVINGSTON*

ABSTRACT

The natural capabilities of wetlands imply that they could be incorporated into a stormwater management system to enhance the removal of pollutants commonly found in urban stormwater. However, the design of such systems must consider many factors to assure that the wetland filter is not damaged nor is the wetland ecologically harmed. Design variables for natural and constructed wetland stormwater treatment systems will be discussed. Criteria used in Florida for natural wetland treatment systems and criteria used in Maryland for constructed wetland systems will be reviewed. The review of this "art" will include a discussion of the many unknowns that must still be determined concerning the long-term ability of using wetlands for stormwater management.

INTRODUCTION

Wetlands are an essential part of nature's stormwater management system. Important wetland functions include the conveyance and storage of stormwater thereby dampening flooding impacts; the reduction of flood flows and velocity of stormwater which reduces erosion and increases sedimentation; and the assimilation of pollutants typically carried in stormwater. Accordingly, there is a great amount of interest in the incorporation of natural wetlands into stormwater management systems, especially wetlands which have been previously drained, or constructed wetlands. This concept provides an opportunity to use, not abuse, one of nature's systems to mitigate the urban runoff impacts associated with urbanization. In addition, by using wetlands for stormwater management, drained wetlands can be revitalized and landowners and developers have an incentive to preserve or restore wetlands.

Unfortunately, the use of wetlands for stormwater management involves a large degree of uncertainty (1). Little scientific information is available concerning the short or long term impacts on wetlands, their natural functions or associated fauna from the addition of stormwater. Therefore, the use of wetlands for urban
-----------------------

*Environmental Administrator, Nonpoint Source Management Section, Florida Department of Environmental Regulation, 2600 Blair Stone Road, Tallahassee, Florida 32399-2400

stormwater management should not be considered the panacea solution to our nation's stormwater problems. Much remains to be learned. This paper will review the current state-of-the-art and discuss the design and performance standards being used for wetland stormwater treatment systems.

STORMWATER MANAGEMENT

In an undeveloped area, stormwater runoff is managed by nature through the hydrologic cycle. This cycle, together with the many components of the natural stormwater management system, is effective in accommodating even severe rainfalls. Unfortunately, land use changes associated with urbanization causes four interrelated effects on the natural hydrology of on area: changes in peak flow characteristics, changes in total runoff, changes in water quality and changes in the hydrologic amenities (2). These changes create a need for the construction of stormwater management systems.

Today, however, stormwater management is far more comprehensive with objectives of providing: 1) surface drainage; 2) flood control; 3) erosion and sedimentation control; 4) reduction of pollutants in runoff; and 5) aesthetic amenities including open space, recreation and water front property. To accomplish these objectives, it is necessary to ensure that the volume, rate, timing and pollutant load of runoff after development is similar to that which occurred prior to development. Therefore, a stormwater management system must be an integral part of the site planning process for every project. The natural site attributes (e.g., soils, geology, slope, water table) will influence the type and configuration of the stormwater system. For example, sandy soils imply using infiltration practices such as retention areas integrated into a development's open space and landscaping while natural low areas and isolated wetlands offer opportunities for detention/wetland treatment. A stormwater management system should be viewed as a "treatment train" in which the individual best management practices (BMPs) are the cars. The more BMPs that are incorporated into the system, the better the performance of the train.

Stormwater Pollution

The need for stormwater pollution control has received increasing attention since the late 1970's when the Federal Section 208 program funded numerous studies on the water quality impacts of runoff. The type and amount of pollutants carried in stormwater is highly variable with pollutant characteristics closely related to land use and rainfall characteristics (3). Other characteristics of stormwater pollution include:

* Higher pollutant concentrations are

associated with more intensive development and higher amounts of imperviousness.

* Erosion and sedimentation during construction can result in very high loadings of suspended solids.

* Stormwater pollutant levels are comparable to secondarily treated wastewater effluent for many constituents.

Of primary importance to water quality is the "first flush". This term describes the washing action that stormwater has on accumulated pollutants in the watershed. In the early stages of a storm the land surfaces, especially impervious surfaces such as streets or parking lots, are flushed clean by the rainfall and resulting runoff. This flushing creates a shock loading of pollutants. Extensive studies in Florida have determined that the first flush equates to the first one inch of runoff which carries 90% of the pollution load from a storm event (4). Treatment of the first one inch of runoff will help ensure that the water quality impacts of stormwater are minimized.

## WETLANDS AS STORMWATER TREATMENT SYSTEMS

As is discussed throughout these proceedings, considerable attention has been devoted toward the use of wetlands for wastewater treatment and nutrient assimilation. However, stormwater hydrology and pollutant characteristics are greatly different from the treated wastewater that is typically placed into wetlands for additional pollutant removal. For example, nutrient concentrations in typical wastewater effluent and stormwater may easily differ by a factor of 10 or more. Therefore, studies of the treatment efficiency of wetlands for wastewater effluent may not be applicable to treatment of stormwater.

Field investigations and research concerning the use of wetlands for treatment of stormwater have been very limited (5,6). The few studies undertaken exhibit great dissimilarities in the type of wetland and stormwater characteristics examined, contain a very slim data base from which to draw conclusions and encountered numerous complications in determining hydrologic components (7). These studies indicate that: (1) a wide disparity exits in the capability of wetlands to remove stormwater pollutants, especially nutrients; (2) the nature of flow and seasonal factors are major influences on pollutant removal capabilities of certain wetlands; and (3) the greatest consistency in pollutant reduction appears to be for BOD, suspended solids and heavy metals (7,8).

Due to the relatively young state-of-the-art, the inherent variations in stormwater characteristics and the complex interactions within a wetland that provide for pollutant removal, it is impossible to predict the treatment efficiency of a wetland stormwater system. There is simply too little monitoring data to develop confident relationships between the many design variables and system pollutant removal. Review papers on the use of wetlands for stormwater treatment are available which summarize the treatment efficiency and performance of wetland stormwater systems (5,6,7,8). Essentially, a wetland system must be designed as a BMP treatment train which accentuates the numerous assimilation mechanisms that are present in each type of wetland to maximize stormwater pollutant removal.

Wetland Assimilation Mechanism

Wetlands may be viewed as "nature's kidneys" since several internal processes allow wetlands to transform some elements (e.g., nitrogen, sulfate), act as a sink or source for others (e.g., carbon) and function as a sediment filter (9). Ultimately, pollutants can be removed by a wetland system through loss to the atmosphere by volatilization, incorporation into the sediments or biota or by degradation. The initial pollutant removal mechanisms in wetlands are physical and chemical processes which are followed by biological processes.

The internal chemical and physical assimilation mechanisms include:

1. Volatilization - Pollutants enter the atmosphere primarily by evaporation but they may also enter by aerosol formation under windy conditions. Common stormwater pollutants removed by volatilization includes oils, chlorinated hydrocarbons, and mercury (8).

2. Sedimentation - This is one of the principal mechanisms by which particulate pollutants are removed from the water column. Deposition is greatly affected by the nature and velocity of flow patterns (slow sheet flow is best). The rate of sedimentation is directly related to the size of the particle, the hydrology, flow velocity and path and storm sizes (10). Sedimentation is important for removal of suspended solids, particulate nitrogen, oils,

chlorinated hydrocarbons and most heavy metals.

3. Adsorption - Dissolved pollutants adhere on to suspended solids, bottom sediments or vegetation thereby removing them from the water column. Adsorption mechanisms are enhanced by shallow water depths and long residence times which increase contact opportunities with soil. Adsorption is an important process for removing ammonium ions, phosphate, and heavy metals. Adsorption is also the primary virus removal mechanism as stormwater percoaltes through soil (11).

4. Precipitation, Dissolution, Complexation - Many ionic species, particularly metals, dissolve or precipitate in response to changes in the chemistry of the wetland environment. Metals will form insoluble sulfides under reduced conditions commonly found in wetlands and will form insoluble oxides and hydroxides under oxidized conditions (8).

5. Filtration - This occurs as particulate pollutants are filtered through sediments, vegetation and biota in the wetland. Sheet flow, reduced velocities and dense vegetation promote greater pollutant removal. Removal of pollutants by filtration through soils (infiltration) is effective in removing organic matter, phosphorus, bacteria and suspended material.

Vegetated wetlands offer high pollutant adsorption and biological assimilation potential and provide an excellent substrate for microbial activity. In general, wetland systems are characterized by: (1) high plant productivity and nutrient uptake; (2) high decomposition activity; (3) large adsorptive areas in sediment substrates; and (4) low oxygen content of the sediments. Each of these properties provide vegetative systems with the ability to remove pollutants. Through interaction with the various soil layers, water and air interfaces, plants can increase the overall capacity of a system to retain or remove pollutants.

Additionally, plant uptake of pollutants, particularly from the sediments, frees more exchange sites for further pollutant adsorption, by the soil. Plants also provide surfaces for bacteria growth and promote filtration and absorption. The primary biological pollutant removal processes in wetlands are (12):

1. Uptake through plant soil interface, via below ground roots, rhizomes, holdfasts and buried shoots and leaves.

2. Uptake through plant-water interface, via submerged roots, stems, shoots and leaves.

3. Translocation through plant vascular system, from roots to stems, shoots, leaves and seeds during the growing season.

4. Differential pollutant uptake, such as preferential storage of trace contaminants in specific plant parts and preferential uptake and accumulation of certain trace elements.

5. Non-specific pollutant uptake, occurring primarily as plants absorb large quantities of nutrients from water and sediments.

6. Uptake and immobilization by plant litter zones, where dead, but not decomposed, plant litter sequesters pollutants through chemical interactions.

### Wetland Design Factors

In designing a wetland stormwater system, one wants to maximize the pollutant assimilative processes which are greatly influenced by various design factors which will subsequently be discussed.

### Pretreatment

It is essential that the wetland and its assimilative processes are not adversely impacted by the addition of untreated stormwater. Therefore, pretreatment is recommended to remove heavy sediment loads and other pollutants such as hydrocarbons which can damage the wetland. Pretreatment is also needed to attenuate stormwater volumes and peak discharge rates so as to not adversely alter the hydroperiod of the wetland and to reduce scour and erosion. Erosion and sediment control practices during construction are essential to prevent heavy sedimentation of the wetland.

## Hydrology

A clear understanding of the wetland's hydroecology is crucial if the wetland treatment system is to perform effectively over a long period of time. The relationships between hydrology and wetland ecosystem characteristics must be recognized and incorporated into the design of a wetland stormwater system. Factors such as the source of water, velocity, volume, renewal rate and frequency of inundation have a major bearing on the chemical and physical properties of the wetland substrate which in turn influence the character and health of the ecosystem as reflected by species composition and richness, primary productivity, organic deposition and flux, and nutrient cycling (7).

Hydrology controls pollutant removal in wetlands through its influence on processes of sedimentation, aeration, biological transformation and soil adsorption. Critical hydrologic factors that must be evaluated in the design of a wetland stormwater system include:

* velocity and flow rate
* water depth and fluctuation
* detention time
* circulation and distribution patterns
* seasonal and climatic influences
* groundwater conditions
* soil permeability

After gaining an understanding of the hydrologic factors ongoing in the proposed treatment wetland, the information can be used to design the system. The type of wetland, (e.g., peatland, cypress dome, marsh meadow, etc.) will greatly influence the wetland's suitability as a component of a stormwater system and the final design and system effectiveness. Of primary importance is the establishment of the hydroperiod of the wetland since this ultimately determines the form, nature and function of the wetland. Hydroperiod is comprehensively defined as the depth, as well as duration, of inundation measured over an annual wet or dry cycle (13). Acceptable high and low water elevations within the wetland will determine the stormwater treatment volume that can be stored in the wetland and the elevation of the discharge structure and its associated bleed-down orifice. The depth of water and period of inundation can result in changes in the plant community which may or may not be beneficial to either the wetland or the removal of stormwater pollutants.

## Vegetation

The vegetation in a wetland is a function of the climate, hydrology and availability of nutrients. In a natural wetland, the vegetation will already be established, and is a factor in the selection of loading rates and expected pollutant removal. In a constructed wetland, plant species can be selected based on considerations such as climate, hydrology, and the response of the plants to pollutants. In either case, a diverse mixture of floating, emergent and submerged plants is desirable, especially plants that form dense stands of submerged stems and leaves or plants with heavy, thick floating root mats since these will help increase filtration and provide many suitable sites for micro-organisms.

In addition to the efficiency of pollutant removal, wetland plants also have specific tolerances to various levels and kinds of pollutants. The addition of polluted stormwater represents a new source of increased nutrients, which will alter the ecological balances of the wetland and may result in a change in the plant community. Since the newly dominant plant species presumably have emerged because of their ability to use the new nutrient source or because of their tolerance to these pollutants, the change in plants should prove beneficial in removing stormwater pollutants (7).

## Maintenance

Very little information is available about maintenance of wetland stormwater management systems. Maintenance is very important since accumulation of organic matter and sediment can alter the pollutant removal effectiveness of the system by decreasing storage volume and altering the sediment. For example, organic accumulations can create an oxygen demand and lower pH which will favor the release of phosphorus and metals from the sediments into the water column. Typical maintenance will likely include removal of sediment accumulations, repair of the conveyance system and discharge structure and vegetation harvesting or burning. Vegetation harvesting is needed to renew the ability of the plants to remove pollutants and to prevent accumulations of plant litter. The timing of harvest is important since it should be performed prior to storage of nutrients in underground plant parts for maximum nutrient removal. In Europe, reeds are harvested annually and rushes every two to three years with the harvested material used for wicker furniture (5).

## MARYLAND CREATED WETLAND DESIGN STANDARDS

In 1982 Maryland enacted state legislation that required the development of stormwater management regulations to assure that stormwater from new development was treated to reduce the pollutant load discharged to receiving waters.

As part of its program the state has developed the guidelines that will be subsequently discussed for the construction of shallow wetland stormwater systems (4).

Water Budget

The most essential ingredient for the successful establishment of a shallow wetland is adequate water to maintain a permanent pool of water. The inflow of water from stormwater, base flow and groundwater must be greater than the outflow of water via infiltration and discharge. If the basin does not intersect the groundwater then the infiltration rate must be less than the inputs from stormwater and base flow to maintain a permanent pool. If the infiltration rate is too high then either a natural (e.g., clay) or synthetic liner should be used.

Wetland Size

To provide the greatest benefits to wildlife and pollution control the wetland surface area should be maximized. If an existing stormwater basin is modified to enhance pollutant removal, the wetland should be created throughout most of the basin. A detention time of 24 hours for the one year storm is recommended to enhance pollutant removal and allow the storage volume to be recovered before the next storm. If extended detention is not possible then the surface area of the wetland should be a minimum of three percent of the contributing drainage area.

General Wetland Design

Table 1 lists the recommended design standards for the shallow wetland stormwater basins while Figure 1 illustrates three potential configurations. Approximately seventy-five percent of the wetland area should have depths less than 12 inches to promote the growth of emergent aquatic vegetation. The outlet structure should be placed in the wetland's two to three feet deep area which covers the remaining twenty-five percent of area in which submerged aquatic vegetation will thrive. A sediment forebay at least three feet deep and containing ten percent of the total volume of the normal pool should be established at all inflow points. Energy dissipation devices also should be used to reduce velocities thus lessening scour and damage to plants. A 2:1 length to width ratio is recommended to reduce short circuiting and maximize the flow path. If inlet and outlet structures are close together, baffles, islands or peninsulas should be used to redirect the water and lengthen the flow path.

Extended Detention Outlet Structure

The outlet structure must dam up the water needed for the

TABLE 1. MARYLAND CONSTRUCTED WETLAND DESIGN GUIDELINES (14)

- Water flow into pond must exceed infiltration rate of basin or a liner should be installed
- 25% of the wetland area should be 2-3 feet deep
- 75% of the wetland area should be under one foot deep
    - 25% should be 6-12 inches deep
    - 50% should be under 6 inches deep
- Locate discharge outlet in deep area of wetland
- Include a 3 foot deep forebay having at least 10% of the total basin volume
- Wetland perimeter should be bordered by a 10-20 foot wide zone that is temporarily flooded at most storms
- Incoming stormwater should flow into the shallow, vegetated area of the wetland
- Length to width ratio should be at least 2:1
- Wetland should be able to detain the one year storm for 24 hours

wetland's permanent pool, detain the treatment volume for an extended time period and allow water to slowly and reliably discharge from the wetland. The discharge orifice is set at the elevation of the permanent pool and sized to meet the desired discharge rate. Stormwater entering the wetland will be detained increasing the volume and depth above the orifice elevation. The small diameter orifice will be vulnerable to blockage from plant material or debris making some form of protection necessary. A wire mesh cube, sphere or other three-dimensional structure with a large surface area should be securely attached to the outlet structure.

Wetland Substrate

Wetland soils which contain a pool of wetland vegetation propagules are the preferred soil. However, it is unlikely that a site will contain a large amount of wetland soil or that such a soil will be available from other sources. When converting an existing basin to a created wetland, areas that do not drain completely or that are in a flow path will often have wet soils that can be carefully stockpiled and spread over the basin bottom after excavation. At least four inches of soil depth is recommended to provide enough substrate to anchor the plants securely.

Wetland Vegetation

Part of the wetland basin should be planted with appropriate aquatic vegetation to enhance the short and long-term development of the wetland and to reduce successful establishment of undesirable plants, especially aggressive volunteers such as Typha spp. (cattail) and Phragmites australus (common reed). Vegetation planting recommendations are summarized in Table 2. A minimum of five wetland species (Table 3) are recommended to take advantage of the wetland's heterogeneity in soil type, depth, water circulation, etc. which create habitat variations. A mix of species also maximizes the presence of vegetation throughout the growing season thus enhancing nutrient uptake.

A minimum of two "primary species", which are aggressive growers, should be planted in four widely separate areas totaling thirty percent of the shallow (less than 12 inches) area. Each area should have site conditions conducive to the growth of the species it will contain such as the depths specified in Table 3. Monospecific planting, to reduce competition within areas, should be done with a spacing of three feet between individual plants. An additional forty clumps per acre of each primary species should be planted in areas conducive to growth throughout the rest of the shallow zone. Each clump should contain

TABLE 2. PLANTING REQUIREMENTS FOR MARYLAND CONSTRUCTED WETLANDS

Include Two Primary Species
1. These species should cover 30% of the shallow zone using a three foot spacing
2. Plant in four monospecific areas
3. Distribute an additional 40 clumps/acre of each species over the entire wetland

Include Three Secondary Species
1. Plant fifty individuals of each secondary specie per acre
2. For each species, plant 10 clumps of five individuals each as close to the edge of the pond as possible
3. Space the clumps as far apart as possible, but no need to segregate species to different areas of the wetland

TABLE 3.  EMERGENT SPECIES FOR WETLAND ESTABLISHMENT

| SPECIES | MAXIMUM DEPTH | AVAILABLE COMMERCIALLY |
|---|---|---|
| **A. Primary Species** | | |
| Sagittaria latifolia (duck potato) | 12 inches | Yes |
| Scirpus americanus (common three square) | 6 inches | Yes |
| Scirpus validus (softstem bulrush) | 12 inches | Yes |
| **B. Secondary Species** | | |
| Acorus Calamus (sweet flag) | 3 inches | Yes |
| Cephalanthus occidentalis (button bush) | 2 feet | Yes |
| Hibiscus Moscheutos (rose mallow) | 3 inches | Yes |
| Hibiscus Laevis (Halbered-leaved r. mallow) | 3 inches | No |
| Leersia oryzoides (rice cutgrass) | 3 inches | Yes |
| Nuphar luteum (spatterdock) | 5 feet (2 ft min) | No |
| Peltandra virginica (arrow-arum) | 12 inches | Yes |
| Pondederia cordata (pickerel weed) | 12 inches | Yes |
| Saururus Cernuus (lizards tail) | 6 inches | Yes |
| **C. Unacceptable Species** | | |
| Phragmites australus (common reed) | 3 inches | No |
| Typha Latifolia (common cattail) | 12 inches | Yes |
| Typha angustifolia (narrow-leaved cattail) | 12 inches | Yes |

one or more individuals of a single species. The number of individuals of each species to be planted is a function of the total area and shape of the site to be planted. In general, the planted areas should be as square as possible to decrease the perimeter area and help preserve relatively homogenous populations of planted species by reducing colonization.

Planting Wetland Vegetation

Site preparation for planting the wetland plants requires that the substrate be soft enough to permit easy insertion of the plants. If necessary, disking the upper six inches of substrate should suffice. Site preparation also involves the flooding of the wetland. If planting is to be done before the wetland is flooded, no more than 24 hours (for bare root plants) to 72 hours (for peat potted plants) should elapse between planting and flooding. If flooding cannot occur within these times, alternatives such as base flow diversion, fire hydrant or irrigation water should be used to keep the plants healthy.

The use of nursery grown plants, plants transplanted from roadside ditches or other wet areas (where permissible) or transplanted dormant underground plant parts are recommended means of plant establishment. The growth form of the propagule will determine the suitable planting time during the year. Actively growing plant material should be set out during the growing season while dormant material should be planted during the winter.

Proper treatment of plant propagule material is essential for successful plant establishment. Growing plants should be kept out of direct sunlight and their roots kept moist. Dormant below ground plant parts should be stored dry, usually in mulch, at temperatures slightly above freezing.

Submerged Aquatic Vegetation (SAV)

Submerged aquatic vegetation is an important food for waterfowl and helps improve the quality of stormwater. The artificial introduction of SAV is not required because of a lack of commercial sources but is encouraged. It is likely that SAV will be brought to the wetland by migrating waterfowl and will become established in the deeper pool.

Maintenance

It is uncertain how long shallow wetland basins will function, especially in removing nutrients and metals. Solids will accumulate and need removal. However, to avoid disruption of the plant community, it is recommended that the outlet structure occasionally be modified to raise the elevation of the permanent pool when solids accumulate.

The procedure can be repeated until peak storage volume requirements are in danger of being compromised, at which time excavation will be needed.

## FLORIDA VEGETATED STORMWATER SYSTEMS

Florida's rapid growth has been accompanied by the construction of lots of impervious areas causing stormwater to become the major source of pollutant loading to receiving waters. Consequently, in February 1982, the Florida Stormwater Rule, Chapter 17-25, Florida Administrative Code, was implemented by the Florida Department of Environmental Regulation. The rule requires all newly constructed stormwater discharges to use appropriate BMPs to treat the first flush of runoff (Livingston). Vegetated systems, such as wet detention or wetlands, are among the more used BMPs.

### Wetland Systems

The 1984 Wetlands Protection Act authorizes the use of two types of wetlands for stormwater management: those wetlands that are connected to other state waters by either an artificial watercourse or by an intermittent watercourse which flows only when rainfall causes the groundwater to rise above the land surface. In addition to providing more natural stormwater management, the legislation also was seen as a way to provide greater economic value to wetlands, an incentive to developers to use wetlands rather than destroy them and a means of restoring many of the ditched and drained wetlands throughout Florida by routing stormwater into them. However, the legislation also recognizes the potential negative impacts that stormwater could have on wetlands and further states that such use shall only be done in a way that protects the ecological values of wetlands.

Because the use of wetlands for stormwater treatment is a largely untested BMP, the department adopted a specific set of design and performance standards for wetland systems and requires monitoring to evaluate the effectiveness of these standards. This data will be used to modify the current standards to more effectively use Florida's wetland resources for stormwater management. For example, initial monitoring confirms the seasonal nature of treatment efficiency of wetland systems with greater removal during active plant growing seasons and with net export of nutrients during winter and early spring. The design and performance standards for wetland stormwater management systems are:

1.  The wetland treatment facility is part of a comprehensive stormwater management system that uses wetlands in

combination with other BMPs to treat the runoff from the first one inch of rainfall; or, as an option for projects or project sub-units with drainage areas under 100 acres, treats the first one-half inch of runoff. Facilities which directly discharge to Outstanding Florida Waters must treat an additional fifty percent of runoff volume.

2. Pretreatment is required to reduce sediments, oils and greases and heavy metals. Pretreatment is also needed to attenuate stormwater volumes and peak discharge rates so that the wetland's hydroperiod is not adversely altered. Typically, swale conveyances and a lake adjacent to the treatment wetland are used to provide the pretreatment.

3. The use of wetlands for stormwater treatment shall not adversely affect the wetland by disrupting the normal range of water level fluctuation as it existed prior to construction of the wetland stormwater system. Normal range of water level fluctuation is defined as the maintenance of the fluctuating water surface changes between the normal low and normal high water levels of the wetland so as to prevent desiccation or over-impoundment of the wetland.

The normal low pool and normal high water elevations are established using existing biological or hydrological indicators present in the wetland. These indicators include water level data, lichen and moss lines, adventitious root formation, epiphytic plant colonies, root tussocks and hammocks, emergent plant community zonation, hydric soils, the landward extent of wetland vegetation, and recent water stain lines and cast rack and debris lines.

4. The design features of the stormwater management system shall maximize residence time of the stormwater within the wetland thereby enhancing opportunity for the stormwater to come into contact with the wetland sediment,

vegetation and micro-organisms. The outfall structure must be designed to bleed down the treatment volume in no less than 120 hours with no more than one-half of the volume discharged within the first 60 hours.

5. To further enhance contact between the stormwater and the wetland's assimilative processes, stormwater must be discharged onto the wetland via sheet flow so that channelized flow is minimized. Spreader swales, wide flat weirs between the pretreatment lake and treatment wetland and level spreaders are some of the distribution methods that have been used.

6. Erosion and sediment controls must be used during construction and operation of the system to minimize sedimentation of the wetland. Sediment control practices must be used in upland areas, be designed to minimize resuspension and discharge of collected sediments into the wetland and allow for recurring maintenance removal of sediments without harming the wetland.

Wet Detention System Design

In addition to natural wetland stormwater treatment systems, a large number of "constructed wetland" systems have been built in Florida since the implementation of the Stormwater Rule. The most common type of constructed wetland is associated with a wet detention system which consists of a permanent water pool ("lake"), a temporary storage area above the permanent pool in which the stormwater treatment volume is detained, and a littoral zone planted with appropriate native aquatic plants.

Detention systems provide numerous benefits and their design should accentuate multiple uses and aesthetic enhancement. Besides flood protection, peak discharge rate attenuation and water quality protection, wet detention systems can also provide a source of fill to the developer, "lake front" property which brings a premium price, open space, recreation and general aesthetic enhancement of a development.

Stormwater treatment by wet detention systems involves many mechanisms including the wetland mechanisms already discussed together with typical lake processes. Recent investigations of wet detention systems show that they can provide a high level of pollutant removal for many constituents, particularly nutrients and metals (3). However, the exact level of pollutant removal will depend upon a number of factors including detention time, amount of littoral zone, aquatic vegetation establishment, pond geometry, pond depth, area ratio, volume ratio and the incorporation of other BMPs with the detention system into a "treatment train" (16).

The current design and performance standards for wet detention systems in Florida include:

1. Treatment volume varies from a minimum of one inch of runoff up to 2.5 times the percent of impervious area.

2. The treatment volume is slowly discharged in not less than 120 hours with no more than half the volume discharged within the first 60 hours following a storm.

3. The volume in the permanent pool should provide a residence time of at least 14 days to promote biological uptake of nutrients (17).

4. At least 30% of the surface area should be a shallow littoral zone with appropriate native aquatic vegetation. The littoral shelf should be gently sloped (6:1 or flatter) and extend out to a point two to three feet below the normal water level. A layer of wetland soil at least six inches thick should be incorporated into the littoral area to promote plant establishment and growth. This soil will provide a source of wetland plant propagules, including nuisance species such as cattails which will need to be removed. At least one-third of the littoral area should be planted with a variety of aquatic plants suitable for the depth conditions. This "aquascaping" will enhance pollutant removal, aesthetics and wildlife habitat and value of the system. Within one year, a minimum of eighty percent of the littoral area must be covered by desirable plants with maintenance undertaken to remove cattails and possibly transplant desired species to areas where plant survival is poor.

The littoral zone should be concentrated near the discharge structure to promote assimilation of dissolved pollutants before they exit the facility. Littoral zones and their plants should also be located on shallow sills separating inline tandem ponds or forebays and near inlet areas to promote filtering before the runoff enters the main impoundment.

5. A maximum depth of eight to ten feet below the invert of the discharge structure should be planned for the permanent pool. If deeper depths are planned, then considerations must be made to prevent anaerobic bottom waters and sediments which will promote release of pollutants from the sediment.

6. Pond geometry should assure a minimum 3:1 length to width ratio. Inlet and outlets should be located as far apart as possible to provide a long flow path. The use of baffles, islands or peninsulas can be used to increase the length to width ratio and flow path.

7. The discharge structure should include a skimmer or other device to prevent oils and greases from leaving the system.

8. Erosion and sediment control practices must be used to retain sediment on-site during construction. Side slopes of detention systems should be sodded to stabilize them. A sediment sump or forebay should be included at the inflows to the wet detention system to facilitate sedimentation and ease of sediment removal.

9. Pretreatment is highly desired especially if the wet detention system is being planned as a real estate lake to enhance property values and promote aesthetic value. By incorporating the treatment train concept, the life expectancy of the system can be greatly extended and maintenance requirements reduced. Swale conveyances, small off-

line landscape retention areas, and a perimeter swale and berm system adjacent to the lake all will greatly reduce sediment, oil, grease and nutrient loadings to the system.

Wet Detention Regional Systems

One of the greatest stormwater management problems facing Florida and the nation is how to reduce pollutant loads from stormwater systems built solely for flood protection which rapidly convey polluted stormwater to the nearest water body. The use of regional wet detention systems which serve large drainage areas with multiple property owners will be heavily relied upon to reduce such pollutant loadings.

Regional facilities offer many advantages including management of stormwater from existing and future land uses; economies of scale in construction, operation and maintenance; and enhanced recreation and open space opportunities within a community. Disadvantages of regional facilities include the need for advance planning and funding to locate and construct systems and institutional difficulties in administering a master stormwater planning approach.

As part of a Clean Lakes Project in Tallahassee, a regional stormwater system was constructed in 1983 to reduce pollutant loads to Megginnis Arm and Lake Jackson. The 2,230 acre watershed consists of a rapidly developing area with commercial, highway and residential land uses. The stormwater system consists of a 28.4 acre detention pond, a 4.4 acre intermittent underdrain sand filter, and a 6.2 acre artificial marsh (figure 3). The marsh has an average depth of 18 inches, except for a 7 foot deep pool near the outfall, and is divided into three sections. The individual cells of the marsh were planted with the emergent macrophytes *Typha* (cattail), *Scirpus* (bulrush), and *Pontedaria* (pickerelweed), respectively.

Studies on the performance of the Megginnis Arm facility reveal that the system is very effective in removing pollutants from urban stormwater. Under normal operating conditions, about ninety-five percent of the suspended solid load is removed, primarily by settling and filtration. Total nitrogen (75%), ammonia (37%), nitrate (70%), nitrite (75%), total phosphorus (90%), unfiltered phosphorus (53%) and filtered phosphorus (78%) were also effectively removed by the system (18). Ammonia and nitrate oxidizing bacteria growing on the surface of the filter's rocks are responsible for much of the nitrogen removal and may assist in removal of phosphorus. Pollutant removal in the marsh system, however, is seasonal with good removal of nutrients and dissolved substances during the growing season (30% - 60%) but net export during winter when plants have died back and are decaying (Tuovla, et.al.). Maintenance to remove accumulated biomass is recommended each fall to prevent release of nutrients.

Recently, the City of Orlando has initiated a watershed wide program to reduce stormwater pollutant loadings to the city's lakes. The Southeast Lakes Project will ultimately lead to the modification of existing stormwater systems currently discharging polluted stormwater to fifteen lakes and fifty-eight drainage wells. While a variety of BMPs will be used, major emphasis is being placed on the use of wetland systems. Urban wetlands will be created or restored throughout parts of the watershed to help reduce stormwater pollution. The first phase of this project is the Lake Greenwood Urban Wetland which is currently under construction. The system will include a series of BMPs within a circuitous wet detention lake system. Wetland vegetation will be planted in littoral zones throughout the system and will also be grown in cages that can be easily removed to harvest the aquatic plants. Monitoring will be performed to determine the efficiency of the overall system and its individual components.

CONCLUSION

Wetlands have great potential to help solve our nation's stormwater management problems. However, much more information is needed to ascertain the possible impacts on wetlands and their fauna from the addition of untreated stormwater. Little is known about the potential for bioaccumulation of heavy metals or other toxics typical of stormwater. Monitoring of wetland stormwater systems is also essential to further determine relationships between design variables and pollutant removal efficiency.

REFERENCES

1. Lakatos, D.F. and L.J. McNemar. "Storm Quality Management Using Wetlands", paper presented at Freshwater Wetlands and Wildlife Symposium, Charleston, South Carolina, March 24-27, 1986.

2. Leopold, L.B. "Hydrology for Urban Planning - A Guidebook on the Hydrologic Effect of Urban Land Use", Circular 554, U.S.G.S., Washington, D.C., 1968.

3. "Results of the National Urban Runoff Program (NURP)", U.S. EPA, December 1983.

4. Miller, R.A. "Percentage Entrainment

of Constituent Loads in Urban Runoff, South Florida", USGS WRI 84-4319, 1985.

5. Stockdale, E.C. "The Use of Wetlands for Stormwater Management and Nonpoint Pollution Control: A Review of the Literature", report submitted to the Washington State Department of Ecology, October 1986.

6. Stockdale, E. C. "Viability of Freshwater Wetlands for Urban Surface Water Management and Nonpoint Pollution Control: An Annotated Bibliography," report submitted to the Washington State Department of Ecology, July, 1986.

7. Chan, E., et.al. "The Use of Wetlands for Water Pollution Control", Municipal Environmental Research Laboratory, U.S. EPA Report 600/2-82-036, 1982.

8. Harper, H.H., et.al. "Stormwater Treatment by Natural Systems", report submitted to the Florida Department of Environmental Regulation, December 1986.

9. Richardson, C.J. "Freshwater Wetlands: Transformers, Filters or Sinks" FOREM (Duke University) 11(2):3-9 (1988).

10. Brown, R. G. "Effects of Wetlands on Quality of Runoff Entering Lakes in the Twin Cities Metropolitan Area, Minnesota," USGS WRI 85-4170, 1987.

11. Silverman, G.S. "Seasonal Freshwater Wetlands Development and Potential for Urban Runoff Treatment in the San Francisco Bay Area", Ph.D. Dissertation, UCLA, Los Angeles, CA, 1983.

12. Kurasy, W. "Effectiveness of Wetland Stormwater Treatment - Some Examples" in *Stormwater Management: An Update*, pp. 145-157, University of Central Florida Environmental Systems Engineering Institute Publication 85-1, Orlando, Florida, July 1985.

13. Knight, R.L. and J.S. Bays "Florida Effluent Wetlands - Hydroecology" WTRDS

No. 2, $CH_2M$ Hill, Gainesville, Florida, 1986.

14. Maryland Department of Natural Resources, Water Resources Administration. "Guidelines for Constructing Wetland Stormwater Basins", Annapolis, MD, March 1987.

15. Livingston, E.H. "Stormwater Regulatory Program in Florida", in Urban Runoff Quality - Impact and Quality Enhancement Technology, pp.249-256, Proceedings of an Engineering Foundation Conference, Henniker, NH, June 23-27, 1986.

16. Livingston, E.H., The Florida Development Manual: A Sound Land and Water Management, Department of Environmental Regulation, Tallahassee, FL, June 1988.

17. Hartigan, J.P. "Regional BMP Master Plans", in Urban Runoff Quality-Impact and Quality Enhancement Technology, pp.351-365, Proceedings for an Engineering Foundation Conference, Henniker, NH, June 23-27, 1986.

18. Tuovila, B.J., et.al. "An Evaluation of the Lake Jackson Filter System and Artificial Marsh on Nutrient and Particulate Removal from Stormwater Runoff", in Aquatic Plants for Water Treatment and Resource Recovery, pp.271-278, Magnolia Publishing, Orlando, FL.

## Discussion of Mr. Livingston's Paper

Question:

How well is the artificial wetlands working?

Answer:

That's mixed. It produces very good outlet water quality at some times, some periods of the seasonal growth cycle. At other times, there appears to be essentially a mass transport through the wetlands. Soluble phosphorus seems to be consistently lowered from about 0.1 mg/l to 0.02 mg/l, though.

Question:

Aren't wetlands transitional places that are filling-in or changing, actually going away in the long-term?

Answer:

Yes, but in many centuries or eons; not in the sort of "design periods" relevant to the uses we want to make of them.

# SUBJECT INDEX
Page number refers to first page of paper.

Analysis, 145
Automobiles, 421

Benefit cost analysis, 84
Bridge decks, 421

Clean Water Act, 49, 100, 109
Combined sewers, 359
Construction methods, 290
Corrosion, 421
Cost effectiveness, 84, 349, 356

Deicing, 421
Design criteria, 49, 122, 214, 239, 290, 307, 359, 467
Detention basins, 14, 122, 145, 164, 180, 214, 239, 258, 268, 280, 451

Environmental engineering, 14
Environmental impacts, 421, 451, 467
Environmental planning, 14
Environmental Protection Agency, 111
Environmental quality, 180

Federal laws, 100
Financial feasibility, 340, 356
Florida, 49, 467

Grease, 403

Hydraulic design, 180
Hydrology, 180

Infiltration, 145, 239, 290, 307
Inflow, 258
Institutional constraints, 340, 356

Management engineering, 68
Manpower, 340
Maryland, 467
Mixing, 164
Monitoring, 224

Nonpoint pollution, 100, 109, 203, 224, 268, 378

Oils, 403
Outflows, 258
Overflow, 359, 390

Percolation, 307
Phosphorus removal, 203, 239
Planning, 324
Pollutants, 180
Pollution abatement, 280
Pollution control, 111, 122, 145, 359, 436

Rainfall frequency, 214
Rainfall-runoff relationships, 214
Recreational facilities, 203
Regulations, 100, 111, 224
Research, 421
Reservoirs, 203
Residence time, 164
Runoff, 68, 122

Salts, 421
Salts effects, 421
Sediment, 268
Separation techniques, 390
Separators, 390
Sewage effluents, 451
Sewage treatment, 451
Solid waste management, 390
Solid wastes, 390
Solids, 390
Sources, 436
Storage facilities, 180
Storage tanks, 359
Storm drains, 436
Storm runoff, 164, 203, 268, 307, 359, 378, 403
Stormwater, 145, 180, 324
Stormwater management, 49, 68, 84, 100, 109, 111, 164, 224, 258, 290, 340, 356, 359, 378, 403, 451, 467
Streamflow, 29
Streams, 29
Suspended sediments, 268
Sweden, 307

Texas, 224
Toxic wastes, 436

Urban areas, 84, 111, 164, 214, 224, 258, 403
Urban runoff, 14, 29, 100, 145, 258, 280, 324, 349, 356, 378, 451, 467
Urbanization, 29

Wastewater treatment, 390
Water management, 349
Water pollution, 378
Water pollution control, 68, 378, 403, 436
Water quality, 14, 29, 100, 109, 111, 145, 258, 280, 324, 349, 356, 359, 403, 421, 436
Water quality control, 68, 145, 324, 356
Water quality standards, 49, 68
Water treatment, 239
Watershed management, 29, 224
West Germany, 359
Wetlands, 180, 203, 239, 280, 451, 467

# AUTHOR INDEX
Page number refers to first page of paper.

Brombach, Hansjoerg, 359

Dorney, John, 280
Driscoll, Eugene D., 145

Ellis, J. Bryan, 14

Gallup, James D., 100

Harrington, Bruce W., 290
Hartigan, John P., 122
Heaney, James P., 84
Helfrich, Mike, 180
Herricks, Edwin E., 29
Hess, Larry G., 258
Holman, Bob, 280
Horner, Richard R., 45
Hubbard, Thomas P., 56
Huber, Wayne C., 66
Hvitved-Jacobsen, T., 214

Jaquet, Neil G., 349
Jennings, M. E., 224
Jones, D. Earl, 356

Livingston, Eric H., 49, 467
Llewelyn, Michael T., 109
Lord, Byron N., 421

Martin, Edward H., 164
Maxwell, Mark, 239
Mulhern, Patrick F., 203

Osborne, Lewis L., 29

Parrish, J. F., 224
Pisano, William C., 390
Pope, Larry M., 258

Samr, Timothy E., 436
Schueler, Thomas R., 180
Schultz, Nancy U., 324
Silver, H. Earl, 340
Silverman, Gary S., 403
Stahre, Peter, 307
Steele, Timothy D., 203
Stenstrom, Michael K., 403

Tucker, L. Scott, 111

Urbonas, Ben, 239, 307

Veenhuis, J. E., 224

Wanielista, M. P., 214
Washtenaw County Statutory Drainage Board and the Michigan Water Resources Commission, 378
Weiss, Kevin, 100
Wenk, William E., 239
Wu, Jy S., 280
Wulliman, James T., 239
Wycoff, Ronald L., 324

Yousef, Y. A., 214

Zarriello, Phillip J., 268

ENVIRONMENTAL S/E
533 W. LIBERTY DRIVE
WHEATON, ILLINOIS 60187
708/690-4090
FAX 708/690-4167

**ENVIRONMENTAL S/E**
933 W. LIBERTY DRIVE
WHEATON, ILLINOIS 60187
708/690-4090
FAX 708/690-4167